ASIALAND Workshop on

The Establishment of Soil Management Experiments on Sloping Lands

Papers and Reports of a Training Workshop
held at Chiang Mai, Thailand

7-18 November 1988

IBSRAM Technical Notes no. 3

The workshop was organized by

The International Board for Soil Research and Management (IBSRAM)
Department of Land Development (DLD), Thailand
The Faculty of Agriculture, University of Chiangmai
The Australian Centre for International Agricultural Research (ACIAR)
Soil Management Support Services (SMSS/USAID)
The International Institute for Aerospace Survey and Earth Sciences (ITC)

Science Editors
E. Pushparajah
and
Samarn Panichapong

Publication Editor
Colin R. Elliott

Correct citation: ASIALAND Workshop on the Establishment of Soil Management Experiments on Sloping Lands. IBSRAM Technical Notes no. 3.

ISBN 974-86952-4-7

Printed in Thailand

CONTENTS

The Context of the Workshop

IBSRAM recognizes that refresher courses and group dissussions on specific subjects are increasingly relevant to present-day research. This is particularly important in relation to the need for constant adjustments to the new knowledge which is regularly evolved. Thus IBSRAM's training activities have dealt essentially with refresher-course elements combined with discussion on methodological guidelines.

The workshop on Site Selection, Characterization and the Establishment of Experiments for Soil Management Networks was part of IBSRAM's general endeavour to promote training activities. The workshop was held in Chiang Mai, Thailand, from 7 to 18 November 1989, and was attended by all the front-line scientests and coordinators involved, or prospectively involved, in the ASIALAND network on the management of sloping lands (ASIALAND - Management of Sloping Lands).

The workshop was organized by IBSRAM in collaboration with the Department of Land Development (DLD), Thailand, the Faculty of Agriculture, University of Chiangmai, the Australian Centre for International Agricultural Research (ACIAR), Soil Management Support Services (SMSS/ USAID), and the International Institute for Aerospace Survey and Earth Science (ITC). Twenty-eight trainees (from nine countries) and twenty-two resource persons participated in the workshop (Appendix IV). The participants played an active role in group discussions, field exercises, classroom lectures and field trips.

The choice of approach and the contents of the training courses were made with the idea of bringing network participants to a measure of reciprocal understanding and of achieving mutual endorsement of methodologies. The topics - site selection and characterization, soil management practices, and experimental monitoring - were of importance and immediate use to the network cooperators (Appendix III).

A multidisciplinary approach was adopted with regard to the subject matter in order to create awareness of areas of concern outside specialized interests. The practical work in the field involved the use of different techniques, not always familiar to the trainees, and complemented the theoretical background by relating academic training to field work in experimental areas. The discussions that followed, and the use made of the knowledge gained in drafting the methodological guidelines at the end of the training workshops, revealed the complexity of the issues invloved in improving soil management.

One of the important objectives of this workshop was in fact to formulate a common methodology to be followed by all participating cooperators in conducting research for the different projects in the network. Guidelines for carrying out network projects were established in connection with site selection, site characterization, erosion and runoff, and for monitoring chemical, biological, and physical parameters (Appendix I).

Some of the lectures, in particular those pertaining to quality control of soil and plant analysis, site variability, experimental design, and statistical packages have been excluded, because they have already been published in IBSRAM Technical Notes no. 2. The lectures are presented in this volume with some stylistic changes, but not necessarily with regard to the formal requirements of papers intended for journals and conference proceedings.

In addition to the classroom lectures, participants were involved in field exercises in site characterization. The site characterization training was concerned with the preparation of a soil - mapping exercise at a detailed level (Appendix II), interviewing farmers with a view to preparing a socioeconomic survey, monitoring infiltration, and measuring resistance to rooting using penetrometer readings. Visits were made to experimental sites to see erosion studies, and to a commercial farm and development projects to see management and conservation practices being implemented in the field.

A factor contributing to the success of the workshop was the careful selection of the participants. However, there was a diversity in the specialization of the trainees - soil chemists, agronomists, soil surveyors, and soil physicists. Although this made the task of the lecturers difficult, the diversity enriched and widened the scope of the discussions, and made it possible to have an in-depth look at some topics which might otherwise only have been covered superficially, or have been overlooked.

Another feature which contributed to the success of the course was that it was conducted in Chiang Mai - an area similar to those in which the projects are to be implemented. This fact, combined with a multidisciplinary array of lecturers, helped sustain the interest of the trainees. The ratio of trainees to lecturers was almost 1:1, which ensured easy access to the available experience and expertise.

INTRODUCTION

Samarn Panichapong

Thai Adviser, International Board for Soil Research and Management

Deputy permanent secretary of the Ministry of Agriculture, deputy governor of Chiang Mai, deputy director-general of the Department of Land Development, distinguished guests, and participants:

On behalf of IBSRAM, I would like to express our gratitude to the deputy permanent secretary for sparing time in his busy schedule to preside at the official opening of this training workshop. Unfortunately, Dr. Marc Latham, director of IBSRAM, cannot be with us because he is now on an official trip abroad. Dr. Pushparajah will make the opening statement on his behalf. Khun Sanan Kimwanich, director-general of DLD, is also on an official trip to the USA. Khun Sitilarp Wasuvat, the deputy director-general, will give the formal address on his behalf.

IBSRAM feels honoured to have been involved in the organization of this training workshop on Site Selection, Characterization and the Establishment of Experiments for Soil Management Networks, and is pleased to record the cooperation it has received in this task from the Department of Land Development, the Asian Development Bank, the Centre for International Agricultural Research, the International Training Centre of the Netherlands, the Department of Agriculture, Chiang Mai University, and Kasetsart University.

In October 1986, IBSRAM organized the First Regional Seminar on Soil Management under Humid Conditions in Asia in Khon Kaen to discuss the possibility of establishing research networks in Asia. The seminar reached an agreement that IBSRAM and the cooperators would prepare research proposals under the title "Land Development and Soil Management in Asia and the Pacific (ASIALAND)". It was agreed that this project should include three research networks: (i) tropical land development for sustainable agriculture, (ii) the management of acid tropical soils, and (iii) the management of Vertisols.

After two years, it was eventually agreed that ADB and the Swiss government would support the research network on the management of sloping lands for sustainable agriculture. Five countries - Indonesia, Malaysia, Nepal, Philippines, and Thailand - joined the network. This training workshop is part of the network's activities to bring all cooperators together to discuss and formulate a standard methodology for establishing experiments for this soil management network. In this way, the results of each experiment will be transferable among the countries involved, and probably transferable to other countries within the same agroenvironmental conditions.

There are 60 participants from 12 countries attending the training workshop. It will cover a wide range of subjects, including soil, climatology, agronomy, experimental design, and socioeconomics. On the last two days, there will also be a meeting of the network's Steering Committee to discuss the technical, administrative, and financial aspects of the network.

In organizing this training workshop, IBSRAM has received valued assistance from many organizations and individuals. On behalf of IBSRAM, I would like to take this opportunity to express our sincere thanks to all of them. Fortunately, too, we have been able to organize this training workshop in Chiang Mai, one of the most beautiful and culturally interesting cities in the North, during the best season of the year. I am sure that your stay in this city will be a pleasant one.

OPENING STATEMENT

E. Pushparajah

Programme Officer, International Board for Soil Research and Management

Deputy permanent secretary of the Ministry of Agriculture, distinguished guests, and participants:

On behalf of IBSRAM, I would like to thank our distinguished guests for their kind presence at this opening ceremony of the training workshop on Site Selection, Characterization and Establishment of Experiments for Soil Management Networks, and to convey the apologies of the director of IBSRAM for his inability to be present today. This is because he has to attend the Centres' Week in the USA and meet IBSRAM's donors.

Let me just introduce IBSRAM in a few words and relate the activities of IBSRAM to this workshop. A detailed presentation on IBSRAM and its role in improving soil management will be given immediately after the break following the opening ceremony.

The International Board for Soil Research and Management was formed in 1983, but did not start to implement its programme until mid-1985 - about 3 years ago. IBSRAM's headquarters are at the Department of Land Development in Bangkok, Thailand. Its primary objective is to promote and assist *applied* soil research into the identification, development, use, management and protection of soils for food production and other agricultural and agroforestry purposes in order to enhance and increase *sustainable* production in developing countries.

IBSRAM's role is to act as a catalyst for the validation of technologies at the farmers' level. It takes a *multidisciplinary* approach to agricultural problems, and encourages the invlovement of national agencies in the application of appropriate technologies on farms in order to ensure sustained production.

IBSRAM has collaborated with national agencies in Indonesia, Malaysia, Nepal, Philippines, and Thailand to form a network called' ASIALAND - Management of Sloping Lands'. The objective of this network is to find ways and means of managing sloping lands on an agriculturally sustainable basis. The trials have been planned and the projects are ready to be implemented.

Many such plans in the past have failed because they were not initiated or monitored properly, and IBSRAM is firmly convinced that the collaborators have to be brought to a level of reciprocal understanding and mutual endorsement of methodologies before any project is launched if we are to achieve a successful outcome.

This training workshop should therefore be viewed as the final preparation for the implementation of ASIALAND - Management of Sloping Lands. We hope that in the next 12 days of lectures, discussions, and field exercises, we will be able to make use of the rich mixture of expertise and the diversity of disciplines represented by the participants, and arrive at a detailed consensus of views on the metholologies to be adopted. This will facilitate the successful implementation of IBSRAM's ASIALAND network on the management of sloping lands for sustainable agriculture.

FORMAL ADDRESS

Sitilarp Vasuvat

Deputy Director-General, Department of Land Development

Deputy permanent secretary, honoured guests, distinguished participants:

It is an honour for me and my department to welcome you all to the premises of the Department of Land Development at Chiang Mai and to provide the venue for IBSRAM's training workshop.

I am happy to note that you have selected Chiang Mai as the venue for the workshop, as this area will enable you all to see for yourselves the problems that are faced in northern Thailand. The Soil Survey Division of my department has reported that 35% of the country is hilly and mountainous, and most of this type of land is concentrated in the northern and western part of the country. As will be clearly evident from your field visits, a large part of these hilly lands have been brought under agriculture, often with zero input, and under shifting cultivation. The pressure for land is so great that the period of fallow, which used to be three to six years in 1984, has now been reduced to one to four years.

The consequence of such practices is soil erosion losses of 50 to 300 tonnes per hectare per year. This not only results in on-site degradation of soils but also in off-site damage due to siltation. The lack of replenishment of organic matter and nutrients further contributes to the declining fertility of the soil. Further, with zero inputs and shorter fallow periods, competition from weeds has increased. The net result of this is a dramatic decline in yields from many of these areas, which is against the principle of trying to increase the productivity per unit area.

Various agencies in the country have been trying to address this problem. This has culminated in a proposal to combine appropriate cropping techniques (such as those recommended by the TAWLD project, DOA, and DOF), and soil conservation measures recommended by DLD. These proposals have to be tested and validated.

We believe that the ASIALAND network on the management of sloping lands for sustainable agriculture would be a useful base for such validation, and have therefore joined the network. This training workshop should enable the cooperators to finalize their metholologies and give them the spurt to move ahead to find the solutions.

It is now my pleasant duty to call on the deputy permanent secretary of the Ministry of Agriculture to officially open this workshop.

INAUGURAL ADDRESS

H.E. Udorn Tantisunthorn

Deputy Minister, Ministry of Agriculture and Cooperatives*

Ladies and gentlemen:

It is an honour for me to preside at the official opening of this training workshop for soil management networks.

Thailand is basically an agricultural country. More than 70% of its population depend directly or indirectly on agriculture, while at the same time a large portion of our export earnings come from agriculture. Thus agriculture and its developement is a priority area of concern for the government.

The expansion of the population, and hence the demand for more food for consumption or for producing agricultural products for cash income, has resulted in a tremendous pressure on land. The consequence is that not only undulating lands, but also steeplands, have been brought into cultivation. The cropping of such lands creates a real concern for our government. The concern arises not only in terms of environmental conservation due to factors such as accelerated soil erosion, but also with regard to the readily noticeable decline in the sustainability of the land.

In the past, when such lands became unproductive the farmers gave up the land and moved further into the forest and opened new land. The consequence is that few forests are now left in Thailand. We not only have to preserve these forested lands, but would also like to see other denuded areas reforested.

The approach to solving these two problems, I believe, would be firstly to find a way of improving and sustaining soil management for agriculture with an input system appropriate to the farmers' capabilities. This would necessarily mean that cash inputs would have to be low to medium inputs. With improved productivity and sustained agriculture, particularly on the undulating terrain and on the lower slopes, the pressure on land requirements could be reduced. This would then allow for reforestation, particularly on the steeper slopes.

Problems of this nature are not peculiar to Thailand. Our ASEAN friends - in particular, Indonesia, Malaysia, and Philippines - also have similar experiences in the management of soils on sloping lands.

---

* Delivered by Mr. Chaisup Supsari, deputy permanent secretary of the Ministry of Agriculture and Cooperatives.

Some of you here may recollect that IBSRAM organized a regional seminar on Soil Management under Humid Conditions in Asia in Khon Kaen in October 1986. At that seminar, my predecessor, Mr. Prayuth Siripanich, called on the participants for something more than mere recommendations which could easily be forgotten. I am glad to note that the participants, working in cooperation with IBSRAM, have taken appropriate action. Five countries with similar problems, i.e. Indonesia, Malaysia, Nepal, Philippines, and Thailand will now, through a network approach, be looking into solutions for the sustained management of soils on sloping lands for food production.

I am told by my officers that they would be addressing the problem of managing sloping lands not only from a basic agricultural perspective, but also with a view to the use of agroforestry, where both food crops and tree crops (e.g. fast-growing legume trees, chinese plum, and coffee) would be grown together. Currently these are mainly proposals which have not yet been implemented. This country and its neighbours who are cooperators in the IBSRAM network want an early solution.

I believe that this training workshop should equip the cooperators sufficiently to enable them to implement their research projects. I therefore have great pleasure in declaring this training workshop open.

Section 1: IBSRAM's network approach

IBSRAM's role in improving soil management

M. LATHAM and E. PUSHPARAJAH*

Abstract

The role of IBSRAM and its approach towards fostering international research on soil management for sustainable production is reviewed. Emphasis is given to a multidisciplinary approach, and to applied research in the farmers' environment. The evolution and mode of formation of networks, with the active participation of the collaborators at all stages of development, is described. Attention is also drawn to the emphasis IBSRAM gives to its training workshops, this workshop provides an example of workshop constituents.

Introduction

IBSRAM was launched in September 1983, or about 5 years ago, but only became really effective in mid-1985, or about 3 years ago. Its headquarters are in Bangkok, Thailand. IBSRAM's primary objective is to promote and assist applied soil research into the identification, development, use, management and protection of soils and lands for food production and other agriucltural and agroforestry purposes, so as to enhance and increase economically sustainable production in developing countries (IBSRAM, 1988).

* Director of IBSRAM and IBSRAM Programme Officer, IBSRAM Headquarters, PO Box 9-109, Bangkhen, Bangkok 10900, Thailand.

The specific objectives are to:
- promote and assist aspects of research on soil and soil management so as to increase sustainable food production in developing countries; this includes validation and transfer at the farm level of existing and improved technologies of agricultural production;
- to provide technical advice, guidance, and practical training to national agricultural systems (NARS) for adaptive research;
- to foster and assist coordinated interdisciplinary activities in order to develop, test, and promote on-farm applications of improved technologies (this could extend into characterization, interpretation of soil maps, soil classification, soil management, and soil conservation);
- to promote the involvement of national agencies of developing countries in soil-related applied research and testing activities;
- to foster cooperation in applied soil research between the developed and developing regions for their mutual benefit; and
- to encourage the coordination of applied soil management and related research for the benefit of developing countries.

IBSRAM's approach to soil management research

IBSRAM readily recognizes and accepts that national research organizations should have the major role in adaptive on-farm soil research. Often limited training staff and/or financial resources are major constraints. The network concept (Greenland *et al.*, 1987) could help to minimize such constraints. Generally networks are initiated by first informing interested NARS of existing knowledge and the potential for its adaptability to local environments. The networks so formed lead to sharing of the knowledge and new findings by one NARS with another, working on the same problems, in a coordinated manner. To achieve this objective, IBSRAM has chosen a collaborative research type of network. In this approach, participating countries collaborate in joint planning, implementation, testing, and validation of the research. The results are not only shared amongst the participating countries but also with nonparticipating countries. We feel confident that this approach will assist in developing NARS capability.

In order to share results in a network, a common approach is essential. Recommendations at IBSRAM's workshops (especially training workshops) on aspects of site selection, characterization, design of experiments, and monitoring of on-farm experiments are being developed into guidelines. This training workshop will discuss and adopt these guidelines.

Establishing priorities and the evolution of networks

IBSRAM's collaborative network approach has evolved from discussions at different seminars and workshops attended by scientists from developing and developed countries. Such joint decisions have been encouraged and supported by donor agencies and recipient countries. The first workshop, held in 1983, decided on the formation of IBSRAM and identified four priority areas for research on soil management (Figure 1).

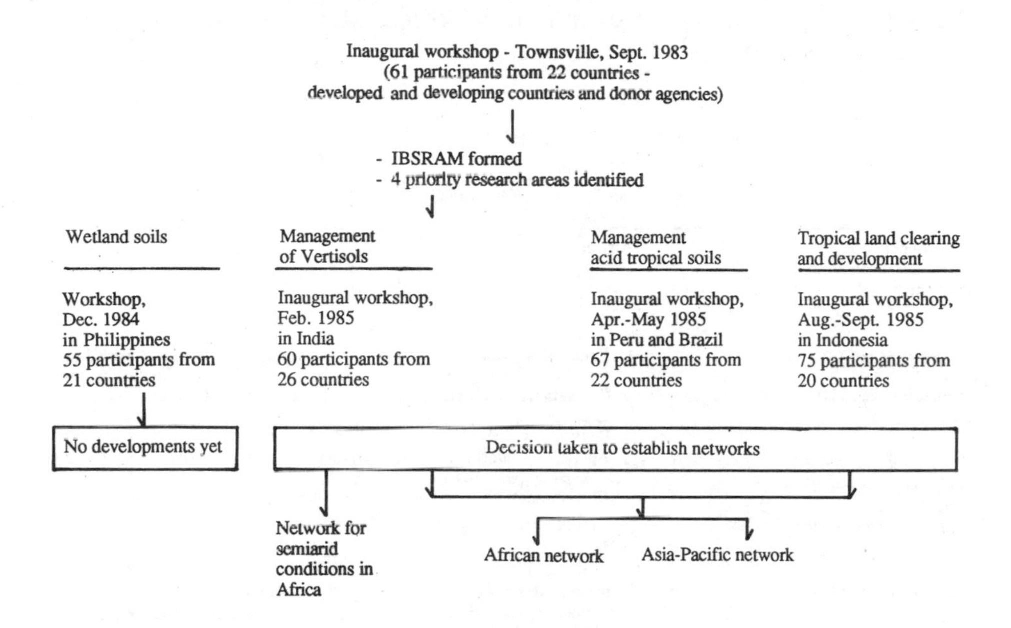

Figure 1. Evolution of IBSRAM's networks.

The subsequent workshops, which were on specific priority areas, involved prospective participants from NARS (particularly scientists working on such problems in their respective countries), scientists from international agricultural research centres (IARCs), research organizations, and donor agencies. Participants at these inaugural workshops were enthusiastic about establishing collaborative soil management networks as a joint effort between NARS and IBSRAM. In identifying the priorities, the participants felt that the problems selected were important areas of research for their countries. The existence of promising improved technologies with development potential offered the best chance of providing beneficial results within a short period.

Originally, global networks were envisaged. However, on subsequent reappraisal it was felt that regional networks provided a more appropriate and realistic approach. Thus regional workshops were organized to develop the proposed regional networks (Figure 2). The regional seminars also considered in greater detail the research areas and the common-core experiments. Some NARS with additional technical capabilities of their own, or in collaboration with IARCs, may carry out support or 'satellite' trials intended to fill gaps in the present knowledge. These latter trials could, for example, be screening for acid-tolerant varieties, or levels of lime or P - and though they are often designed to meet the needs of individual cooperators, they may eventually benefit other cooperators in the network.

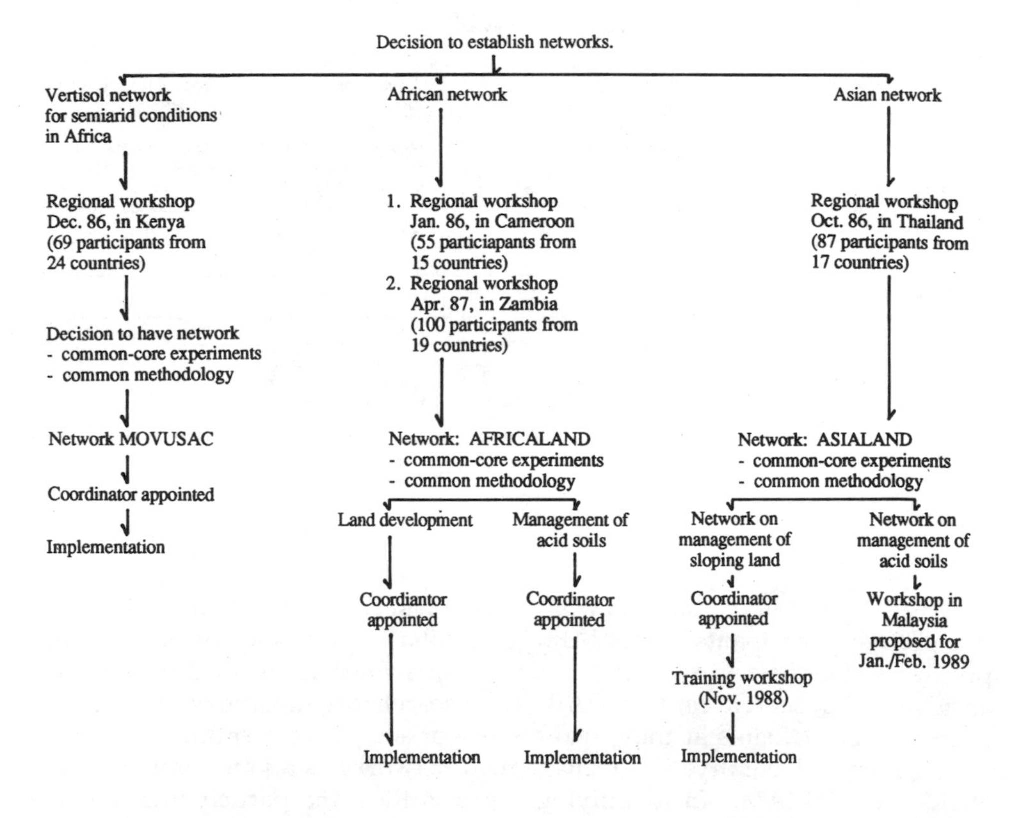

Figure 2. Evolution of regional networks.

Core experiments: The standard core experiment has the three following basic treatments:

- the normal practices of the farmers (this is used as a reference treatment);
- low cash input technologies (including the use of improved acid-tolerant varieties, improved practices, and some inputs of fertilizers, e.g. Ca and P;
- medium to high cash inputs (this involves the use of high-yielding varieties, amelioration for pH, use of fertilizers and pesticides, and crop rotation).

These treatments have many adaptations. The main objective is to measure whether the newer systems are sustainable, and ultimately whether they are acceptable to the farmers. In this context, it need to be emphasized that such trials should be carried out on farmers' fields, or at least in the farmers' environment.

The network concept

A simple and conceptual view of a network to achieve the objectives is indicated in Figure 3. This shows that a few NARS can be involved in a network, all having a common-core experiment, while some have one or more 'satellite' trials. In some cases, the NARS may collaborate with IARC's in one or more of the 'satellite' trials.

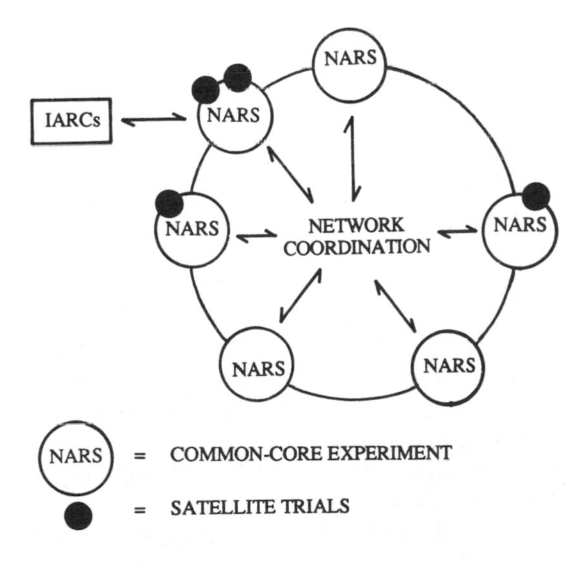

Figure 3. Organization of network: A simple framework.

The above approach for a network is only a conceptual one. In practice, a network involves not only the NARS but a catalyst (in this case IBSRAM), a coordinator (appointed by IBSRAM), and a donor or a group of donors. Typical interactions in a network are shown in Figure 4.

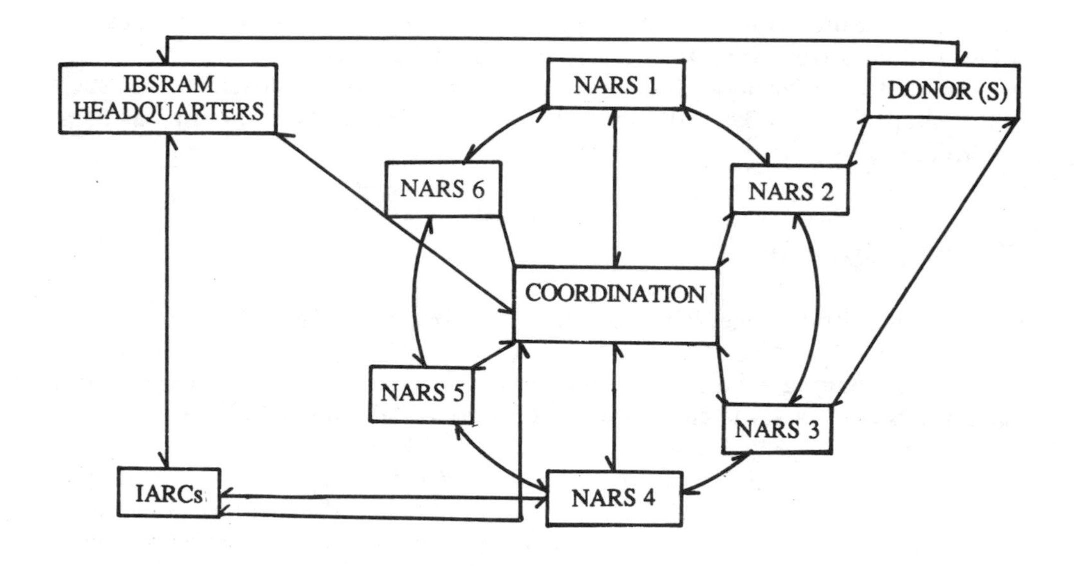

Figure 4. IBSRAM network organization.

An important feature often overlooked in forming networks is the need for a 'network' within the country itself. As indicated earlier, any project proposal for acceptance in a network must satisfy the following criteria:-

- the investigation should be considered important for the country concerned, and the NARS should be already involved in similar research (or be willing to invest in such research);
- the country proposal should fulfil the network objectives;
- the project should be technically acceptable; and
- the project should be considered to be economically acceptable.

IBSRAM feels that in order to ensure that the country proposal is more meaningful, a small country 'network', as indicated in Figure 5, would be useful.

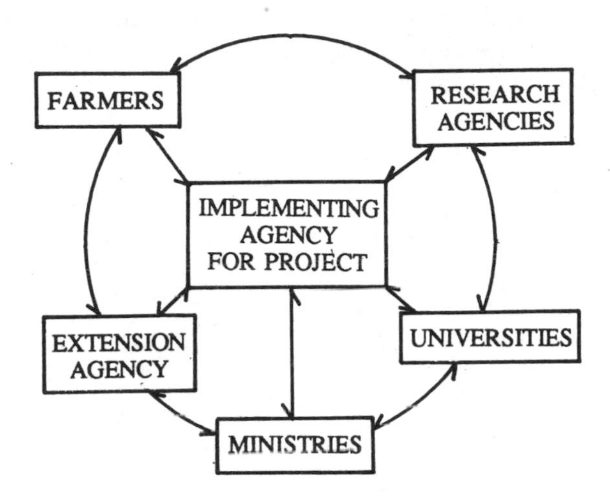

Figure 5. A network within a country.

The acceptance of projects and network implementation

The national institution(s) formulate a project proposal based on its priorities and submits it to IBSRAM. The proposal (s) are reviewed by a Network Coordinating Committee, which may suggest amendments to enable greater conformity to the network, but without altering the main objectives. The proposals are then considered and endorsed by IBSRAM's board. On approval, each national cooperator may seek donor support individually or IBSRAM will assist in obtaining donor support.

IBSRAM, through a programme coordinator assisted by a Review Committee, will help to crystalize ideas, and coordinate and assist the cooperators; the actual research is done by the cooperators themselves. The coordinator is the key link between the collaborators and IBSRAM. Through regular visits and consultations the coordinator helps to strengthen cooperators' programmes in the following areas:

- site characterization;
- exchange of control soil samples and analytical methods;
- design of experiments, analysis, and interpretation;
- regular meetings, during which progress will be reviewed, and invloving all collaborators in the network;
- monitoring; and
- advice on the need for short training courses, etc. for the collaborators.

Clearly, there will be a time lapse from the inaugural workshop to the time the network is implemented (Figure 6). After the initiation of the idea, formal workshops and training courses need to be held so that the cooperator will be well equipped to commence the field trials in a coordinated manner.

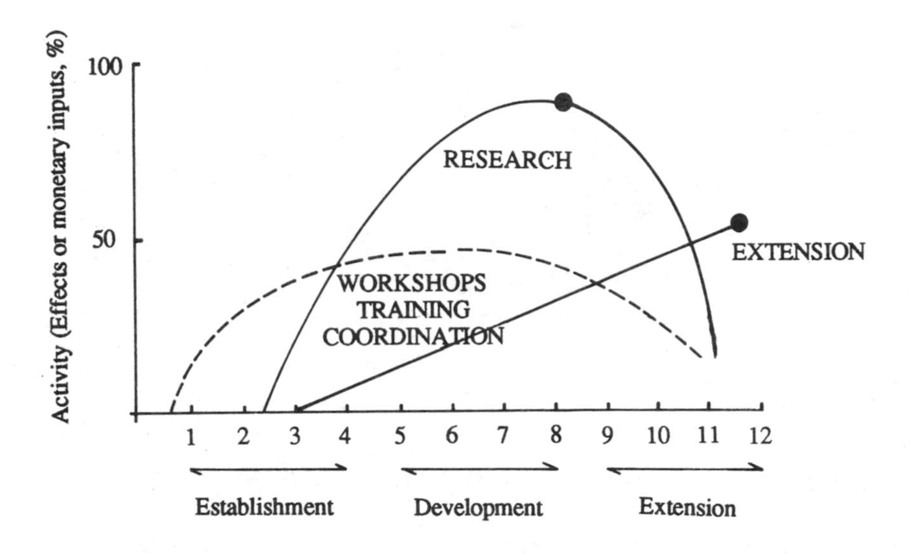

Figure 6. Progress of a successful network.

The time lapse could be longer or shorter depending on how soon donor agencies respond. Though most networks are initially envisaged to run for three years, it is IBSRAM's belief that the peak of research activity will be around the sixth to seventh year; this research would however incorporate a large element of field demonstration on the most promising economical and sustainable system likely to have farmer acceptance. The concurrent extension input, which will be increasingly promoted with time, cannot be overemphasized.

Training - also an objective of this seminar

Training is one of the major activities of IBSRAM. It is essential in supporting the activities of the soil management networks. Training for site selection and characterization and for monitoring experiments is especially emphasized, as these activities are crucial operations in the network. A proper characterization of the sites and monitoring of the experiments with a common methodology takes on a more important function in a network, since they allow a comparison of results and an exchange of information. A failure to consider these aspects could mean a loss of valuable time and investment, as evidenced by the failure of many agronomic experiments in the past.

Since its networks have become operational, IBSRAM has become actively engaged in training activities. Two training workshops have already been held, and this training has been fruitful. The current workshop has been specially formulated for the ASIALAND network on the management of sloping land.

The purpose of this training workshop is to define and develop a common methodology for site selection, characterization, monitoring, and evaluation. The topics covered include:

- the network approach;
- the agroecological approach to agriculture;
- farming systems;
- site selection;
- site characterization;
- measurements (erosion, runoff, sampling, etc.);
- experimental design and data processing; and
- field demonstrations and exercises.

We believe that this workshop will be of immediate benefit to the collaborators. Both the participants and the lecturers are from different disciplines to allow for a multidisciplinary approach. This should elicit a wider range of discussion and an in-depth look at some aspects of the topics which may otherwise be overlooked. Further, the ratio of trainees to lecturers has been kept low (almost 1:1), and it is hoped that this will result in more active discussions.

IBSRAM is convinced that training workshops have to be conducted in areas similar to those where the network projects are to be implemented. The similarity of the environment is a key factor in the field exercises, and Chiang Mai has been selected as the venue for the training workshop as it is the location of one of the sites for the project on sustained management of sloping lands in the ASIALAND network.

It is hoped that this training workshop will give the participants some measure of reciprocal understanding and that it will result in a mutual endorsement of the methodologies to be used in the ASIALAND network.

References

GREENLAND, D.J., CRASWELL, E.T. and DAGG, M. 1987. International networks and their potential contribution to crop and soil management research. *Outlook on Agriculture* 16(1):42-50.

IBSRAM. 1988. *IBSRAM's strategy, programmes and budget plans.* IBSRAM: Bangkok. 58p.

LATHAM, M. 1987. The IBSRAM land development and soil management network program in monsoon Asia. In: *Soil management under humid conditions in Asia and the Pacific (ASIALAND)*, 13-22. IBSRAM Proceedings no. 5. Bangkok: IBSRAM.

Land development and management in Asia: sloping lands

ADISAK SAJJAPONGSE*

Abstract

The network on the management of soils for sustainable agriculture has three areas of priority concern, i.e. soil management on sloping lands for sustainable agriculture, the management of acid soils, and crop management on Vertisols after rice. After a consideration and evaluation of various projects proposed by different countries at the regional ASIALAND seminar in Khon Kaen, it was decided to make the management of sloping lands for sustainable agriculture the first priority. The network involves five countries, namely Indonesia, Malaysia, Nepal, Philippines, and Thailand, and is funded by ADB and the Swiss government. This paper describes the progress of the projects in the different participating countries. Although the projects are still in their infancy stage, site selection and the treatments to be tested in all the countries have been chosen and finalized.

Introduction

The network on the management of soils for sustained agriculture in Asia has three priority areas of concern, i.e. the management of soils on sloping lands, the management of acid soils, and crop management on Vertisols after rice crops. These soils, which are considered to be marginal soils in Asia, have

* ASIALAND Coordinator, IBSRAM Headquarters, PO Box 9-109, Bangkhen, Bangkok 10900, Thailand.

become increasingly important for agricultural production in recent years. Areas on sloping lands and acid soils, which were once covered by forests, are being progressively cleared. Such lands can support only few crop cycles before they become degraded and unproductive. Vertisols occur in lowland areas, where they are usually used for one paddy crop per year, after which they are not used for the rest of the year due to a lack of water. However, growing a second crop on the Vertisols is possible if proper crop management is introduced.

In September 1986, IBSRAM organized a regional seminar in Khon Kaen, Thailand. The seminar decided on the need for an Asian network on land development and management (on sloping lands), the management of acid soils, and post rice soil and crop management on Vertisols. Initially priority was given to the establishment of a network on postclearing management of sloping lands, with five participating countries - Indonesia, Malaysia, Nepal, Philippines, and Thailand. Subsequently, a detailed consideration and evaluation of the various project proposals were submitted to ADB for consideration for funding. The network is now funded by ADB/Swiss Aid. The various project-implementing agencies are: the Centre for Soil Research (CSR) of the Agency for Agricultural Research and Development (AARD), Indonesia; the Rubber Research Institute of Malaysia (RRIM); the Division of Soil Science and Agricultural Chemistry (DSSAC), Nepal; the Philippines Council for Agriculture, Forestry and Natural Resources Research and Development (PCARRD), Philippines; the Department of Land Development (DLD), Thailand.

Project development

During the period 19 September to 6 October 1988, a team consisting of the ASIALAND coordinator and a consultant visited the different network participating countries to finalize the project proposals and to visit the experimental sites. An account of the development of the various projects is given below.

Indonesia: Two sites located in the transmigration area about 60 km south of Muaro Bungo on Sumatra Island were selected. One site is on degraded land and the other one is on newly cleared land. Both sites are easily accessible by car, and are on slopes of less than 10%.

Treatments which will be tested include alley cropping, legume cover, residue management, and farmers' practices - in combination with different rice-based cropping systems. A total of 12 treatments will be tested in a randomized complete block design in three replications.

Five research officers and administrators will be involved in the project. All aspects of the project, including technical and financial considerations, have been finalized.

Malaysia: The experimental site is about 70 km from Kuala Lumpur. It is easily accessible by car and located on an old rubber plantation. The area has a slope of about 20%.

The main objective of this experiment is to determine soil fertility changes and the sustainability of intercropping under rubber. Five treatments related to farming systems under rubber will be tested. They are farmers' practices (rubber with minimum maintenance), low input (rubber + legume), high input I (rubber + maize and groundnut in sequence), high input II (rubber + banana and papaya), high input III (rubber + banana + maize and groundnut).

Five research officers will be involved in the project. All the details of the project, including the implementation time schedule, have been finalized.

Nepal: Two sites on farmers' fields located in Naldung Village, Panchayat, about 30 km east of Kathmandu, have been selected. Both sites are 1900 m asl and have a slope of about 50%. One site is an old terrace field and the other one is on newly cleared land.

The treatments which will be tested for reducing erosion and ensuring sustainable agriculture on these two sites are as follows: farmers' practices (conventional tillage), farmers' practices + mulching, minimum tillage + grass cover, and farmers' practices + recommended doses of fertilizer for the terraced site; farmers' practices (terracing), alley cropping, grass strips, hillside ditches, agroforestry, and intercropping for the newly cleared land.

Five research officers will be involved.

Philippines: Two sites with slopes greater than 30% have been selected. One site is in Batangas Province, and the other is in Tanay, Rizal Province, both located in southern Luzon.

Four treatments will be tested. These include farmers' practices (ploughing up and down the slope), alley cropping (*Glyricidia* and napier as hedgerows), alley cropping with high input, and alley cropping (banana + *Fapothilla* [chicu] as hedgerows). The layout will be in a randomized complete block design with three replications.

The Philippines Council of Agriculture, Forestry and Natural Resources Research and Development (PCARRD) will coordinate the work. Six research officers will be involved.

Thailand: Two sites have been selected, one in Chiang Mai and the other in Chiang Rai province. The two sites are easily accessible by car. The site in

Chiang Mai is 600 m asl and has a slope of about 30%, and the site in Chiang Rai is 900 m asl and has a slope of about 40%. The treatments which will be tested on the two sites are traditional cropping, alley cropping, a combination of perennial cash crops and annual subsistence crops, hillside ditches, and agroforestry. All the details of the treatments and the implementation schedule have been discussed and finalized. The details of disbursement of funds have also been finalized.

Seven research officers will be involved at the Chiang Mai site, and another three at the Chiang Rai site.

Conclusion

The project proposals for all the countries are now in good order. The fund disbursement schedule has been arranged, and the implementation timetable has been agreed. The selected sites will be fenced and cleared before the end of December, and the planting will commence early next year - except in Malaysia and Indonesia, where the planting will commence in December 1988.

Section 2: Agroecological approach

Agroecological approach to tropical highland research and development

H. HUIZING*

Abstract

Agroecology deals with agroecosystems which are parts of farming systems. Agroecosystems are evaluated in terms of productivity, stability, sustainability and practicability. For soil management research, the present agroecosystems are the starting point. Agroecological knowledge provides the basis for the selection of experimental sites, the design of experiments, and the prediction of experiment results. Current methods dealing with agroecosystems are 'diagnosis and design' methods as used in farming systems research and 'land evaluation' methods based on A framework for land evaluation *(FAO, 1976). Both methods have their strenghts and weaknesses. In this paper, a combination of the two methods is proposed as a systematic procedure for soil management research in tropical highland areas.*

Introduction

Human management changes natural ecosystems into agroecosystems. In these agroecosystems, solar energy, water, nutrients, labour, machine power, and material inputs are transformed into food, fibre, or fuel. Agroecosystems consist of two interacting components: crop or livestock on the one hand, and

---

* International Institute for Aerial Survey and Earth Sciences (ITC), 350 Boulevard 1945, B.P. 6, 7500 AA Enschede, The Netherlands.

their biophysical environment on the other. The biophysical environment includes all characteristics of the earth's surface that affect crop and/or animal growth, i.e. sunshine, rainfall, landform, soil, runoff/run-on, groundwater, flora, fauna, etc. Agroecology deals with the understanding of agroecosystems and with the prediction of their performance.

Agroecosystems are components of more comprehensive and complex agricultural systems. The analysis of agricultural systems is facilitated when they are treated as a hierarchy of subsystems (Fresco, 1986). Figure 1 is an example of such a hierarchy. It shows a regional system, farming systems, and agroecosystems. Each lower level system is part of a higher level system. Agroecosystems are placed at the lowest level. They are part of and receive inputs from the farming system. Farming systems generally contain more than one agroecosystem.

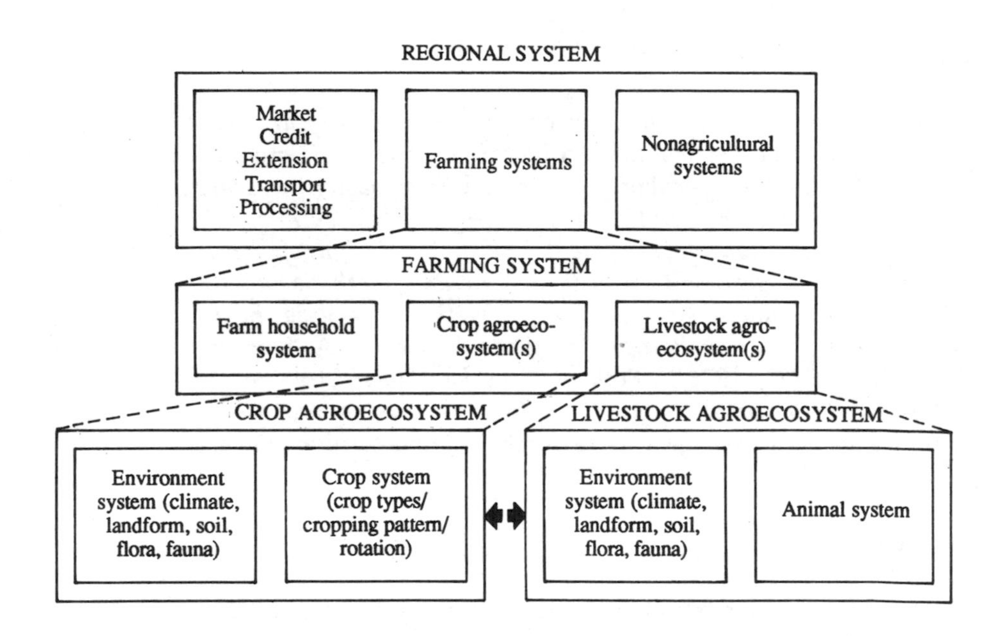

Figure 1. Agriculture as a hierarchy of systems (adapted from Fresco, 1986)

Man's major interest is in the "performance" of agroecosystems. The evaluation of the current or expected performance can be based on four aspects (Conway, 1985; Young, 1986):

- productivity,
- stability,
- sustainability, and
- practicability.

Practicability as used here takes into account constraints posed by the farming or regional system which affect the adaptability and acceptability of agroecosystems by farmers. For example, the productivity of maize-based cropping systems in parts of the Central Highlands in Thailand could be increased substantially by hand cultivation instead of downslope ploughing and by better weeding; but the practicability of these practices is constrained by on-farm labour availability: weeding, for instance, is required at a time when the same farmers plant their rice crop in adjoining valleys.

Agroecology is a necessary component of farming systems, cropping systems research, and agricultural (land-use) planning.

Agroecology and soil management research

The overall goal of the ASIALAND network is to develop productive, sustainable and practicable soil management technologies for newly cleared lands (Latham, 1987). This will be done by selecting, testing and evaluating improved soil management technologies. The evaluation will be based on productivity, stability, and sustainability criteria. The practicability of the technologies also needs to be assessed because it will affect the success of the introduction of the technologies in existing farming systems. In addition, practicability will be one of the requirements for the funding of soil management research projects (Nangju, 1987). The research projects should therefore be of a multidisciplinary character.

Basic agroecological considerations in the selection, design, and evaluation of technologies are:

- performance constraints of present agroecosystems;
- knowledge acquired by soil management experiments of national and international research centres/agencies; and
- prediction of the performance of agroecosystems after the introduction of improved or new crop and soil management technologies.

Present agroecosystems and their problems should be the starting point of soil management research (Latham, 1987). The development of improved technologies should reduce or eliminate these problems. Improvements based on adaptations of current farming practices, rather than the development of new technologies, will generally contribute to the practicability of the technologies. The prediction of the performance of agroecosystems must be based on a good understanding of agroecological processes. Prediction is an important aspect in (a) the design of improved technologies, and (b) guiding the application of the results of the research to environments where they are applicable.

Current methodologies that include assessments of agroecosystems and their performance are:

- the diagnosis and design (D&D) method as used in farming systems research; and
- agroecological characterization and land evaluation (LE) methods.

Diagnosis and design method

A farming systems research and experimentation programme is usually preceded by a "diagnosis" phase in which the structure and functioning of present farming systems is analyzed (Norman and Collinson, 1985; Norman, 1982). In the diagnosis, the performance of agroecosystems is assessed in relation to constraints at the regional and farming systems level (Figure 1). Characteristics of the biophysical environment are taken into account in the diagnosis. The diagnosis leads to the identification of groups of farmers with similar problems or constraints. The diagnosis is the basis for the design of on-farm experiments to test improved technologies that may reduce or eliminate these problems.

The diagnosis includes socioeconomic and biophysical aspects. In practice, however, the biophysical aspects are mostly covered in a generalized way only. The performance of agroecosystems is assessed in terms of inputs and outputs, but generally does not include an assessment of biophysical limitations of the site where the outputs are obtained.

Table 1 shows the ICRAF D&D methodology for the development of agroforestry systems that can help to solve current problems in annual crop agroecosystems (Young, 1986). The methodology includes three phases:

- A prediagnostic phase, which is mainly based on an analysis of existing information.
- A diagnosis phase, which includes field surveys to identify land-use problems. The surveys include farmers' interviews as well as site observations. The diagnostic phase includes an analysis of the causes of the problems (declining yields, declining soil fertility, low material inputs due to low cash income, etc.).
- A design phase which (i) identifies priorities with respect to interventions, (ii) specifies interventions and technologies, and (iii) designs a (mostly on-farm), research programme.

The results of successful experiments may be extended to farming systems with similar problems through extension services, programmes, or projects.

Table 1. Procedures in diagnosis and design for agroforestry.

Steps	PREDIAGNOSTIC PHASE
1	**Environmental description**
2	**Identification of existing land-use systems**
3	**Description of existing land-use systems**
	DIAGNOSTIC PHASE
4	**Diagnostic survey** Information needed for diagnosis of land-use problems and potential
5	**Diagnostic analysis** Analysis of land-use problems and potential
6	**Derivation of specifications for appropriate land-use systems**
	DESIGN PHASE
7	**Identification of feasible interventions** - listing of feasible interventions - selection of priorities
8	**Technology specifications** Derivation of technical specifications for each intervention, and arrangement in order of importance
9	**Technology design** Selection of plant species and management practices that best satisfy technical specifications
10	**Design evaluation** Analysis of selected land-use systems

Agroenvironment characterization and land-evaluation methods.

In the characterization of agroenvironments, only those properties of the environment that have a distinct impact on the performance of agroecosystems need to be assessed. This performance strongly depends on the type of crop system. The better the crop system is defined, the more specific must be the criteria for agroenvironmental characterization.

Small-scale agroecological characterizations are available for most Southeast Asian countries. Some are limited to climate (Eelaart, 1973; IRRI,

1974; Oldeman and Frère, 1982). Others also include soil aspects (FAO, 1980). Most of these characterizations are for broadly defined agricultural activities, i.e. annual upland cropping, wetland rice growing, etc. They indicate the scope for these activities for very large areas, without considering, for instance, climatic hazards, yield stability due to drought hazards, and local variability in soil distribution.

More detailed characterizations in the form of land capability or land suitability assessments are often available for provinces and project areas (scales 1:250 000 and larger). These characterizations are based on soil (and landform) surveys. Production and sustainability assessments in these characterizations are generally based on empirical information for the areas concerned.

Systematic land evaluations using principles and procedures of *A framework for land evaluation* (FAO, 1976) have been conducted for many areas in Indonesia, the Philippines, and Thailand. The generalized land evaluation procedures are shown in Figure 2. The evaluation takes into account government development objectives and socioeconomic aspects. The main procedure, however, is the comparison of present or improved land-utilization types (specified in technical and management terms) with tracts of land (described in terms of environmental attributes) and the assessment of the suitability of land in terms of productivity and sustainability. This assessment requires an understanding of present agroecosystems and/or a prediction of the performance of the systems after the introduction of improved/new technologies.

In most land evaluation studies, land-use types and their agronomic and management attributes are described in a generalized way only. A constraint in many land evaluation methods is that the knowledge on land-use performance in relation to different land attributes is insufficient.

An agroecological approach for soil management research in tropical highlands: a proposal.

The following stepwise procedure for soil management research in tropical highlands contains elements of the diagnosis and design methods, land evaluation methods, and methods applied by CIAT to define microregions relevant to research on cassava production (Carter, 1986). It also includes socioeconomic aspects which are not part of agroecology proper. Socioeconomic aspects, however, have a great influence on the practicability of the agrotechnologies to be tested. A summary of the procedure is shown in Table 2. The procedure is considered as an interactive process: in each following phase, (preliminary) assessments made in previous steps need to be adjusted based on the new information that has become available. The procedure contains the following phases (steps): outlined in Table 2, which are discussed in more detail in the subsequent sections.

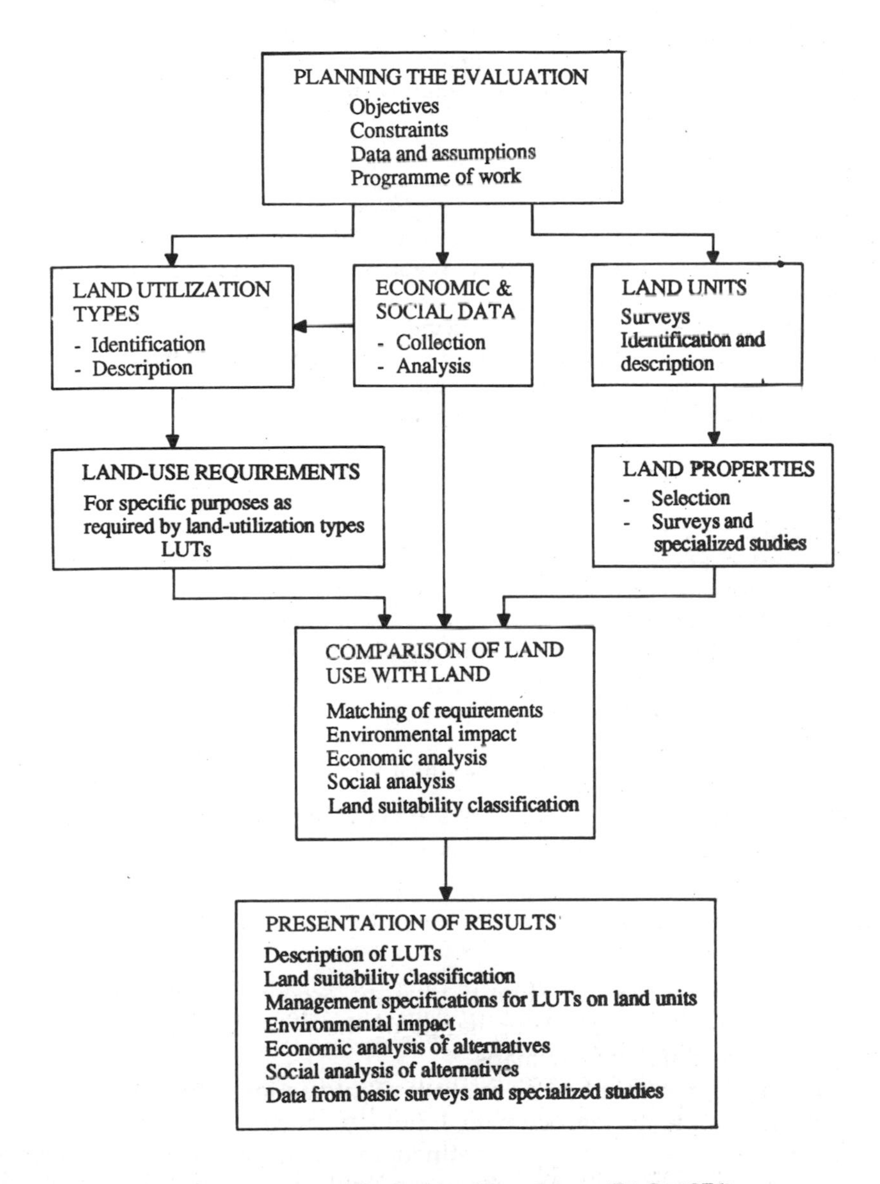

Figure 2. Procedures in land evaluation (adapted from FAO, 1976).

Table 2. An agroecological procedure for soil management research in tropical highlands: a proposal.

Regional description

1. Description of the main present agroecosystems in the region under consideration:
   i. description of the agroenvironment
   ii. description of the agricultural land use, management, and recent trends
   iii. interactions between (i) and (ii).

Diagnosis

2. Identification of the key constraints (emphasis on crop and soil management)
3. Field check on constraints, their severity and the extent and scope for improvements
4. Preliminary list of possible soil management technologies
5. Selection of priority areas for research within the region concerned
6. Field survey on cropping and farming systems in priority areas
7. Selection of relevant, practicable technologies

Design

8. Site selection for experiments and/or trials
9. Site characterization
10. Design of experiments/trials
11. Conducting the experiments/trials

Evaluation and implementation

12. Evaluation of performance
13. Planning and implementation of the transfer of successful technologies to other areas.

Regional description

The regional description should give an account of the main present agroecosystems in the tropical highlands of the region concerned, based on existing statistical data and/or maps.

1. This description is based on an analysis of interactions between:
   - agroenvironments: agroclimate, lithology, landform and soil, and
   - agricultural land-use types defined in terms of crops, technical and management attributes and changes over the years.

The description can be achieved by combining available data on agroclimate and agroecological zones, detailed reconnaissance soil and landform maps, and land-cover/land-use data by overlaying and identifying land-use trends (planted areas, yields). Satellite and airphoto data may be used for land-use change analysis and updating of land-cover/-use data when available data is inadequate.

Diagnosis

The diagnosis should include:

2. Identification of key constraints (with emphasis on soil management), e.g. soil acidity, soil fertility decline, soil erosion, clearing and postclearing technology, etc. that significantly affect the productivity, stability, or sustainability of the present agroecosystems. Key constraints should be ranked according to extent and expected severity. If possible, agroecosystems with different levels of constraints on a map should be indicated and delineated.
3. Field check, including (i) discussions with local, knowledgable persons (soil scientists, extension staff, etc.) on the key constraints identified in step 2 and the scope for their improvement, followed by (ii) cross-checking with a limited number of village heads, farmers, and site observations in areas with agroecological constraints.
4. Identification and listing of possible crop and soil management technologies that can reduce or remove the constraints identified in steps 2 and 3, e.g. clearing and tillage methods, cropping patterns that will provide a better cover in periods with high rainfall intensities, introduction of tree (crop) components in agroecosystems, use of mulches, water conservation techniques, control of burning, etc. The listing of improved crop and soil management technologies should be based on (i) performance prediction on the basis of known agroecological processes; (ii) results of research conducted in the area or in other areas with a comparable agroenvironment; and (iii) local information and opinions (see step 3).
5. Selection of priority areas for research should be based on the severity and extent of production/stability/sustainability problems. Areas where the agroenvironment is such that technologies identified in step 4 are evidently physically or economically not feasible (e.g. very steep slopes, highly erodible soils, shallow stony or gravelly soils, etc.) should be avoided.
6. Multidisciplinary field surveys in priority areas by a team consisting of soil scientists, agronomists and agroeconomists on cropping and farming systems covering on-farm resources (land, labour, capital); current agro-

ecological, on-farm, and regional constraints; farmers' needs and priorities; and response of farmers with respect to acceptability/practicability of possible soil and crop management technologies (identified in step 4).

7. Selection of technologies for experimentation that are physically relevant and socioeconomically practicable (based on step 4 and step 6).

Design

The design should include:

8. Site selection for experimentation. Existing maps and airphotos should be used to select representative sites based on landform/soil relationships.
9. Site characterization based on agroenvironmental properties that are expected to influence the productivity, stability, and sustainability of the technologies to be tested.
10. Experimental design appropriate to the needs of the treatments and site.
11. Conducting experiments. This aspect of the project should preferably include both experiments under the control of the researchers and trials under the control of the farmers.

Evaluation and implementation

12. Evaluation of experiments and trials should be made in terms of productivity, stability, sustainability, and practicability.
13. Planning and implementation of the transfer of successful technologies to other areas should be an important component.

Conclusion

The main conclusions are:

- Present farming and its constraints should form the starting point of any soil and crop management research.
- In order to arrive at technically feasible and practicable soil management technologies, all factors and processes that are current constraints to the performance of agroecosystems have to be considered. Many factors may be involved. However, understanding and further analysis of a limited number of key constraints is generally sufficient.
- Both diagnosis and design methods and agroecological characterization and land evaluation methods can support soil management research. Both

methods have their strengths and weaknesses. In D&D, the agro-environment is mostly not taken into account in sufficient detail. In land evaluation, technology and management are often considered in generalized terms only.

- Any approach to soil management research should be iterative: preliminary ideas on relevant research options have to be supported by further data collection; the analysis of new data, however, often makes it necessary to reformulate the preliminary ideas.

References

CARTER. 1986. Collecting and organizing data on the agro-socioeconomic environment of the cassava crop: case study of a method. In: *Agricultural Environments* ed. A.H . Bunting. Wallingford, UK: CAB International.

CONWAY, G.R. 1985. Agricultural ecology and farming systems research. In: *Proceedings of an international workshop,* (Richmond, N.S.W., Australia).

EELAART, A.L.J. van den. 1973. *Climate and crops in Thailand.* Rep. SSR-96. Bangkok: DLD.

FAO, 1976. *A framework for land evaluation.* FAO Soil Bulletin no. 32. Rome: FAO.

FAO, 1980. *Report of the Agroecological Zones Project.* World Soil Resources Report 48/2. Methodology and results for S.E. Asia. Rome: FAO.

FRESCO, L.O. 1986. *Cassava in shifting cultivation. A systems approach to agricultural technology development in Africa.* Amsterdam: Royal Tropical Institute.

IIRI. 1974. *Agroclimate map of the Philippines.* Los Baños, Philippines.

LATHAM, M. 1987. The IBSRAM land development and soil management network programme in monsoon Asia. In: *Soil management under humid conditions in Asia and the Pacific,* 13-22. IBSRAM Proceedings no. 5. Bangkok: IBSRAM.

NANGJU, D. 1987. Soil management research: a suggested approach. In: *Soil management under humid conditions in Asia and the Pacific,* 69-95. IBSRAM Proceedings no. 5. Bangkok: IBSRAM.

NORMAN, D.W. and COLLINSON, J. 1985. Farming systems research in theory and practice. In: *Agricultural Systems Research for Developing Countries.* ed. J.V. Remeni. ACIAR Proceedings no. 11. Canberra, Australia: ACIAR.

NORMAN, D.W. 1982. *The farming systems approach to research.* FSR Paper Series no. 3, Kansas, USA: Kansas State University.

OLDEMAN, L.R. and FRERE, M. 1982. A study of the agroclimatology of the humid tropics of S.E. Asia. Rome: FAO.

YOUNG, A. 1986. Land evaluation and agroforestry diagnosis and design: towards a reconciliation of procedures. *Soil Survey and Land Evaluation* 5(3).

Agroecosystem analysis for on-farm research in the Chiang Mai valley

METHI EKASINGH*

Abstract

Agroecosystem analysis was developed to serve as a framework for a multidisciplinary team of researchers, planners, and agricultural development workers to interact in a series of seminars and workshops. This system of analysis was designed to characterize the target area and to generate key questions or guidelines which would lead to relevant research and development, and bring about a significant improvement in productivity, stability, sustainability and equitability of agricultural systems. The procedure for agroecosystem analysis begins by setting the objective of the workshop, and explaining system definition and pattern analysis, which are the main activities carried out by the participants.

Multidisciplinary interactions during the workshop will generate well-focused key questions or guidelines that can be used to formulate testable hypotheses. Since the analysis cuts across various system hierarchies, farm-level research initiated from the key questions can be linked to the broader system, and the research results will have a significant impact on the target agroecosystem as a whole, and not only on the individual farm.

This paper briefly describes such procedures and provides some of the examples from the agroecosystem analysis of the Chiang Mai valley. It illustrates how the outcome of the analysis was used to formulate on-farm yield constraints to soybean production.

* Department of Soil Science and Conservation, Chiang Mai University, Chiang Mai, Thailand.

Introduction

The formulation of an agricultural research programme that aims at developing technology to suit specific environments requires a knowledge of existing farming systems in order to prioritize research efforts. In recent years, a number of techniques for site description have been employed. They range from quick, qualitative and low cost to time-consuming, extensive and expensive techniques. Among them is agroecosystem analysis (AA), which was first developed and used in Chiang Mai (Gypmantasiri *et al.*, 1980), and later implemented in many research and development projects both in Thailand (KKU-Ford Cropping Systems Project, 1982) and elsewhere (KEPAS, 1984; KEPAS 1985; Conway and Sajise, 1986). The paper briefly discusses the procedure for agroecosystem analysis with examples drawn from the Chiang Mai case study. Details on the concept and methodology of AA for research and development have been elaborated in Conway (1985) and Conway (1986). This paper also illustrates how the output from AA could be used to help formulate an on-farm experiment on soybean yield constraints in an irrigated area in the Chiang Mai valley.

Concept of agroecosystem analysis

Agroecosystem analysis is based on the concept that agroecosystems were modified from natural ecosystems through the process of agricultural development for the purpose of food and fibre production. These systems are organized in a hierachical level with well-defined boundaries. An example of hierachical organization of an agroecosystem as illustrated by Hart (1986) is shown in Figure 1.

An agroecosystem is assembled by combining numerous elements and evaluating those elements and their interactions. The behaviour or the output of the system can be differentiated into four distinct system properties. They are:

- *Productivity:* the quantity of products per unit of resource, usually measured in terms of yield or net income per hectare.
- *Stability:* the variability of productivity when the system is subjected to normal fluctuation of climate or economic conditions.
- *Sustainability:* the ability of the system to maintain productivity in face of repeated stress and perturbation.
- *Equitability:* a measure of how evenly the productibility is distributed among the human components of the system.

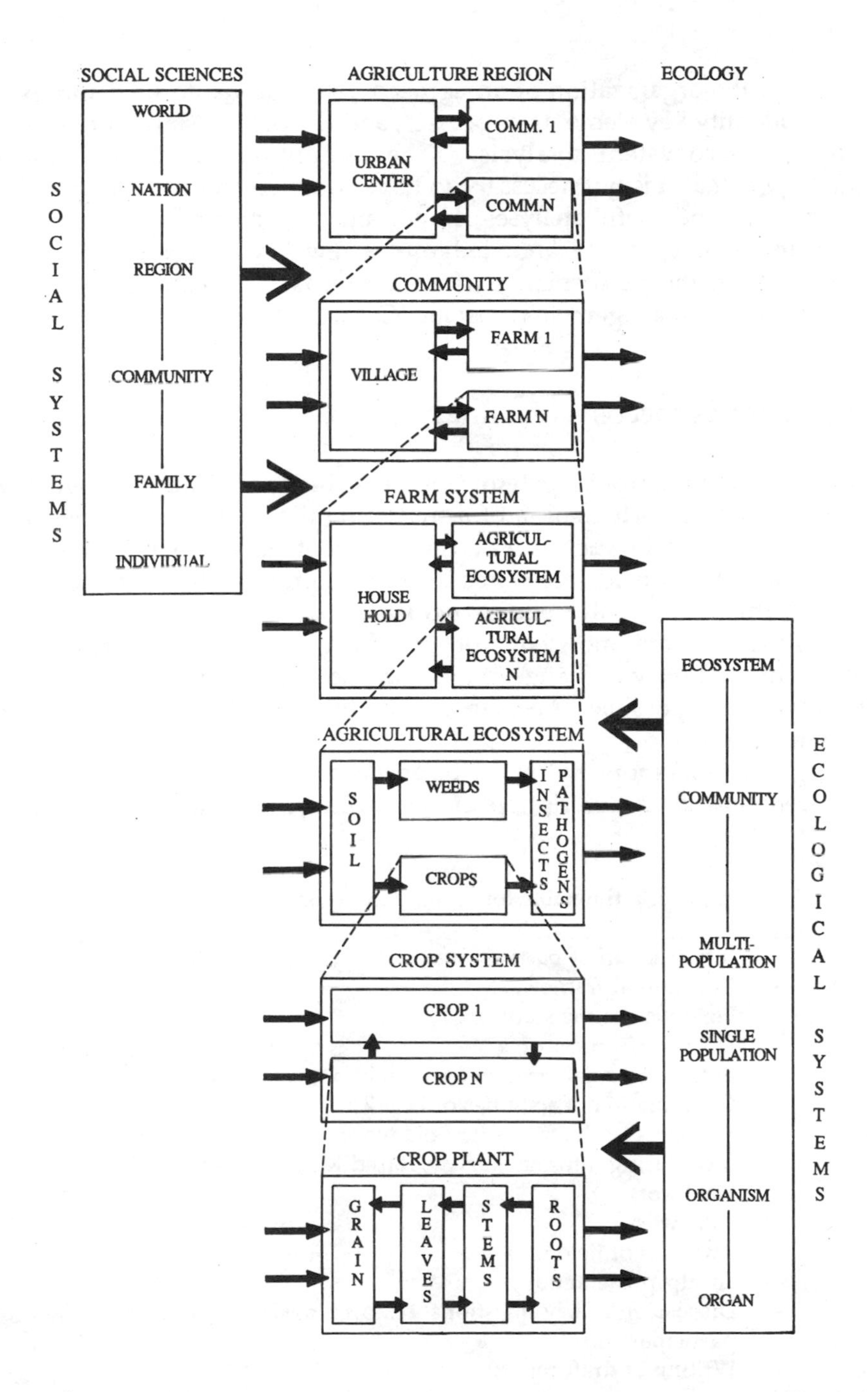

Figure 1. Conceptual framework for a hierarchy of agricultural systems (Hart, 1986).

Although the organization of an agroecosystem seems to be complex, it is possible to identify key elements, processes, and the performance of the systems by using agroecosystem analysis. The assumption necessary for this methodology is that it is not necessary to have complete information in order to produce realistic and useful analyses. Important properties of an agroecosystem could be understood by the knowledge of a few key processes. Significant improvements in the performance of an agroecosystem can be achieved by changing a few key management decisions (Conway, 1985).

Procedure for agroecosystem analysis

Analysis of an agroecosystem can be achieved through a workshop involving the active participation of a multidisciplinary team. The workshop should be organized in such a way that relevant data and information are prepared and transformed into maps, transects, graphs, diagrams, and other graphical forms. This will help participants express key relationships and issues derived from the knowledge of their own disciplines to participants of other disciplines. Lively exchange of opinion and better communication among disciplines are then possible. An example of a timetable for a 10-day workshop is shown in Table 1.

During the workshop, the steps in agroecosystem analysis as shown in Figure 2 are followed. A description and examples of each step are given below.

Table 1. An example of a timetable for an agroecosystem workshop.

Day 1	-	Introduction of participants
	-	Conceptual framework
	-	Field visit to the study area
Day 2	-	Details of procedures
	-	Break into small groups for pattern analysis of secondary data
Day 3	-	Continuation of activities of Day 2
Day 4	-	Summary of the topics for field work
	-	Introducing informal survey, Rapid Rural Appraisal (RRA)
Day 5	-	Field work
Day 6	-	Field work
Day 7	-	Analyses of field data
Day 8	-	Group presentation
Day 9	-	Discussion of key questions, research design, and implementation (all participants)
Day 10	-	Writing of draft report

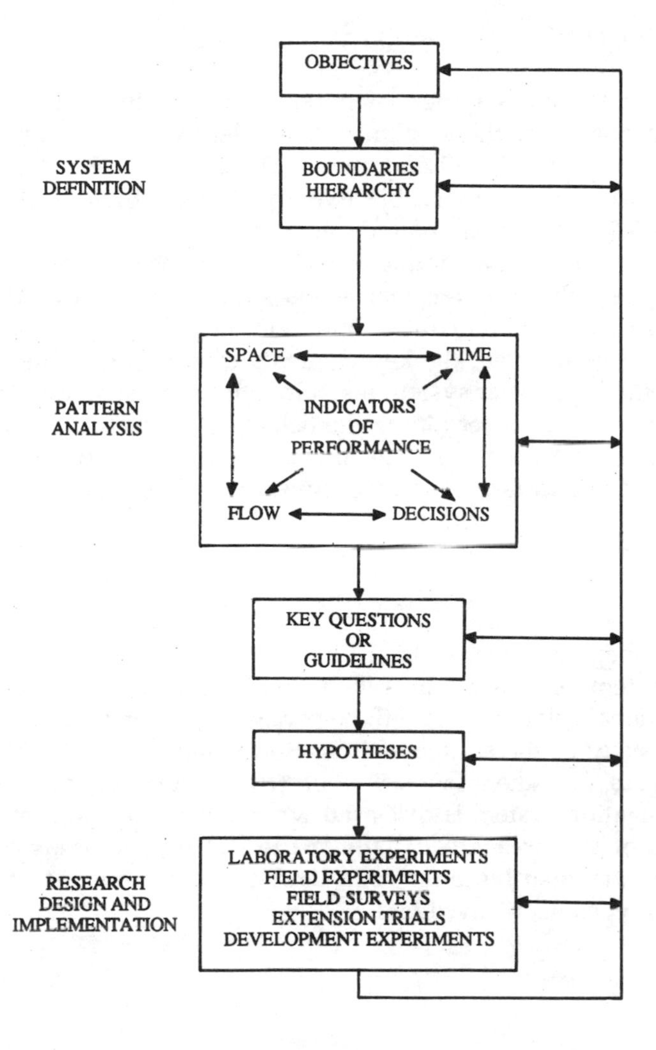

Figure 2. The procedure for agroecosystem analysis (Conway, 1985).

Objective

The first step in the analysis is to set an objective. The objective has to be as precise as possible to direct the analysis to the goal. An example of the objective of a workshop is:

"To identify research priorities for improving productivity of irrigated farms in the Mae Taeng Irrigation Project".

System definition

Once the objective is set, the system boundary and hierachy can be defined. In the above objective, the system can be clearly defined by the physical boundaries i.e. head work, tail end, main irrigation and drainage canals of the irrigation project, etc. Other objectives may need biological or economic conditions to define the system boundary.

Apart from the system boundary, the hierachy of the system must be defined. At least three hierachical levels should be identified: the system level of interest, and the one above and below it. The analysis of the higher system hierachy will reveal the key elements that influence the system being focused on, while the lower system hierachy will increase understanding of the mechanism involved in determining the behaviour of the target system. In the above objective, the irrigation system is the focus of the analysis, while Chiang Mai valley and farms are respectively the levels above and below it.

Space

Thematic maps and transects are efficient ways of displaying the spatial pattern of system components. When thematic maps are overlayed, some obvious functional relationships will be revealed. Thematic maps showing the cropping intensity of the source of irrigation water in Chiang Mai valley are given in Figure 3. The association of triple cropping systems with the traditional irrigation system is disclosed when these maps are overlayed.

Transects of the area are usually made during the analysis, because it displays the spatial distribution of soils, land use, and problems encountered on different land types (Conway, 1986).

Time

Graphs showing the seasonal and long-term changes in climatic parameters, crops, labour, prices, soil fertility, and other dynamic components are helpful in detecting potential problems and the changing trends in the system. The pattern of flow rate in the main canal of Mae Taeng irrigation system is shown in Figure 4. It demonstrates the typical variation in the availability of irrigation water of most irrigation systems in northern Thailand. This information suggests that the irrigation canal is closed twice a year. The system was designed for full support of paddy rice during July to December, and partial support for the second crop(s) grown after rice during January to April. Irrigation water is not adequate for growing a third crop during May to July.

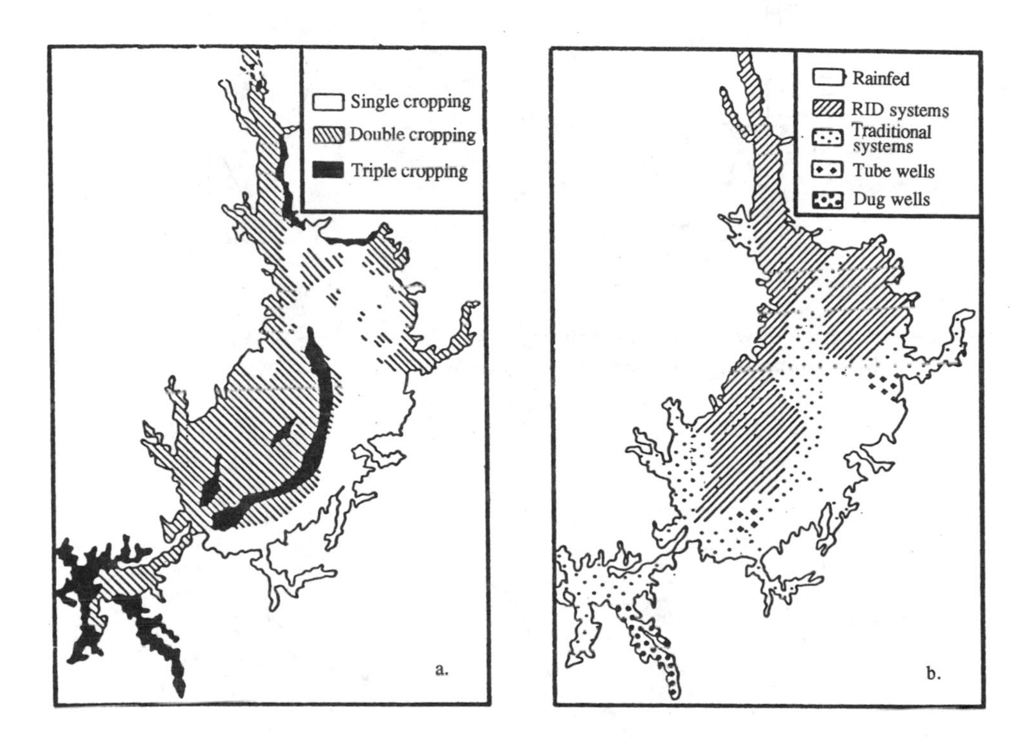

Figure 3. Thematic maps of (a) cropping intensity (b) government (Royal Irrigation Department) and nongovernment irrigation systems in the Chiang Mai valley (Gypmantasiri *et al.*, 1980).

Long-term changes in the cultivated area of the second crops grown after rice in the same irrigation system are shown in Figure 5. This graph can stimulate discussion among the participants in a workshop on the reasons behind the decline in the cultivated area of the second rice crop, and the substitution of soybean in that area while the planted area of the other crops remained the same.

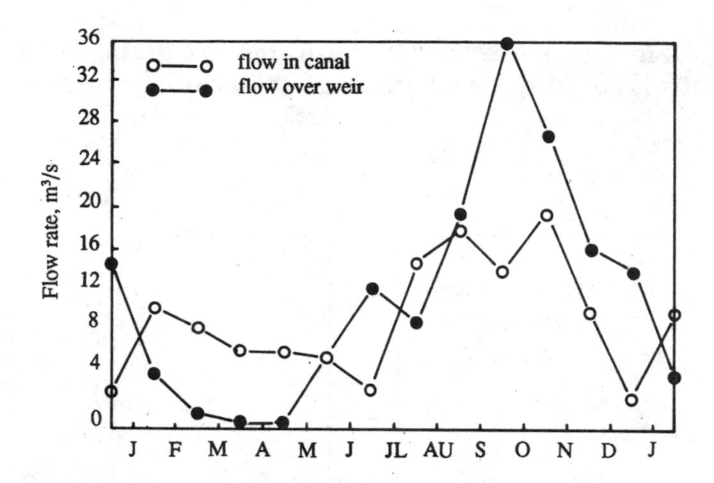

Figure 4. Average flow rate in the main canal and over the weir at the head work of the Mae Taeng irrigation system, Chiang Mai.

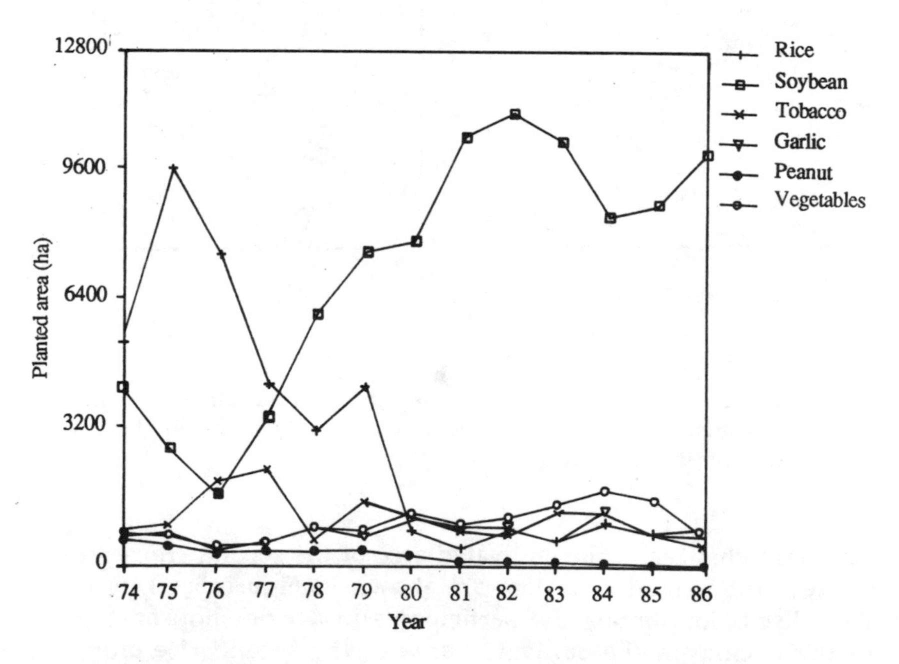

Figure 5. Changes in the cultivated area of major crops grown after rice during the dry season in the Mae Taeng Irrigation Project.

Flow

The flow of materials and information in the system are useful for tracing the bottleneck of input supplies and marketing channels, or the constraints in the process of information and technology transfer. The flow diagram should be simple, and include only important components (Figure 6).

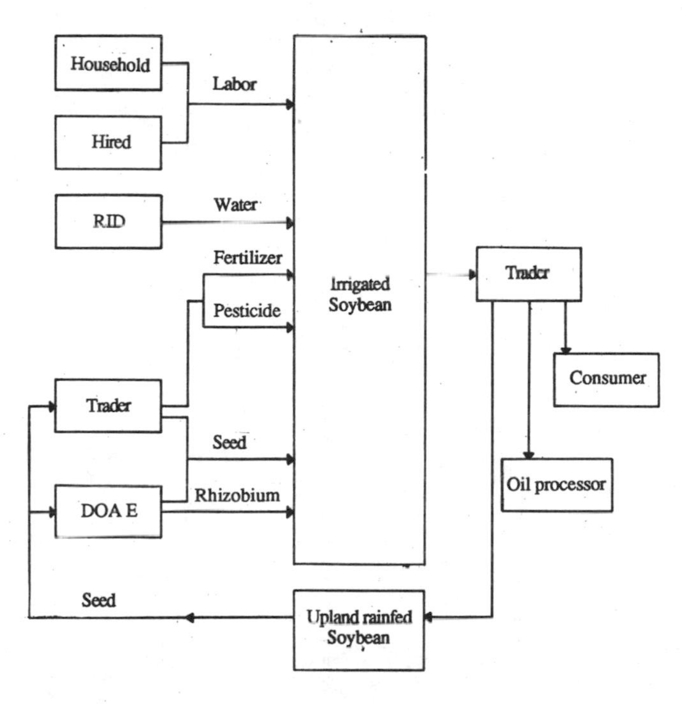

Figure 6. Flow of materials in the soybean production system in Chiang Mai.

Decisions

Since productivity may be improved by changing the key decision making of those involved in the system, the pattern of decisions should be analyzed and shown as a decision tree (Conway, 1986), and as indicated in Figure 7. Farmers in the Chiang Mai valley chose to grow soybean as a second crop mainly because this crop requires less cash inputs, labour and water than other crops grown during the same period. The resource requirements of soybean are compatible with the available resources of the farmers in the area. The farmers' decision on the types and quantity of input for a particular crop can be further analyzed, and the results will be useful in improving the productivity of the farm.

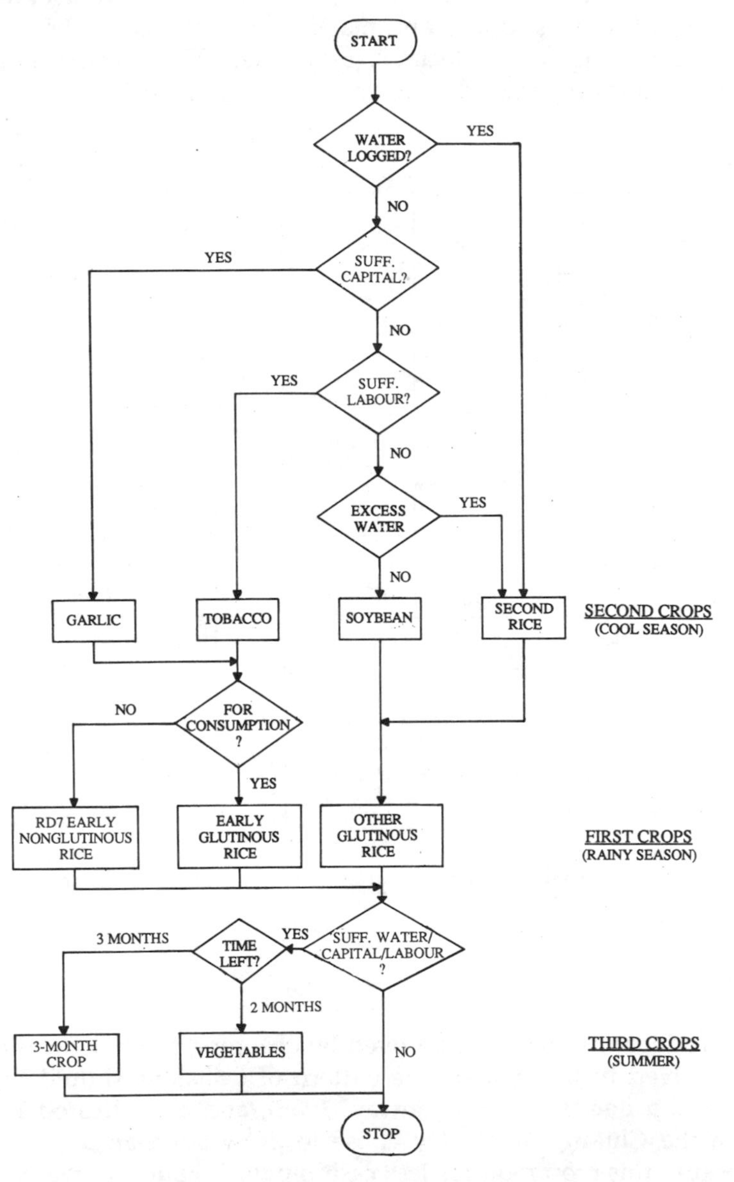

Figure 7. A simple farmers' decision-making model in selecting crops and cropping systems in the Mae Taeng irrigated area.

Key questions

During the discussion in each step of the procedure, many questions may emerge. Participants will have to select the key questions that lead to further research or guidelines for significant improvement in the productivity, stability, sustainability, or equitability of the system. Key questions should be well focused and lead to the formulation of hypotheses which can be tested. Some of the key questions which arose from an analysis of the agrosystem in Chiang Mai valley (Gypmantasiri *et al.*, 1980) were: "What is the best time to close irrigation systems for maintenance so as to improve cropping system options?" and "How can the yield constraints of the rice-soybean subsystem be alleviated in order to improve the productivity and stability of the system?". Subsequent research activities initiated from the latter question will be used as an example, and are briefly described in the next section.

Research design and implementation

To resolve a key question, a hypothesis is formulated and further field surveys, experiments, or laboratory investigations are conducted by a multidisciplinary team of scientists. For example, the key question on the productivity of the rice-soybean subsystem in the Chiang Mai valley was pursued. A multidisciplinary team was formed to conduct an exploratory survey, followed by a formal survey using questionnaires which examined key agronomic and socioeconomic issues such as water availability, cultivation methods, utilization of inputs, the past performance of soybean, and common problems in soybean production. A sampling of the crops from the farmers' fields was also conducted to verify data on plant population, yields, and the yield components of soybean. The study was confined to the Mae Taeng Irrigation Project, where rice-soybean was found to be a predominant system from earlier agroecosystem analysis.

The results from the exploratory and formal survey revealed that most farmers did not plough their land prior to planting. The most common varieties of soybean grown were SJ4 and SJ5, although SJ5 is more dominant in the Hang Dong and Sanpatong areas. Row planting and rice-stubble planting were practiced equally in the study area. The seeding rate and plant density were significantly higher than those recommended by the Department of Agricultural Extension. Inappropriate seed quality also gave rise to poor stands, which in turn contributed to the high seed and production costs. Infestation by narrow-leaved weeds was found to be a serious problem for soybean production, but only half of the farmers controlled the weeds by using herbicides, or by hand weeding, or both.

Almost half of the farmers inoculated seeds with rhizobium, and a large percentage used either organic or chemical fertilizers. The most common fertilizer used was grade N:P_2O_5 of 16:20, but the rate of N-P_2O_5-K_2O applied was significantly less than that recommended. The efficiency of fertilizers was also markedly decreased by the incorrect placement of fertilizers on the soil surface by most farmers.

Insects caused moderate to severe damage to soybean, and most farmers used insecticides - although the amount and frequency of spraying varied widely from one farmer to another.

From the results of this study, it is believed that weed control, fertilizer use and plant density were factors that contributed to the yield gap between farmers' fields and experimental plots. Although this gap was evident from the yield obtained from crop sampling of soybean in 57 farmers' fields, subsequent on-farm experiments were set up in order to quantatively measure the contribution of each of those factors to the yield gap.

A design for an on-farm experiment involved twenty farmer cooperators. Two levels of technology, i.e. improved technology and farmers' practices were tested for plant density, fertilizer use and weed control. The physical and biological environment of the farms, as well as current farm practices and socioeconomic data, were monitored. The yields were measured for each plot, and the yield gap and the contribution of each factor (including their interaction) were analyzed using the method described by De Datta *et al.* (1978). The marginal benefit-cost ratio over the farmers' practices was also evaluated for each factor. The evaluation of the contribution of each tested factor will help improve recommendations for soybean production to suit the farm resources in this area.

References

CONWAY, G.R. 1985. Agricultural ecology and farming systems research. In: *Australian systems research for developing countries*, ed. J.V. Remenyi, 43-59. Canberra: ACIAR.

CONWAY, G.R. 1986. *Agroecosystem analysis for research and development.* Winrock International Institute for Agricultural Development.

CONWAY, G.R. and SAJISE, P.E. 1986. *The agroecosystems of Buhi: problems and opportunities.* Los Baños, Philippines: PESAM, University of the Philippines.

DE DATTA, S.K., GOMEZ, K.A., HERDT, R.W. and BARKER, R. 1978. *A handbook on the methodology for an integrated experiment survey on rice yield constraints.* Los Baños, Philippines: IRRI.

GYPMANTASIRI, P., WIBOONPONGSE, A., RERKASEM, B., CRAIG, I., RERKASEM, K., GANJAPAN, L., TITAYAWAN, M., SEETISARN, M., THANI, P., JAISAARD, R., ONGPRASERT, S., RADANACHALESS, T. and CONWAY, G.R. 1980. *An interdisciplinary perspective of cropping systems in the Chiang Mai valley: key questions for research.* Chiang Mai, Thailand: Faculty of Agriculture, University of Chiang Mai.

KEPAS. 1985. *The upland agroecosystems of East Java.* Jakarta, Indonesia: Kelompok Penelitian Agro-Ekosistem, Agency for Agricultural Research and Development.

HART, R. 1986. Ecological framework for multiple cropping research. In: *Multiple cropping systems,* ed. C.A. Francis, 40-56. New York: Macmillan.

KEPAS. 1984. *The sustainability of agricultural intensification in Indonesia: A report of two workshops of the Research Group on Agroecosystems.* Jakarta, Indonesia: Kelompok Penelitian Agro-Ekosistem, Agency for Agricultural Research and Development.

KEPAS. 1985. *The upland agroecosystems of East Java.* Jakarta, Indonesia: Kelompok Penelitian Agro-Ekosistem, Agency for Agricultural Research and Development.

KKU-FORD CROPPING SYSTEMS PROJECT. 1982. *An agroecosystem analysis of northeast Thailand.* Khon Kaen, Thailand: Faculty of Agriculture, Khon Kaen University.

Land evaluation and soil management research

H. HUIZING*

Abstract

In land evaluation, the suitability of land is determined for land-use types that are specified in terms of technology (crops, crop varieties, inputs) and socioeconomic attributes (e.g. labour, capital, and infrastructure requirements). Agroecological knowledge as well as socioeconomic information is required for such a suitability assessment. Land evaluation and soil management research are complementary. Land evaluation can contribute to soil management research by:

- *the identification of soil management problems;*
- *the seclection of priority areas and sites for soil management research; and*
- *the finding of areas where improved or new technologies developed by this research are applicable.*

Introduction

Various land evaluation systems exist. A widely used system is the USDA land capability classification system (Klingebiel *et al.*, 1961) which is based on the interpretation of climate and soil survey data for broadly defined land uses (cultivation, grazing, forestry). Local versions of this system have been

---

* International Institute for Aerial Survey and Earth Sciences (ITC), 350 Boulevard 1945, B.P. 6, 7500 AA Enschede, The Netherlands.

developed in several S.E. Asian countries to suit their specific environmental conditions.

In the early 1970's, an attempt was made to develop a more universal approach to land evaluation (Brinkman and Smyth 1973). This attempt resulted in a *Framework for land evaluation* (FAO, 1976). Guidelines for the evaluation of land for rainfed agriculture, forestry, irrigated agriculture and extensive grazing have been developed since then (FAO, 1983, 1984, 1985, and 1988 respectively). The FAO approach to land evaluation differs from previous land evaluation systems in several aspects. A major difference is that land, according to the FAO framework, is evaluated for current or improved uses that are specified in terms of technology, management, and socioeconomic attributes.

Land evaluation based on principles and procedures of the FAO framework is currently applied by various government agencies (CRS, Indonesia, BS Philippines, DLD Thailand) and consultant groups in Southeast Asia. The results of the land evaluations are used for land-use planning and regional agricultural planning. One of the aims of the land evaluations is often to guide results of agricultural research (e.g. intensified cropping patterns based on the use of improved, early-maturing crop varieties) to tracts of land where these results are applicable.

In the following sections,the FAO approach to land evaluation will be discussed in relation to crop and soil management research.

Principles and procedures of land evaluation

Land evaluation is the process of assessing the suitability of land for alternative, specified land-use types. This process includes: (i) selection and description of current or improved land-use types (LUTs) that are relevant to the area under consideration; (ii) land resource surveys; and (iii) the actual suitability assessments.

The basic principles fundamental to suitability assessments (FAO, 1976) are:

- the selected LUTs are relevant to the local, physical, and socioeconomic context;
- the LUTs are specified in technical and socioeconomic terms;
- the evaluation involves the comparison of two or more LUTs;
- land suitability refers to use on a sustained basis;
- land suitability includes a comparison of yield and inputs; and
- land evaluation requires a multidisciplinary approach.

Procedures involved in land evaluation adapted from FAO (1984) are given in Figure 1.

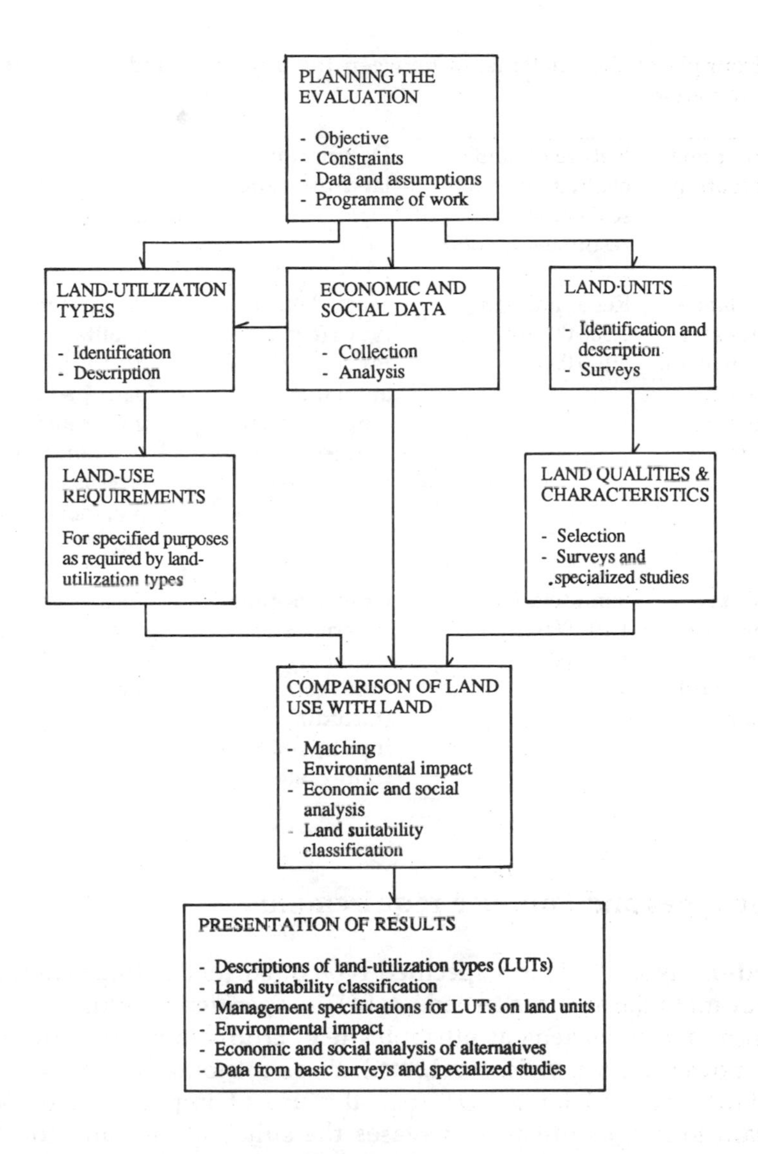

Figure 1: Land evaluation procedures .

The detail and accuracy of information provided by land evaluations depend on the objectives of the land evaluation. Depending on these objectives, land evaluation may be of an exploratory, reconnaissance, semidetailed or detailed nature (for reconnaissance and semidetailed land evaluations, see Table 1).

Table 1. Example of the relationship between the objectives and the nature of land evaluation.

Objective/context of land evaluation	Nature of land evaluation and the scale of land resources map(s)	Description of land-use types	Terms in which assessments are made
Selection of land-use priorities and promising areas for development in (larger) regions; choice of project location	Reconnaissance; 1:500 000 to 1:100 000	Major land-use types (e.g. forestry for timber; rainfed annual crops); or crops with broadly defined management levels	Performance in qualitative terms (good, moderate, poor); sometimes inputs and outputs in quantitative physicial terms (kg, mandays, etc. in ranges per ha)
(Pre-) feasibility study for districts, projects, programmes; project identification	Semidetailed; 1:100 000 to 1:25 000	Crops/cropping paterns, specified technology and management (including. levels of inputs, practices, timing, etc.)	Performance in quantitative physical and/or financial terms

Land-use types and land use requirements

A land-use type (LUT) is a specified type of current or improved land use that is relevant to the area concerned. A LUT is specified in terms of:

- technical and management attributes (e.g. crop/cropping pattern, inputs used, power source, yield level, etc.); these attributes often influence the suitability of land for a LUT (e.g. the use of improved crop varieties resistant to fungus diseases increases the suitability of land for the crop species in more humid areas at higher elevations; contour planting and limited tillage reduce erosion and thereby may improve the sustainability of the use); and
- socioeconomic attributes (e.g. labour and capital use, farm size, infrastructure) which describe the farming system(s) and the facilities provided by the regional system (e.g. transport, input supply, extension, credit) for which the LUT is considered.

Land-use requirements are the specific land conditions required for the proper functioning of the LUT. General requirements for crop production, are shown in Table 2.

Table 2. Major land-use requirements for rainfed cropping.

1. Ecological (growth) requirements
   - suitable climatic regime
   - foothold for rooting
   - availability of soil moisture
   - availability of oxygen in root zone
   - availability of nutrients
   - absence of hazards such as damaging floods or storms
   - absence of salinity and alkalinity
2. Management requirements
   - easy clearance
   - soil workability
   - accessibility to fields
3. Conservation requirements
   - resistance to erosion

A major limitation in land evaluation is the lack of adequate quantitative knowledge with respect to the mentioned requirements, particularly in tropcal countries. These include questions such as: How much moisture, nutrients, or foothold is required for the LUT to be productive? What should be the resistance to erosion of a tract of land for sustainable use? A proper assessment of land-use requirements must be based on knowledge of "yield-management-land quality" relations as obtained through experiments and trials, farmers' knowledge and experience, and field observations; and/or through modelling approaches of crop growth and land degradation.

Land resources

The term "land" in land evaluation refers to aspects of the biophysical environment that significantly influence the performance of the use. "Land" as used in land evaluation, and "environment" as used in (agro-) ecology are therefore synonyms. The main types of information on land resources needed for land evaluation for agricultural purposes are on agroclimate, surface and groundwater resources, landforms, soils and land cover/land use.

Land resources are generally described in terms of land characteristics, i.e. biophysical properties of the land that can be estimated or measured. Land characteristics are also used to assess "land qualities", i.e. properties of the land that have a known (or supposed) effect on the performance of a land use. Land qualities are expressed in the same terms as land-use requirements (Table 2): land-use requirements express the conditions required by the LUTs; land qualities show the actual conditions provided by the land. For the assessment of a land quality, one land characteristic or a combination of characteristics is used. The combination of land characteristics (e.g. rainfall, soil texture, soil structure, organic matter, and soil depth) for the assessment of the land quality - in this case "available soil moisture" may be done empirically or through "transfer functions" (Bouma and Van Lanen, 1987); transfer functions are mathematical expressions relating different land characteristics to a land quality.

Land characteristics and qualities vary in time and space. With respect to temporal aspects of land characteristics, subdivisions can be made into: (i) relatively stable land characteristics (e.g. landform, soil texture, subsoil pH and CEC, available water capacity), (ii) characteristics that fluctuate over the years (e.g. start of rainy season, drought periods, incidence of pests and diseases). Field surveys, generally supported by the interpretation of airphotos and/or satellite images, are needed to describe the spatial variability. The delineation of mapping units during these surveys is mainly based on relatively stable land characteristics.

Land suitability

Land suitability is the fitness of a tract of land for a specified LUT. It is based on the comparison of the requirements of a LUT with the qualities of the land (Figure 1). The term "land suitability" suggests that only land is evaluated. The actual assessment of suitability, however, is based on a prediction of the performance of a LUT (crop system with defined management) on a tract of land with known characteristics or qualities. In land suitability assessments, LUT/land combinations are the subject of the assessments, rather than tracts of land alone. A LUT/land combination is called a "land-use system" (Beek, 1978); when the land use is agricultural, a land-use system can be considered as a synonym of "agroecosystem".

Agroecological knowledge is required for such assessments (see also section on assessment of land-use requirements). Land suitability classification is thus a form of "applied agroecology" in which on-farm resources and regional, physical, and socioeconomic constraints are also taken into account.

Limitations of current land evaluations in Southeast Asia

The land evaluation methodology described above consists of a number of concurrent and/or subsequent activities, which include the collection, analysis and integration of different data sets. The proper application of the methodology requires a close cooperation of natural resources scientists, agronomists and agrosocioeconomists. In practice, many land evalutions carried out in Southeast Asia are not integrated multidisciplinary evaluations, despite claims that the FAO methodology is followed. In addition, the reliability of the land suitability assessments is often difficult to judge. These shortcomings are caused by logistic and technical limitations, which include:

- institutions dealing with natural resources often do not have qualified personnel in the fields of agronomy and socioeconomics;
- multidisciplinary cooperation involving the cooperation of various institutions is difficult to organize; and
- the current knowledge on tropical agroecosystems is mostly inadequate, which makes the evaluation of the performance of land-use systems difficult.

Land evaluations and soil management research

Existing land resource information and land evaluation procedures can be used for soil and crop management research in its regional analysis and diagnostic phases for:

- the identification of the severity and extent of current or potential soil management problems;
- the selection of priority areas and sites for research; and
- site characterization.

In some cases, existing land evaluations may be used directly. Mostly, however, the land resource information provided by the evaluations has to be reinterpreted in the light of the soil management and technology considered for experimentation. Additional field observations will be required for this.

Another role of land evaluation can be to guide the results of soil management research to new areas where the improved or new technologies are applicable. Examples of such evaluations can be found in FAO (1971) - areas suitable for new, high-yielding rice varieties in Bangladesh in three seasons; Rondal (1982) - suitability of land for double-cropping patterns, based on early maturing IR36 rice varieties and other crops for an area in the Philippines traditionally used for single cropping; and Taffesse Asres (1986) - suitability of highland and mountain areas in Chiang Mai Province, Thailand, for a rust-resistant *Coffea arabica* variety.

The complementarity of land evaluation and agricultural research is shown in Figure 2 based on Young, (1986). In comparison with the FAO (1984) system (Figure 1) in the diagnosis of land-use problems, a "research loop" ("R") has been included in the land evaluation process of Young (1986) shown in Figure 2.

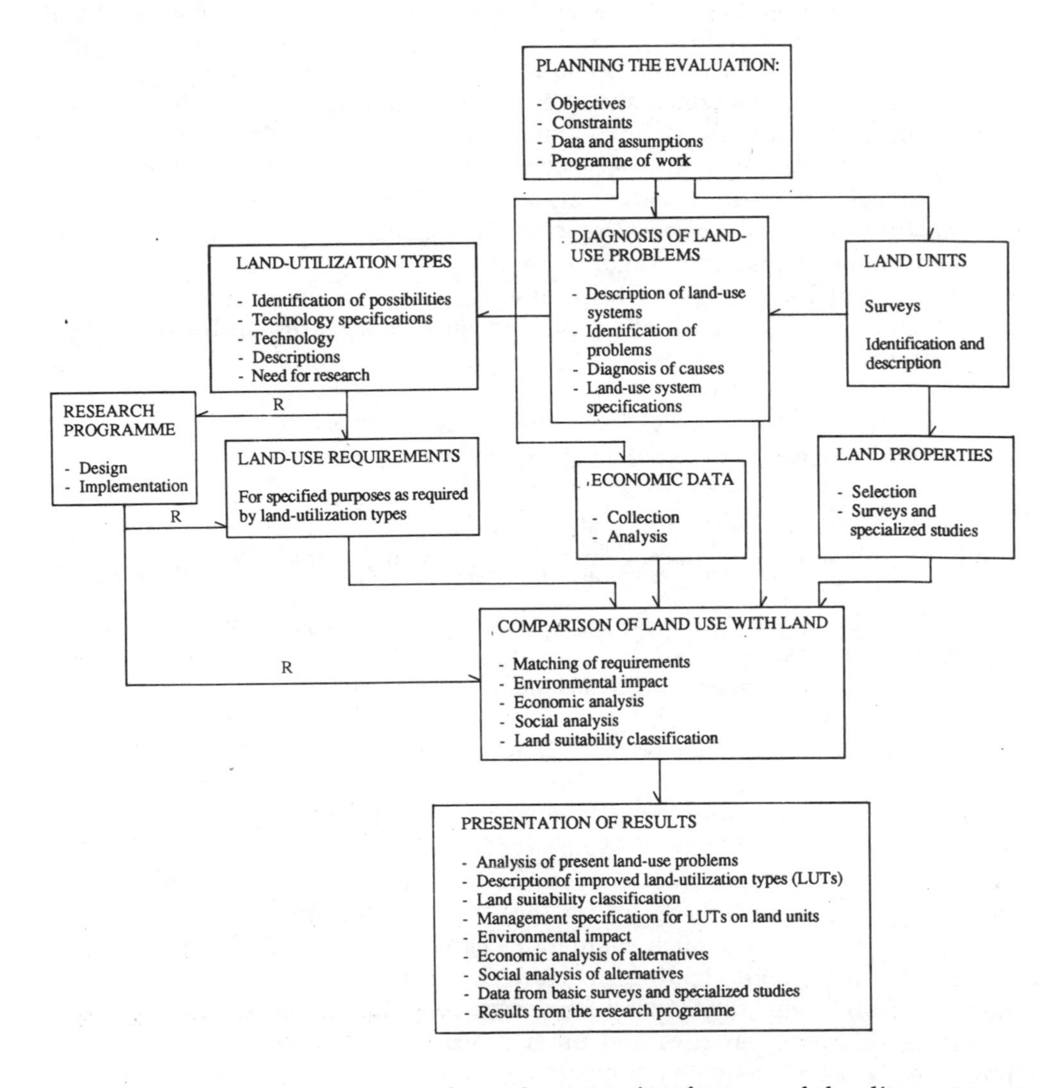

Figure 2. Land evaluation procedures, incorporating features of the diagnosis and design approach used in agricultural research (Young, 1986).

Conclusion

In conclusion it can be said that:

- Land evaluation is conceptually an applied form of agroecology that takes into account the resources of current farming systems and the socioeconomic setting of these farming systems.
- Land evaluation and soil management research are complementary. Land evaluation may provide data for the diagnosis phase and for the design of experiments; it can also contribute to finding areas where improved or new technologies developed by the research are applicable.

References

BEEK, K.J. 1978. *Land evaluation for agricultural development.* ILRI Publication no. 23. Wageningen, The Netherlands: ILRI.

BOUMA, J. and VAN LANEN, H.A.J. 1987. Transfer functions and threshold values: from soil characteristics to land qualities. In: *Quantified land evaluation procedures.* ITC Publication no. 6. Enschede, The Netherlands: ITC.

BRINKMAN, R. and SMYTH A.J., eds. 1973. *Land evaluation for rural purposes.* ILRI Publication no. 17. Wageningen, The Netherlands: ILRI.

FAO, 1971. *Agricultural development possibilities.* Soil Survey Project, Bangladesh. AGL: SF/PAK 6, Technical Report no. 2. Rome: FAO.

FAO, 1976. *A framework for land evaluation.* Soil Bulletin no. 32. Rome: FAO.

FAO, 1983. *Guidelines: land evaluation for rainfed agriculture.* Soil Bulletin no. 52. Rome: FAO.

FAO, 1984. *Land evaluation for forestry.* Forestry Paper no. 55. Rome: FAO.

FAO, 1985. *Guidelines: land evaluation for irrigation.* Soil Bulletin no. 55. Rome: FAO.

FAO, 1988. *Guidelines: land evaluation for extensive grazing.* Soil Bulletin no. 58. Rome: FAO.

KLINGEBIEL, A.A. and P.H. MONTGOMERY. 1961. *Land capability classification.* Agricultural Handbook no. 210. Soil Conservation Service. U.S. Department of Agriculture. Washington DC: U.S. Government Printing Office.

RONDAL, J.D., 1982. Land evaluation in an area with a long dry season and limited potential for irrigation development in the Philippines. Two case studies in the Ilocos Region. Unpublished M.Sc. thesis. Enschede, The Netherlands: ITC.

TAFFESSE ASRES, 1986. Agro-ecological zones for growing *Coffea arabica* in Chiang Mai Province, Thailand. Unpublished M.Sc. thesis. Enschede, The Netherlands: ITC.

YOUNG. 1986. Land evaluation and agroforestry diagnosis and design: towards a reconciliation of procedures. In: *Soil Survey and Land Evaluation* 5(3).

Section 3: Farming systems

Rainfed farming systems for upland soils in Thailand

CHANUAN RATANAWARAHA*

Abstract

The paper reviews agricultural development in Thailand since the initiation of Five-Year National Economic and Social Development Plans in 1961. Now the country is under the Sixth Plan (1987-1991), which has placed more emphasis on the diversification of agricultural production to overcome the risks faced by farmers. Strategies for achieving the goals set within the Sixth Plan are outlined, notably integrated farming systems, multiple-cropping systems, ley farming, appropriate cultural practices, and agroforestry farming.

Introduction

Thailand is one of a few developing countries which not only produces enough food for her own consumption, but also has a surplus for export. The agricultural GDP, which is worth about US$8,000 million annually, is equivalent to 26% of the total national GDP. The revenue obtained from the export of agricultural products was about 56% of the national total export revenue, and amounted to US$3772 million in 1985. Two-thirds of the labour force of 26 million is engaged in agricultural production, and about 30 million of the total 50 million population depend on agriculture for a living. During the past two-and a half decades (since 1963), the rate of increase in agricultural

* Farming Systems Research Institute, Department of Agriculture, Bangkhen, Bangkok 10903, Thailand.

production has been remarkably high (6-7% in the early period, but has declined to 2-3% at present).

The increase in agricultural production in Thailand during the past decades is mainly attributable to the expansion of crop production in former upland forest areas. The upland forest area has decreased sharply from 23 million ha in 1961 to 14.9 million ha in 1985 (equivalent to 50% and 29% of the total land area of the country respectively). The most crucial effect of the deforestation of this vast upland area has been the frequency of drought and the irregularity of rain currently observed compared to that in the past. There is evidence that the average annual rainfall in the last six years has been 8-14% less than that recorded in the past thirty years (Table 1).

Table 1. The decline in the amount of rainfall in Thailand during 1981-1986.*

Region	Annual rainfall (mm)		Differences
	Average during past 30 years	Average during past 6 years (1981-86)	
Upper northern	1,357	1,234	-125
Low northern	1,277	1,146	-131
Central	1,321	1,151	-170
Eastern	1,948	1,825	-123
Northeastern	1,396	1,279	-117
Western	1,093	984	-139
South (west coast)	3,307	3,037	-270
South (east coast)	2,293	1,972	-321

* Source: Agroclimatogy Section, FSRI.

Of the total 17 million ha of arable land, 81.6% or 13.8 million ha is under rainfed agriculture. Almost all the economically important field crops such as corn, cassava, kenaf, and food legumes are grown in the upland deforested rainfed area. The growth of these crops is subject to the adverse effects of deterioration of the natural environment, and notably erratic rainfall, infertile soil, pests, and outbreaks of disease. These are therefore the main causes of unsustainability, instability, and a decrease in production in these areas. The yield per unit area of economic crops is decreasing (Table 2). The overall rainfed rice yield is low (1.3 t/ha) compared to that of irrigated rice (3.2 t/ha).

Table 2. Yield of cassava, kenaf, and sugarcane during a 5-year period in northeast Thailand.

Crops	Yield (kg/ha/yr) 1983	1984	1985	1986	1987
Cassava	17,762	14,094	13,662	12,000	13,800
Kenaf	969	1,106	1,119	1,106	1,094
Sugarcane	45,380	42,375	45,300	41,156	40,910

* Source: Office of Agricultural Economics (1987).

Among the five main regions of the country - the central, eastern, northern, southern and northeastern regions - the last is the most important for rainfed agriculture. This is because the region occupies one-third of the total land area and accounts for one-third of the population of the country, the soils present are low in fertility, and the area experiences greater frequency of moisture deficits. The latter characteristic is due to its location further inside the mainland and almost surrounded by mountain ranges which results in the interception of rain. The northeastern region has the lowest rice yield per unit area (1.2 t/ha for rainfed paddy and 2.6 t/ha for irrigated paddy).

Out of the total cultivated area of four million ha, only 190 000 ha are served by an irrigation system, and hence most of the agriculture is rainfed. The sandy soils and unsuitable terrain constitute a serious constraint to the performance of irrigation systems. Since the irrigation systems can only serve a small proportion of the cultivated area, groundwater is an important resource for water supply to agriculture. At present there are at least 250 000 usable boreholes, supplying a total of about 310 000 m^3 of water per day (Ramnarong, 1985). However, the salinity of available groundwater is a constraint for agriculture in some areas. Cassava has been found to be one of the most suitable crops to be grown in the dry and infertile soils of the Northeast under rainfed, upland conditions. Cassava production now accounts for two-thirds of the region's farm cash income.

Review of policies, legislation, and strategies

Thailand has been implementing Five-Year National Economic and Social Development Plans since 1961. At present, the Sixth Plan (1987-1991) is in effect. During the First to Fourth Plans (1961-1981), emphasis was placed on the creation and building up of the infrastructure. Irrigation systems, electricity, highways and feeder roads, bridges, schools, and hospitals, were among these infrastructure items. Agricultural production has also increased and expanded, mainly with regard to field crops for export, such as maize, sorghum, cassava, and sugarcane. Much of the increased production was attributable to the expansion of cultivated land into previously forested land.

Though the country's economic growth rate has been favourably high (at an average of 7% per year), resources and income have not been well distributed. The gap of income per capita between the agricultural and nonagricultural population has been steadily widening, from 1:5 in the 1970s to 1:9 in the 1980s. In addition, natural resources and the environment have deteriorated, causing a drastic decline in yield in the agricultural sector.

Migration of labour from rural rainfed areas to towns and foreign countries to seek employment is now considered to be a serious problem. Slums, prostitution, low wages, homeless children, child labour, and criminal activities are among the problems created by overpopulation in the capital city. Many villages in the northeastern region lack male labour because the men have left to find other employment. Most of the poor families in the rainfed areas have sold their lands to meet the expenses and payments to commission agents to obtain employment in foreign countries, mainly in the Middle East. They have been hoping that the wages received from their new jobs will cover their expenses later on, but the experience has generally proved to be disappointing.

The policies and strategies of the Fifth Plan (1982-1986) were devised to overcome the problems evident in the previous plans. Unakul (1985) stated that "the Fifth Plan emphasizes restructuring the production process, improving efficiency in the utilization of land, water, forestry resources, and fishing grounds, and improving the agricultural marketing and pricing system in order to create greater justice for farmers. The expansion of cultivated land will be limited. The plan is also intended to increase the standard of living and income of farmers, and thus reduce the problems of rural poverty".

With regard to rural development, the government has placed much emphasis on developing of the so-called "backward" areas, mainly confined to the rainfed agricultural areas throughout the country. Names, locations, numbers, and constraints of backward villages have been defined and described by a new computerized data collection and analysis centre at Thammasat University. From the beginning of the Fifth Plan, the development administration system has emphasized a bottom-up planning process. Of a total of about fifty thousand villages in the country, 12 587 villages in 38 provinces are categorized as backward, in accordance with criteria relating to basic needs, availability of infrastructure, productivity, and income per capita.

US$3360 million (84 000 million baht) has been spent in implementing the development of backward rural areas in the past seven years (1982-88), with satisfactory results. The official report indicates that the standard of living of the people in the backward areas has improved. More than 10 000 kilometres of feeder roads have been built to facilitate the transportation of agricultural products to the markets. In providing water for both agriculture and human consumption, more than 4000 reservoirs, and 5000 shallow wells and underground deep wells have been built, which makes a total of 1.5 million units of water-resource wells. Of this number, it was estimated that an average of 29 wells, consisting of 7 deep wells and 22 shallow wells, are now available in each village.

For the ongoing Sixth Plan (1987-1991), the government has placed more emphasis on diversification of agricultural production to overcome the risks faced by farmers. Such risks could be due to the decline in prices, and to crop failures due to unpredictable agroclimatic conditions. A list of about 150 potential crops has been identified for a final selection for research, development, and production. The development of backward rainfed areas is an extension of the achievements of the Fifth Plan, with more emphasis on the distribution of income among agricultural and nonagricultural sectors.

Strategies for an increase in agricultural production in the rainfed rural areas have been prepared with regard to past and present experience. Emphasis has been placed on the reclamation of lost sustainability and the stability of the farmers' income, as well as the improvement, or arrest of the deterioration, of the environment. The income of farmers in the rainfed rural areas, which formerly was obtained from growing mainly cash-earning crops, will be modified to the new production scheme, which emphasizes both the need to stabilize income and to improve natural resources at the farm level. To achieve this goal, the strategies outlined below will be implemented in accordance with the potential of each specific locality.

- The land-use intensity will be increased by means of appropriate cropping systems. This can be accomplished through intercropping, relay-cropping or mixed-cropping systems, using drought-resistant varieties of crops grown with the existing main crops. A large amount of research and development on such cropping systems under rainfed conditions is being conducted by several research organizations and projects. Such research and development indicates that there are promising cropping systems which provide a more sustained and stable generation of income. These include combinations such as mungbean-rice, mungbean-rice-soybean, kenaf-rice, groundnut-rice, cowpea-rice, sesame-rice, etc. In the most severe drought-prone areas (rain-shadow areas), such as Chaiyapum province of the northeastern region, kenaf-rice cropping has proved to be the most suitable system in terms of stability and income generation.

An integrated farming system, combining activities which have a positive reciprocal response amongst each other (e.g. growing crops in combination with raising animals and fish culture in paddy fields), will help make full use of natural resources at the farm level, as well as bringing a more balanced diet and income to the farm families. The pilot projects conducted in many locations throughout the rainfed areas of the country revealed that income was increased by not less than 15-20% in the first and second year after the integrated farming system was introduced, and subsequent income increased steadily. Further, labour utilization and income are more evenly spread during any yearly period.

Rehabilitation of the deteriorated natural environment by encouraging farmers to grow more fruit trees, rubber trees, and fast-growing trees for fuel and wood will be implemented earnestly. Annual cash crops to provide cash income during the period before the trees bear fruit are encouraged. These cash crops include groundnut, sweet corn, sesame, and upland rice, which will be grown in the interrows of the trees.

The application of appropriate low-cost technologies, using available local, natural resources such as animal manure, leguminous cover crops, green manure, azolla, blue-green algae, herbicinal pesticides, alley cropping, and ley farming, will be emphasized.

Labour availability is an essential component of the rainfed rural development programme. A continuous and evenly adjusted labour force on the farm during a yearly time span needs to be assured. Small-scale agroindustries and other manufacturing ventures within the villages of rainfed rural areas are important off-farm employment options which can help to solve the labour-drain problem.

At present, there are several ways of putting these strategies into practice. The main objectives are to increase income per capita and thus to improve the livelihood of poor rural farmers in upland rainfed areas. This can be done by:

- multiple-cropping systems,
- ley farming,
- integrated farming,
- appropriate cultural practices, and
- agroforestry farming.

The first three systems are the more important options, and will be discussed in detail.

Multiple-cropping systems

Concept and principle

Multiple cropping is the practice of growing more than one crop on the same piece of land. In terms of space, it can either be mixed or interrow cropping, and in terms of time, it can either be simultaneous or relay cropping. The objectives of multiple cropping are as follows:

- to increase land-use intensity, so that the yield will be increased and more income obtained;
- to rehabilitate the agroecosystem, notably by replenishing soil properties, prolonging the soil moisture availability period, and providing more favourable conditions by growing trees as an intercrop;
- to increase the degree of sustainability and to ensure the stability of production on the farm (both from an agronomical and economical point of view);
- to minimize crop losses due to pests and diseases; and
- to establish a mutual relationship between crops in the cropping system so that they improve each other's productivity - as, for example, by the use of leguminous nitrogen fixing-crops, or of shade-loving crops with tree crops.

However, to gain additional benefit from multiple cropping, some factors need to be considered in arriving at the most appropriate cropping systems. These include:

- the component crops of a cropping pattern must be planned and implemented jointly, because the growing of one component crop greatly influences the productivity of the other;
- the cultivation techniques of each crop needs to be well planned (e.g. it should be decided whether it is to be grown by direct seeding, trans-planting, or sowing);
- the management practice for a component crop in the pattern should be chosen not to maximize production of that particular crop, but rather to maximize production of the whole system (e.g. soybean may produce a better yield if planted on well prepared seed beds, but in present cropping systems it is generally grown under minimal tillage to conserve residual soil moisture).

Characteristics of upland cropping systems

The cropping system will include both the cropping pattern and the crop types.

Cropping patterns

In terms of time and space, multiple-cropping systems can be categorized into four main types, namely:

Intercropping: This is a system where more than one crop is grown on the same piece of land at the same time. The crops can be either on a mixed- or row-cropping pattern, which are characterized as as follows:
- mixed intercropping, where the rows of crops have no distinct arrangement; and
- row intercropping, where at least one crop is planted in rows.

Gomez and Gomez (1983) explained that between any two crops grown simultaneously, intercropping is denoted with a plus sign (+). For example, an intercrop of corn and mung beans is written as corn+mung beans. Thus, an intercrop of corn and mung beans followed by corn will be written as corn + mung beans-corn.

Relay cropping: This is a system where a second crop is planted before the harvest of a maturing crop. Gomez and Gomez (1983) have denoted the relay cropping by a slash between crops. For example, the cropping pattern denoted by rice/mung beans-corn is a relay crop of mung beans after rice which is followed by corn.

Sequential or double cropping: Sequential or double cropping refers to the growing of two or more crops where the planting of one crop follows immediately after the harvest of the previous crop. This can be done by complete land preparation before the planting of the second crop, or by planting of the second crop in the crop stubble after an interrow cultivation to kill weeds.

Ratoon cropping: Ratoon cropping is another form of multiple cropping, which provides an increased yield and continuous harvesting seasons. Sorghum, millet, sugarcane, and cotton cultivars exhibit varying degrees of ratoonability (regrowth after harvest).

Benefits and impact

Economic stability and sustainability

An appropriate multiple-cropping system can be the means of producing the food requirements of poor rural farm households, who form the majority of the population of Thailand. This is because, firstly, they increase the diversity of farm products in such a way that self-sufficiency in the daily food requirements of the family can be achieved. Secondly, in most of the agricultural farming communities the products usually provide for more than one family, which means that the remainder can be either stored for longer periods, or sold in the local markets for cash income. Thirdly, with a good cropping system and suitable crops, sustainability can be achieved.

The success of this approach can be judged by the variation of yield and incomes gained over a period of consecutive years as measured by the coefficient of variation (c.v.) value and the probability of net income. Krisnamoothy (1980) and Rao *et al.* (1979) conducted extensive trials on sorghum-pigeon peas and corn-soybeans. The intercrop combinations had a much lower probability of producing a low income than monoculture cropping. For example, the probability of net income falling below US$350/ha is once in every four years for the corn or soybean monocrop, but once in 13 years for the corn-soybean intercrop (Table 3).

Table 3. Net income, coefficient of variation (c.v.), and probability of net income per hectare falling below US$350 for a monocrop and for intercropping combinations.

	Income		
Cropping pattern	Net income (US$/ha)	c.v. (%)	Probability of net income/ha falling below US$350
Sorghum	361	53	0.48
Pigeon peas	523	47	0.24
Sorghum + pigeon peas	618	35	0.11
Corn	632	65	0.24
Soybeans	506	47	0.26
Corn + soybeans	852	41	0.08

Sources: Rao *et al.* (1979), Krisnamoorthy (1980).

Low cash input

The high risk of rainfed upland agriculture is the most important constraint preventing farmers from adopting high-input technologies such as chemical fertilizers, and chemical pesticides. Locally produced materials, such as organic fertilizers from farm waste products and some herbicidal pesticides, are now used by upland farmers in rainfed areas with great success. This helps to increase yield, stabilize income and decrease the chance of crop failures.

Increase of productivity by land-use intensity

Multiple cropping can increase productivity. Generally, it has been found that the combination of crops in multiple-cropping systems increases productivity by 30 to 60% over monoculture. The land equivalent ratio (LER) is the total land required using monoculture to give a total production of a given crop equal to the yield produced on the same area of intercropping. It is calculated by determining the ratio of the yield of a crop in a mixture to its yield in monoculture under the same treatment levels. The LER is defined as:

LER = summation n i-1 (x/y)

where: x is the yield of crop i in intercropping
y is the yield of crop i in pure stand

To determine the LER ratio when mung beans and corn, for example, are planted in a monocropping and an intercropping system, assume:

- the yield of soybeans in monocropping = 800 kg/ha
- the yield of soybeans in monocropping = 3500 kg/ha
- the yield of soybeans and corn in intercropping = 550 and 2300 kg/ha respectively

Therefore, the LER

= (soybean intercrop/soybean monocrop) + (corn intercrop/rice monocrop)

$$= \frac{550}{800} + \frac{2300}{3500} = 1.34$$

This shows that a monocrop planting area of 1.34 ha is equivalent to 1 ha of intercropping with the same two crops.

Soil improvement through legumes and nitrogen fixation

Intercropping and relay cropping provide opportunities for planting mixtures or sequences of grain legumes and nonlegumes. It is generally accepted that the nitrogen-fixation activity of a legume increases under low soil nitrogen conditions. There have been numerous trials which have proved that the yield of main crops increases by growing them together with legume crops. One of these trials was conducted under a NERAD project in Srisaket Province in the northeastern region. The results are shown in Table 4.

Table 4. Yield and income from rice with and without cowpea prior to rice in Srisaket Province (in the 1986 growing season).*

Treatment	Yield (kg/ha)	Income (US$/ha)	Expense (baht/ha)	Profit (baht/ha)
Green manure	3,390	312	79	233
Ccontrol	2,550	249	79	170
Difference	840	63	-	63

* *Notes:* 1. The variety of rice was KDML 105.
2. Results obtained from 18 trial plots at two different locations.

Source: NERAD (1987).

Prateep *et al.* (1988) reported that the combination of minimum tillage and a leguminous cover crop (*Stylosanthes hamata* cv. Verano) as a living mulch, can increase both the root yield and shoot weight of cassava from 16.6 to 23.2 and 14.2 to 26.2 t/ha respectively, and consequently resulted in a high marginal rate of return (175%) as compared to traditional farmers' practices (Table 5).

Table 5. Effect of four management systems on the fresh weight of roots and shoots and the harvesting index of Rayong-1 cassava cultivar (1987-1988).

Treatment	Fresh weight (kg/ha)		Harvest index
	Roots	Shoots	
Farmers' practices	16.6	14.2	0.54
No tillage + pre-emergence herbicide	20.1	16.7	0.55
Minimum tillage + *Stylosanthes*	14.1	21.0	0.40
Minimum tillage + *Stylosanthes* + mowing	23.2	26.2	0.47

Effect of crop residue and organic waste in intercropping

A by-product of more intensive cropping systems is the production of greater quantities of crop residues and organic wastes. The proper handling of crop residues and organic wastes can provide the necessary organic-matter turnover to improve soil workability and water infiltration, and recycle needed nutrients into the soil. Over two-thirds of the potassium and zinc and about one-third of the nitrogen, phosphorus and sulphur taken up by crop plants are contained in the leaves and stems of plants. If these residues are returned to the field, either directly or indirectly in the form of farmyard manure, the need for applied chemical fertilizer can be reduced.

Petchawee *et al.* (1984) reported that the use of plant residue mulch with fertilizer had a remarkable effect on maintaining a high level of corn yield throughout eight growing seasons. Under no-mulch practice, the corn yield fluctuated to a great extent. Virakornphanich *et al.* (1984) also reported the same effect of mulching with a plant residue to increase the yield of corn production in reddish-brown lateritic soils.

Pest and disease control

There are many combinations of cropping systems where each of the crops can provide reciprocal effects of controlling pests and diseases among each other. For example, Burandy and Raros (1975) have shown that *Plutella* infestation in cabbages by the "diamond-backed moth" is reduced by the presence of tomatoes as an intercrop (Table 6).

Table 6. Intercropping patterns and influence on pest incidence.

Intercropping	Pest affected
Cabbages + tomatoes	*Plutella*
Corn + mung beans	Weeds
Corn + peanuts	Corn borer
Cress + grass	*Phytum irregulare*
Sesame + sorghum	*Pyralid welburn*
Cotton + corn	*Heliothes*

Source: Gomez and Gomez (1983).

Effects of vegetative ground cover

Ground cover is very important in reducing soil erosion, particularly under heavy rainfall in humid tropical countries. A suitable intercropping pattern can cover the land area, resulting in significant reduction in soil erosion. For example, cassava as a monocrop takes 63 days to provide a 50% ground cover as compared to only 50 days for a cassava + corn intercropping (Lal, 1980).

Ley farming

Concept and principle

Ley farming is a means of soil fertility improvement and conservation to enhance agricultural production. The concept involves the planting of pastures as part of a cycle in a rotational cropping system in which livestock is introduced into the system to utilize the pastures economically, and becomes a significant component in the ley farming. This system of farming can provide a sustained agricultural production system as well as a means of natural environmental rehabilitation for rainfed upland areas. Phalaraksh (1987) reported that in addition to the income earned from livestock (between US$79-90/ month), the farmer also obtained a higher yield of crops grown after the pastures, as compared to the income from continuous cropping (Table 7).

Table 7. Comparison of cassava yield from plots with and without pasture in the rotation cycle.

Plot	Fresh weight (kg/ha)	No. of tubers per plant	Height (cm)
A (continuous cropping)	14,481	6.7	147
B (following pasture)	28,900	7.5	304
T-test (0.01)	**	ns	**

Source: Phalaraksh (1987).

Furthermore, Phalaraksh (1987) discovered that other main crops, such as kenaf, also showed a similar trend of production. A technique of ley farming, through which sustained crop production is achieved in rotation with a two-to-four- year rest period when the land is under a grass/legume pasture, is therefore possible. During this period the depleted plant nutrients in soils can be replenished.

The reasons for this are that during the resting stage of soil under the pasture, the soil will not be disturbed by cultivation, enabling the formation of larger aggregates and thus a better soil structure. The grass root materials also add some organic matter to the soil, in addition to the supply provided by dead leaves. Legume plants increase soil nitrogen through their ability to fix nitrogen, and cattle dung on the pasture area contributes a significant amount of organic material to the soil, thereby enhancing its cation-exchange capacity. The water-holding capacity of the soil is also improved, resulting in greater moisture retention.

Integrated farming systems

Concept and principle

"Integrated farming is the integration of appropriate agronomic practices, crops and animals in such a manner that there is a balance between production and allocation of land and resources" (Upawansa, 1975). Intergrated farming is not the same as mixed farming or different kinds of farming coordinated by management. It is a system incorporating microbes, plants and animals, and involves repeated use of waste matter from human consumption, animals, crops, and products from the various activities of a farm. Due to this linkage, it is possible to convert more of the sun's energy into photosynthetic products and extract energy needed in the form of food, animal feed, fuel, fertilizer, and fibre.

This system conserves energy, soil, water, plants and nutrients, while directly or indirectly enriching the soil with water, plant nutrients and microbes, and also storing energy in the form of biomass. It can be a means of overcoming the uncertainties which poor farmers in developing countries have to face, especially in situations where the price of agricultural products is very low and unpredictable, and the cost of farm investments are increasing drastically. Integrated farming can be considered at the micro (farm or household) level, and the macro (national agricultural policy) level.

Integration at the farm level

This leads to:

- stability and sustainability of both economical and environmental conditions in the farming system;
- increased productivity per unit of production (land, labour, and capital);
- improved nutrition and health for the rural population;
- increased efficiency of energy utilization per capita;
- improved microclimate on the farm; and
- a high standard of living, contributing to self-confidence and self-reliance on the part of the farmers.

Integration at the national level

This results in:

- reduced energy consumption in farming;
- reduced seasonal variations in labour requirements;
- creation of employment in the case of intensive farming;
- reduced purchases of inputs for farming;
- improved natural vegetation - at the same time conserving forests, land, labour, and capital;
- improved climate, rainfall, ecological balance, and an efficient use of solar energy;
- conservation of soil, moisture, food, feed, fertilizer and fuel; and
- creation of more self-reliance in the nation, thereby helping to cope with the problem of trade deficits.

There is no well-defined scope for integrated farming, indicating (for example) how many activities should be incorporated on a single farm. It is a versatile system, suitable for any kind of social, economic, physical and biological environment. The system should allow the introduction of any innovation at any time or any location. It can be as small as half a hectare for one family or as large as 25 ha for a commercial-scale farm.

General guidelines for integrated farming

Integrated farming systems for small farmers should be developed along the following lines:

- It should be considered as an intervention which is carried out to improve the existing land-use system. The emphasis should be on improving rather than transforming land use. A rapid appraisal of the agroecosystems should be made so as to diagnose the present land-use situation. The physical, biological, and socioeconomical aspects of the community should be studied and analyzed before intervening in the system.
- It should open up new opportunities for raising the income levels of small farmers, without in any way putting agriculture in jeopardy.
- It should pay particular attention to the sustainability of the system. Trees and shrubs, apart from providing useful basic products (wood, fodder, fruits, etc.) protect the soil against erosion, provide organic matter to maintain soil fertility, bring up nutrients from deeper soil layers, and create a more favourable microclimate.

References

BURANDY, R.P., and RAROS, R.S. 1975. Effect of cabbage tomato intercropping on the incidence and oviposition of the diamond back moth, *Plutella xylostella* (L.). *Philippine Entomologist* 2(5).

FSRI (Farming Systems Research Institute). 1989. Report on the amount of rainfall during the past six years as compared to the average in the previous thirty years. Agroclimatology Section. Unpublished report. Bangkok: FSRI.

GOMEZ, A.A., and GOMEZ, K.A. 1983. *Multiple cropping in the humid tropics of Asia.* Ottawa, Canada: IDRC.

KRISNAMOOTHY, C. 1980. Low-input cropping systems. Paper presented at the Symposium on Potential Productivity of Field Crops Under Different Environments, IRRI, Philippines, September 1980.

LAL, R. 1980. Soil erosion as a constraint to crop production. In: Soil-related constraints to food production in the tropics, 405-423. Los Baños, Philippines.

NERAD (North East Rainfed Agriculture Development Project). 1987. *Proceedings of the Annual Farming Systems Technical Workshop.* Bangkok: Ministry of Agriculture and Cooperatives.

OFFICE OF AGRICULTURAL ECONOMICS. 1987. Division of Policy and Agricultural Development Plan. Report no. 84 (11). Bangkok: Ministry of Agriculture and Co-operatives.

PETCHAWEE. 1984. Long -term effect of mulching with fertilization under cropping corn-legumes on crop yield and improvment of soil chemical-physical properties. Paper presented at the International Seminar on Yield Maximization of Feed Grains Through Soil and Fertilizer Management, 12-16 May 1986, Bangkok, Thailand.

PHALARAKSH, K. 1988. Ley farming on upland areas in Northeast Thailand. In: *Sustainable Rural Development in Asia, 4th SUAN Regional Symposium on Agroecosystem Research* (Khon Kaen).

PRATEEP. 1988. A promising low-input management to sustain high cassava yield in northeast Thailand. Paper presented at 8th International Society for Tropical Root Crops, 13 October - 5 November 1988, Bangkok, Thailand.

UNAKUL, S. 1985. Food policy analysis, ed. Theodore Panayotou, 189-193. Bangkok: Agricultural Development Council.

VIRAKORNPHANICH, P., MASANGSAN, W., MORAKUL, R., CHONGPRADITNAN, P., CHANDRAPANIK, S., SUNGTHADA, O, and INOUE, T. 1984. Improvement of soil productivity by mulching and nitrogen fertilizer for corn products in reddish brown lateritic soils. In: *Dynamic behavior of organic matter and available nutrients in upland soils of Thailand.* A report under the cooperative research work between Department of Agriculture, Thailand and Tropical Research Center, Japan. Bangkok: Department of Agriculture. 377p.

Cropping system experiments

ADISAK SAJJAPONGSE*

Abstract

Due to urbanization, industrialization, and a lack of new areas, increasing crop production by expanding the area planted is difficult. The answer may be to increase productivity. In order to increase crop production by growing more crops with higher yield per year effectively and economically, research in cropping systems is essential. To arrive at a successful cropping system which is acceptable to farmers, three factors need to be considered: environmental conditions, technical feasibility, and economic variability. This paper describes the various cropping systems commonly found in tropical Asia, the factors affecting cropping system design, and design and testing methodology.

Introduction

Crop production can be increased either by expanding the area planted to crops, thereby increasing the yield per unit area, or by growing more crops per year. Due to urbanization and industrialization, increasing crop production by expanding the area planted to crops is rather limited and difficult. However, the use of modern technology and high-yielding varieties enables increased crop production. Knowledge of the climatic conditions of an area allows for more than one crop per year to be introduced and planted successfully. In order

* ASIALAND Coordinator, IBSRAM Headquarters, PO Box 9-109, Bangkhen, Bangkok 10900, Thailand.

to increase crop production by these measures, more effective and efficient research in cropping systems has been initiated.

A cropping system is simply the system in which a combination of crops is grown over a period of time or a combination of crops interplanted over the same area. The objective of this paper is to describe briefly the accepted approach for cropping system experiments.

Types of cropping systems

The major determinants for the selection of a cropping system are the climate, the social structure of the community, crop preferences, and resources. Some of the cropping systems which are commonly found in tropical Asia are described below.

Single cropping. This is a system in which only one crop is grown in one year.

Mixed cropping. This system invloves growing more than one crop species on the same piece of land at the same time. The different species are planted in organized intercropped rows with regular spacing, or they are grown mixed randomly and unequally distributed over the land. The latter procedure is commonly used in subsistence agriculture in tropical Asia.

Relay cropping. In this system, crops are planted between the rows or the plants of an already established crop. Normally this practice is conducted when the established crop is at or near senescence so that shading on the interplanted young plants is not detrimental.

Multistorey cropping. This system involves growing crops under trees such as coconut, rubber, and oil palm. Since the trees are widely spaced, making it possible for light to get through their leaf canopies, it is possible to grow crops underneath them. An example of a three-storey cropping system may be peanuts or sweet potatoes as the layer immediately above the ground, papaya at two to five metres above the ground, and a coconut canopy at five to fifteen metres above the ground.

Sequential cropping. The aim of this system is to grow more than one crop a year on the same piece of land. The different crops occupy the land at different times of the year, one crop being planted after the other has been harvested.

Factors affecting cropping system design

In order to arrive at a successful cropping system which is acceptable to farmers, three factors need to be considered in the design of a cropping system: environmental conditions, technical feasibility, and economic viability.

Environmental conditions

The environmental factors are physical, climatological, and biotic - factors which involve rainfall or precipitation, day length, radiation, temperature, and soil conditions (texture, fertility, and topography). To arrive at a feasible cropping system under a particular environment, it is necessary to match the physical and climatological requirements of the crop (during its growth duration) to the physical and climatological environment (during the year) of the area. The matching can be done as follows:

1. By developing tables for the environmental requirements for each crop. Data showing the sensitivity and tolerance of the crops to environmental stress, such as drought and flooding, are also useful.
2. By expressing the environmental conditions of the area during the year in tables or graphic form (for soil moisture conditions, temperature, day length, solar radiation and major insect and disease incidence).

By matching the information in the above two sets of data, suitable growing seasons for the crops considered can be identified.

Technical feasibility

Cultural management and input levels are the two main factors determining the feasibility of a cropping system in a given area. The levels refer to the type and expenditure for equipment, pesticides, fertilizers and labour inputs. The availability of these resources often determines if a crop can be included in a cropping system. By comparing the estimated resources requirements of a cropping system with the resources available, the feasibility of the cropping system can be determined. Acceptability is another determinant of the technical feasibility. If the crops to be introduced are not socially acceptable to the farmers, the new technology will not be readily adopted - even though the profitability from the system may be high. Some technical considerations which need to be taken into account in designing cropping systems are described below.

Availability of new crops and varieties

Previously, there were only a few crops and varieties available for use in a cropping system. Now, through breeding programmes and the introduction of species from different areas, more crops and varieties with different traits are available - such as early or late maturity, shade tolerance, drought resistance, heat tolerance, and disease and pest resistance.

Irrigation

Water is required throughout the life cycle of crops. Interruption of the water supply to a plant will certainly affect crop yield to some extent. In designing a cropping system, drought-tolerant varieties should be used if drought is anticipated or the water supply is likely to be insufficient or interrupted, especially during the reproductive stage of the crop.

Fertilizer

In general, the rates and types of fertilizer application have a close positive relationship with the yield and quality of crops. Most improved varieties respond to fertilizer application significantly, and therefore require high rates of fertilizers in order to produce a high yield. Complete yield failure can occur if insufficient amounts of fertilizer are applied to an improved variety.

Economic viability

The economic viability of a cropping system is determined by analyzing the cost of the production inputs and the prices of the products produced by the system. This analysis includes costs of labour, pesticides, fertilizers, and all the other operational inputs required by the system and the returns from the system. The profitability and the returns from the system are then compared with those of the existing systems. To be feasible, the profitability of the new cropping system must be higher than that obtained from the existing systems.

Although yield can be assessed in monetary terms, several difficulties are encountered in such an analysis. The main difficulties are the seasonal fluctuation of prices for produce, and estimates of the value of family labour and hired labour, payed for in kind rather than cash.

Design and testing

A framework for cropping systems research is shown in Figure 1. Initially, physical, biological, social environment, and resources are assessed quantitatively. Using this information, and also by drawing on the experience of soil scientists, agronomists, plant pathologists, and entomologtists, a number of potential cropping systems can be designed. The next step is to compare these potential cropping systems with the prevailing system (farmers' practices). Using information gathered from socioeconomic conditions, the feasibility of the potential cropping systems is evaluated, and those which seem more relevant will be tested under research station conditons. When the results of the testing are known, potential cropping systems can be classified according to the degree of technology they require, and then assessed in relation to the technology currently available to the farmer.

When sufficient knowledge about the performance of the different cropping systems being tested under the controlled conditions of the research station has been obtained, the promising systems will be further tested on the farmers' fields and under the control of researchers. Extension workers will be involved at this stage. The next step is to have the systems tested by farmers under the guidance of extension workers, and the last step is the implementation and extension of the recommended cropping system.

In the testing during the first year, many different cropping systems may be studied in a nonreplicated trial. In the second year, the number of systems should be reduced and the number of replications may be increased. During the third year, only the most promising cropping systems should be continued.

Data collection

To evaluate the economic viability of a cropping system and to understand the interaction between environmental conditions and the cropping system, the following data must be collected.

Climate

The climatic factors important for plant growth and performance are precipitation, temperature (minimum and maximum), solar radiation, and sunshine hours. In addition, if possible, information on pan evaporation should be collected.

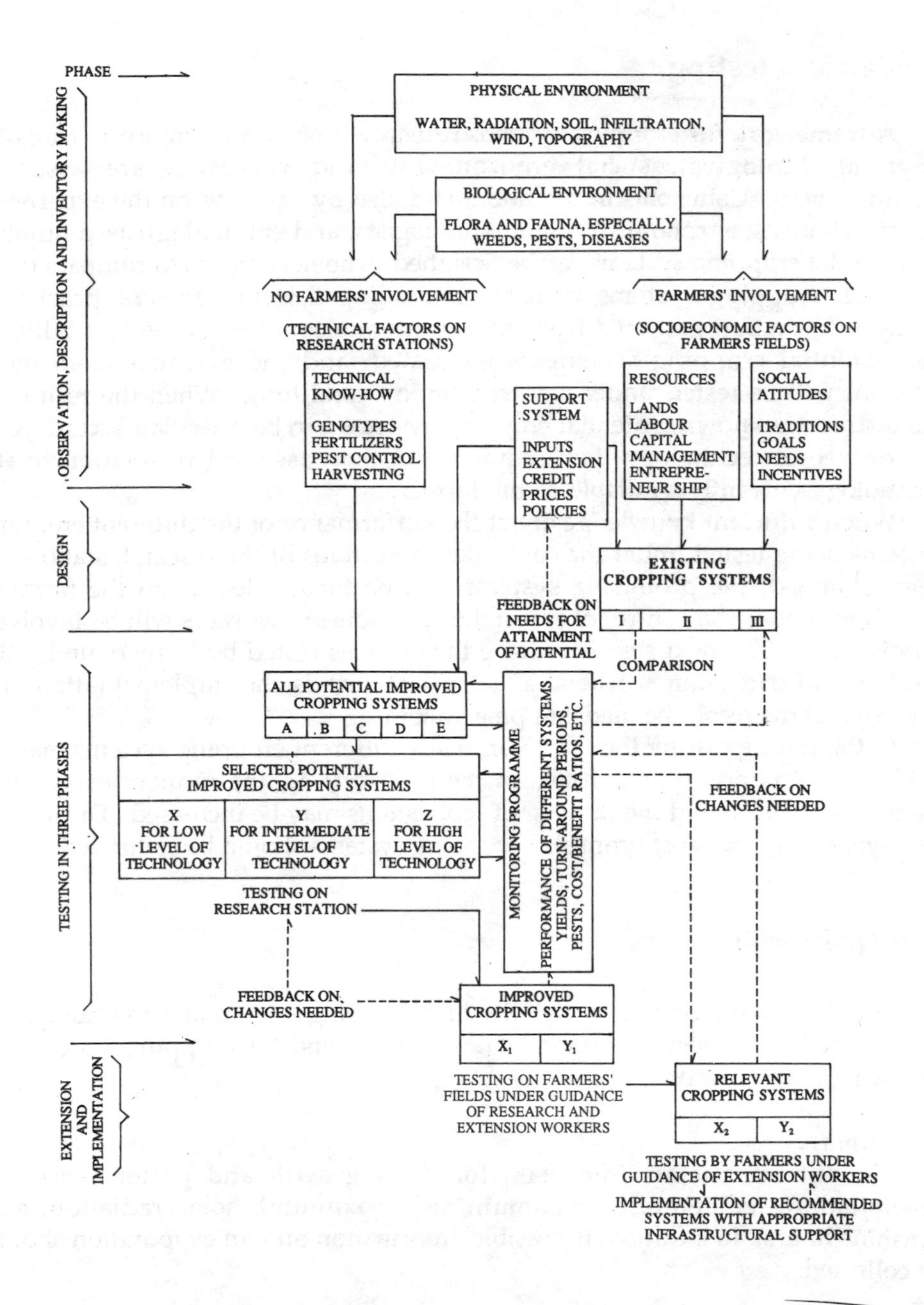

Figure 1. A framework for cropping systems research in the tropics (adapted from Beets, 1982).

Land

The general topography, the underground water table of the field, and the chemical and physical properties of the soil are the most important characteristics of the experimental site. Soil properties such as pH, organic-matter content, nitrogen, phosphorus, potassium, micronutrient contents, texture, and the type of clay minerals should be analyzed. Information on these characteristics of the soil is important for the interpretation of results, for cross -site comparisons, and for the use of the results for extrapolation studies.

Crop husbandry and production operation

Information on the variety, planting methods, seeding rates, plant spacing, irrigation, weeding, and other crop management practices for each crop component in the system should be recorded. At harvest, the yield and its components must be measured. Operational costs such as those incurred by hiring labour, purchasing planting materials, and the price of the products must be collected so that the economical aspects of the system can be determined.

References

ANNONYMOUS. 1974. *Multiple cropping systems in Taiwan.* Taipei, Taiwan: Food and Fertilizer Technology Center. 77p.

BEETS, W.C. 1982. *Multiple cropping and tropical farming systems.* Boulder, Colorado: Westview Press. 156p.

ZANDSTRA, H.G., PRICE E.C., LITSINGER J.A., and MORRIS R.A. 1981. *A methodology for on-farm cropping systems research.* Los Baños, Philippnes: IRRI. 147p.

Socioeconomic evaluation of cropping systems

CHATT CHAMCHONG*

Abstract

The type of socioeconomic evaluation which is required varies with the purpose and nature of the cropping system it is associated with. Cropping systems have been developed with an emphasis on one of the following parameters: seasonal variations, topographical conditions, irrigation facilities, the type of crop combination, industrial outlets, marketing prospects, soil conservation, input efficiency, the cost of production, income distribution, and the specific policy on each cropping system.

The economic evaluation can be made in a number of ways, the choice depending upon the purpose of the cropping system. These analyses include cash flow, gross returns on the farm business or returns per unit of area planted, net returns or net returns per unit of output of each crop, net returns per unit of area planted, net returns per unit of labour, cost per unit of area planted or per unit of output, and returns over variable costs. Estimates of social evaluation are based on the farmers' views, combined with economic evaluation and field observations by the researcher of the factors involved with the particular cropping system concerned. These factors include soil and water conservation, the distribution of labour and other inputs, marketing, and the prices of the crops produced.

---

* Department of Agricultural Economics, Faculty of Economics and Business Administration, Kasetsart University, Bangkok 10903.

Introduction

It is generally accepted that being without food is like being without life, so agriculture is the most significant base for many types of development. More advanced technology is always more systematic in its method of operation. In order to achieve greater productivity, crop production should follow both scientific and economic principles.

In most parts of the world, natural resources are becoming scarcer. High-profit production with an emphasis on the conservation of nature provides the most efficient management system. A good cropping system should maximize the use of inputs, achieve higher profits and maintain a balance of nature and a sufficient supply of food. Cropping systems which meet these objectives have been implemented by many farmers in many countries, often as a result of extension and promotion activities by agricultural agencies. A socioeconomic evaluation needs to be undertaken in order to select a more suitable system under specific enviornments, and to evaluate the benefits which will accrue by introducing the improved system.

The nature of cropping systems

A socioeconomic evaluation will vary according to the purpose and nature of the cropping system, and thus the method of evaluation developed should be aimed at serving some specific system and purpose. Cropping systems have been developed with an emphasis on different parameters, which are outlined below.

Cropping based on season. In this case the cropping system developed in most cases is based on the climatic conditions and the natural environment in a particular area or region. In Asia and in the Pacific region, there are generally two or three seasons (wet, cold and dry), depending on the latitude of the area.. The short period of growth for the crops is dictated by seasonal variations, and they need to be adapted to the changes in the weather conditions in any given region.

Cropping based on topography. In this instance, the cropping system is developed to suit a particular topography and is adapted to hilly, sloping, undulating, or lowland conditions.

Cropping based on irrigation. The cropping system in this case will be either fully irrigated, partly irrigated, or nonirrigated (i.e. rainfed).

Cropping based on the nature of the crop. This system is developed through crop combinations in relation to agricultural environments. The examples of cropping systems in this case are:

- annuals: corn, sorghum, mung bean, black bean, or soybean
- perennial tree crops: durian, rambutan, or mangosteen

Cropping based on industrial outlets. This cropping system is developed in accordance with the availability of facilities for processing the raw produce. For example, a cropping system for the fruit juice industry could be pineapple, passion fruit, oranges, or strawberry, while that for canned vegetables could be asparagus, cabbage, cucumber, young corn, mushroom, or bamboo shoot. Such systems should allow for almost all-year-round availability of produce for processing, and thus these systems may comprise two combinations of crops.

Cropping based on marketing prospects. Crop marketing strategies will determine the system which is adopted. An export-oriented cropping system might involve corn, cassava, or rice; a domestic-oriented system (or import substitution system) could be rice, bamboo shoot, or fruit trees; and a system for both domestic and export markets might be mango, pomelo, or bamboo shoot.

Cropping based on a specific policy. This cropping system is developed to meet the specific policy or strategy of the government, entrepreneurs, or farm groups and is aimed at satisfying their integrated needs. These include:

- demonstration of farm and cultural products
- farming in conjunction with restaurant or fishing park business
- farming in conjunction with reforestration or a community forest
- farming in conjunction with a recreation area or resorts

Cropping with a view to soil conservation. Besides income from crop production, this system aims at soil quality improvement by incorporating crops of the legume family, or by using barrier crops.

Cropping to maximize input efficiency. This system is developed to achieve a higher input efficiency of operation by focusing on the lowest cost per unit of input, the highest yield per unit of input, or a reduction in the period of production. Efficiency is achieved through intensive cropping, reducing labour requirements, reducing fuel consumption, and increasing the area planted per machine hour.

Cropping for better income distribution. This cropping system focuses on the year-round income generated by all crop enterprises, e.g. coconut production for juice, for palm sugar, or for coconut meal.

Cropping to achieve cost reduction. This system is sometimes similar to that of input-efficiency cropping, but in some cases, efficiency is improved through higher costs but increased outputs.

Socioeconomic evaluation

This involves two stages. The first concerns the evaluation of traditional cropping systems to assess if there is a need to change the traditional cropping system to a more sophisticated one. The second stage involves the evaluation of the implemented or recommended system to decide if any modification to the system is required:

The evaluation process

The stages in the evaluation process can be listed as follows:

- review of the purpose of the cropping system in the area to be investigated;
- selection of alternative evaluation methods and techniques to suit the objectives;
- determination of the variables in the evaluation methods;
- compilation of secondary data to support the selection of the recommended cropping system;
- data collection through farm records, multiple-visit surveys, and single -visit, seasonal, and annual surveys;
- data processing and analysis; and
- reporting on the findings, conclusions, and recommendations.

Evaluation methods

Many alternative evaluation techniques and methods are available for the evaluation of the selected cropping system. A brief description of some of the methods is given below.

Cash flow analysis

This method is recommended for well-planned systems where close supervision is provided to farmers to ensure a systematic maintenance of farm records, and multiple visits and surveys are made in each period of the farming process, i.e. land preparation, planting, crop care, harvesting, and selling.

A cash flow analysis for these activities can be recorded as shown in Tables 1 (a) and (b).

Table 1 (a). Cash inflow.

Cash revenue	M	J	J	A	S	O	N	D	J	F	M	A	Total
Season 1													
Crop 1													
Crop 2													
Crop 3													
Crop 4													
Subtotal													
Season 2													
Crop 1													
Crop 5													
Crop 6													
Crop 7													
Crop 8													
Subtotal													
Season 3													
Crop 1													
Crop 7													
Crop 8													
Subtotal													
Grand total													

Table 1 (b). Cash outflow.

Cash expenditure	M	J	J	A	S	O	N	D	J	F	M	A	Total
Labour, machines, service costs													
Land preparation													
Planting													
Crop care													
Harvesting													
Primary processing													
Subtotal													
Material costs													
Seeds													
Fertilizers													
Chemicals													
Fuel													
Tools													
Subtotal													
Other costs													
Repair services													
Interest on loan													
Land rent													
Grading/ packaging													
Transportation													
Tax													
Administration													
Other													
Subtotal													
Grand total													
Balance (Cash inflow - cash outflow)													

Returns on farm business

Gross returns are calculated according to the formula:

Gross returns = $(Pc_1.Qc_2) + (Pc_2.Qc_2) + (Pc_3.Qc_4)$

$$= \sum_{i=1}^{n} Pc_iQc_i + \ldots\ldots\ldots\ldots + (Pc_n.Qc_n)$$

where:

Pc_i = price of crop *i*

Qc_i = quantity of crop *i* produced

Gross returns per unit of area planted with each crop in the system is:

$$\frac{\text{Total returns of crop } i}{\text{Planted area of crop } i}$$

This method is applicable to crops with perfect competition, where price does not vary from one market to another. Comparison on returns among crops in the system can also be made if production costs of each crop are slightly different or are not taken into account.

The net returns can be total net returns or net returns per unit output or land.

Net returns = (total revenue) - (total fixed cost + total variable cost)

= TR - (TFC + TVC)

= TR - TC

The net returns per unit of output of each crop in the system is:

$$\frac{TR - TC}{\text{Total output}}$$

or = output price - average total cost per unit of output

= $Pc_i - ATC_i$

where: Pc_i = price of crop *i*

ATC_i = average total cost of crop *i*

The net returns per unit of area planted/harvested of each crop in the system is $TR_i - TC_i$

$$\text{The net return/unit of area planted} = \frac{TR_i - TC_i}{\text{Area planted of crop } i} = \frac{TR_i - TC_i}{Ha_i}$$

where: Ha_i = area planted of crop i in hectares.

The assessment of net return per head of the family of an active farm family aims at evaluating that the farmer has sufficient income. The basis of the calculation is:

$$\text{Net return per head/year} = \frac{\text{Net return}}{\text{No. of family members working on farm}}$$

Where farmers' families could have income from nonfarm activities, gross returns per head is used. This is based on:

$$\text{Gross return per head/year} = \frac{\text{Total revenue}}{\text{No. of family members working on farm}}$$

In order to compare the cost advantage in introducing a new crop, the following calculations can be used:
Cost per unit of area planted of each crop in the system -

$$= \frac{\text{Total cost}}{\text{Area planted}}$$

Total cost = total fixed cost + total variable cost

Cost per unit of output of each crop in the system -

$$= \frac{\text{Total cost}}{\text{Total output}}$$

When fixed costs are not taken into account or there are some constraints or difficulties in estimating the benefits, the following calculation can be used:

Returns over variable costs = TR - TVC

where: TR = gross revenue from all crops in the system

TVC = total variable costs of all crops produced under the system

Cost components

The cost components are shown in Table 2.

Table 2. Cost components.

Cost items	Family	Hired	Total
1. Variable costs			
1.1 Labour			
- Land preparation			
- Planting			
- Crop care			
hand-weeding			
machine-weeding			
spraying			
pruning			
fertilizing			
wartering			
other			
- Harvesting, drying			
- Threshing, grading, packaging			
- Primary processing			
- Other			
1.2 Materials			
- Seeds			
- Chemicals			
- Fertilizers			
- Fuel			
- Tools with one-year life span			
- Containers			
- Other			

Table 2. (cont'd)

Cost items	Cash	Noncash	Total
2. Fixed costs			
- Renting			
land			
equipment			
- Permanent labour			
- Interest on loan			
- Insurance			
- Repairs*			
- Management			
Total			

Total cost = total fixed cost + total variable costs.

* Cost of repairs can be classified as fixed or variable costs depending on the specific case concerned.

The methods for evaluation can be selected from some of those recommended above, depending on the purpose of the evaluation. These methods can also be developed to suit the case of each specific study.

Social evaluation

The social evaluation will be based on the farmers' views in conjunction with field observations by the researcher, and will be concerned with the following items in the cropping system.

- soil, water, and conservation issues involved in the cropping system;
- labour and other inputs in the cropping activities;
- marketing and prices of crops produced under the system;
- the food supply obtained from the system;
- other intangible benefits; and
- problems concerned with the cropping system, notably:
  * agricultural problems,
  * natural conservation problems,
  * prices of products, and marketing problems,
  * income distribution problems,
  * household food supply problems, and

* infrastructure availability problems.

Social evaluation will need to be made on both the positive and negative aspects of the cropping system.

Recommended reading

CASTLE, E.N., BECKER, M.H. and SMITH, F.J. 1972. *Farm business management*. New York: Macmillan.

DOLL, J.P. and ORAZEM, F. 1978. *Production economics theory with applications*. London: John Wiley & Sons.

HARSH, S.B., CONNOR, L.J. and SCHWAB, G.D. 1981. *Managing the farm business*. New Jersey: Prentice-Hall.

KADLEC, J.E. 1985. *Farm management: decisions, operations, and controls*. New Jersey: Prentice-Hall.

BROWN, M.L. 1979. *Farm budgets - from farm income analysis to agricultural project analysis*. World Bank, Baltimore: John Hopkins University Press.

OSBURN, D.D., and SCHNEEBERGER, K.C. 1978. *Modern agriculture management*. Boston, Virginia: Prentice-Hall.

SHANER, W.W., PHILIPP, P.F., and SCHMEHL, W.R. 1982. *Farming systems research and development: guidelines for developing countries*. Colorado: Westview.

IRRI (International Rice Research Institute). 1987. Training report. Agricultural Economics Department. Nigeria: IRRI.

Physical sustainability evaluation of cropping systems in tropical highlands

H. HUIZING*

Abstract

Evaluation of the physical sustainability of cropping systems is concerned with the maintenance of productivity as well as with the conservation of resources for long periods. A major constraint, however, is the lack of reliable procedures to predict the sustainability of tropical cropping systems. Crop yields, for instance, are generally not directly related to specific soil or land properties. In this paper, two approaches are described: (i) the use of "key land qualities" that act more or less independently on crop growth and yield, a number of which can affect crop growth and can be assessed quantitatively (e.g. the availability of soil moisture and nutrients) by means of crop growth models; and (ii) the use of "site-specific observations" on landform and soils in combination interviews with farmers on land-use history, management and yields - which provides important information on present cropping systems.

Introduction

A cropping system is an agricultural land-use unit comprising crops, weeds, pathogens, soil, and water (Fresco, 1986). The physical sustainability of cropping systems refers to the maintenance of productivity as well as to the conservation of soil resources for long periods.

* International Institute for Airial Survey and Earth Sciences (ITC), 350 Boulevard 1945, B.P. 6, 7500 AA Enschede, The Netherlands.

Cropping systems are the result of human activities that change (semi-) natural ecosystems into crop agroecosystems. In tropical highlands, the population increase and the shortage of land are the major causes underlying this change. Generally the effects of the change are drastic at first. Deep-rooted vegetation is replaced by shallower-rooting crops that are unable to use nutrients released by weathering in deeper layers; bare soil is exposed to rain and runoff for part of the year, which leads to erosion of fertile surface soil materials; soil organic matter rapidly diminishes because of mineralization and a reduced supply of organic materials; infiltration decreases; and by harvesting crops, nutrients are removed from the system. This process of land degradation is associated with declining productivity. The decline in yield is mostly rapid in the first few years after clearing, and usually becomes more gradual afterwards (Figure 1). The rate of decline depends on soil type and management. Improved soil management may reduce the high initial rate of yield decline, and thereafter establish an equilibrium in which yields do not decline further.

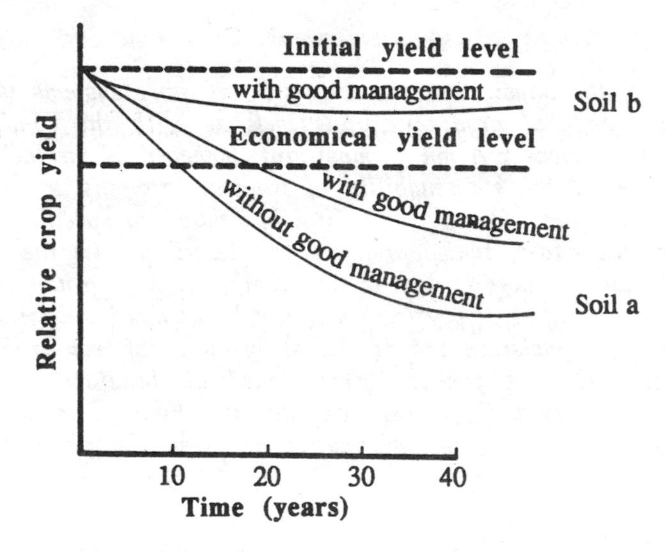

Figure 1. Yield trends on different soils and for different management (tillage) systems (Lal, 1988).

The effects of land degradation processes are easy to identify when changes occur rapidly. Later, when land degradation and yield decline are more gradual, the identification of changes is more difficult and often becomes masked by the effects of weather fluctuations and pest and disease incidence over the years.

Land productivity and land degradation

Evaluation of the sustainability of cropping systems involves the assessment or prediction of changes in yields and land properties over long periods, and consequently systematic long-term observations are needed. Such records are scarce for the tropics and, when available, are mostly incomplete. Evaluation of sustainability has to be based, therefore, on a knowledge of yield decline and land degradation processes obtained through (i) shorter-term experiments, (ii) experiences of farmers, and (iii) systematic field observations covering the effects of land use and land-use management history on similar soils at different sites. Such knowledge is essential for the prediction of the performance of current as well as improved cropping systems at different sites. 'Black box' approaches, ignoring the available knowledge on factors and processes underlying yield decline and land degradation, are likely to result in data sets for different sites that cannot be compared, and that cannot be used for the prediction of the performance of cropping systems at new locations.

Yield decline is related to land degradation. Crop yield and yield decline are integrated responses to many interacting landform, soil and climate parameters, and are therefore difficult to relate to individual parameters (Lal, 1988). Yield decline, for instance, is often not directly related to soil losses (in mm per year) by water erosion (Young, 1986; Lal, 1988).

Understanding a "whole" cropping system and all the complex interactions of its soil and crop subsystems is extremely difficult. A simple approach is desirable. Such an approach can be based on the identification of a limited number of "key land qualities" that have a more direct and better-known influence on productivity. Key land qualities in tropical highlands are:

- soil moisture availability;
- nutrient availability;
- physical conditions of surface soil affecting crop establishment and germination;
- weed population; and
- incidence and severity of pests and diseases.

The five listed qualities act "more or less" independently on crop growth and yield. They incorporate the effects of management (e.g. nutrient availability is based on the combined effect of natural fertility and applied fertilizer). A difficulty, however, is to assess key qualities in quantitative terms, because they are also influenced by the interaction of many land properties. Recent developments in modelling, however, make it possible to predict crop yields or yield levels based on soil moisture availability and the availability of major soil nutrients. In these predictions, the effects of individual qualities on crop yields are assessed separately; the predictions generally assume that other land qualities are not limiting. If several qualities cause yield reduction, the sufficiency of the most limiting quality determines the yield (Law of the Minimum).

Assessing the sufficiency of soil moisture and nutrients by models

The sufficiency of a land quality should be assessed by means of observable or measurable land properties. Because of interactions between land properties, such an assessment is not easy. Another difficulty is that qualities and related soil properties change within a crop season or crop year and between years. Seasonal changes are most pronounced in the quality "soil moisture availability". Also nutrient availability is affected by seasons (e.g. the increased availability of most nutrients early in the rainy season because of rapid mineralization of organic matter). The influence of pests and diseases on crop productivity fluctuates over periods of several years.

Considerable progress has been made in the last decade in the development of models that predict (relative) crop yields in relation to sufficiency levels of some land qualities. Two main types of model are available: (i) empirical (or expert) models and (ii) simulation models.

Empirical or expert models use decision rules based on general and local expert knowledge. Individual soil properties (observed or measured in the field or in the laboratory) are used to assess the sufficiency level of land qualities. Generally decision rules have to be modified when applied to environments that differ from the ones for which the rules were established. Examples of expert systems that include land quality assessments are LECS (Wood and Dent, 1983) and ALES (Rossiter *et al.*, 1988). LECS was designed for conditions prevailing in the outer islands of Indonesia. ALES makes it possible to assess relative yields using decision trees which structure (local) expert knowledge.

Simulation models are based on known plant growth processes and (dynamic) land attributes. It is often claimed that these models are universally applicable. However, simulation models require calibration and validation when applied in new areas. There are complicated models that require detailed data obtained through sophisticated methods as input, and simple models with modest data requirements. Simple models often, but not always, produce less accurate results.

Examples of models that calculate crop yield reductions in relation to soil moisture deficits are the SWATR/CROPR models (Feddes *et al.*, 1978), the IBSNAT model, the FAO model (Doorenbos and Kassam, 1979) and the

WOFOST model (Van Keulen and Wolf, 1986). Yield reductions calculated for experimental sites can be related to long-term records of neighbouring climatic stations to assess their probability of occurrence, i.e. whether they apply to normal or exceptional years. With the WOFOST model (version 4.1) and the QUEFTS model, crop yields can be estimated on the basis of supply and uptake of N, P and K, taking into account interactions between these nutrients. The QUEFTS model bases its estimation on the following topsoil properties: pH-H_2O, organic carbon, P-Olsen, and exchangeable K. The SCUAF model (Young *et al.*, 1988) predicts soil erosion and changes in soil organic matter, together with feedback effects of soil changes on plant growth.

A lack available data generally restricts the use of simulation models to experimental sites. Most soil moisture balance models, for instance, require quantitative data on bulk density, hydraulic conductivity and moisture retention. This data is difficult and/or time-consuming to obtain for a large number of sites. By the development of (pedo-)transfer functions based on regression analysis, this quantitative data can be related to data on land characteristics collected during routine soil surveys (Bouma and van Lanen, 1987). For instance, bulk density values can be calculated using a mathematical expression based on organic-matter content, texture and structure; and soil moisture retention values can be based on organic-matter content, clay content, and bulk density. A simulation model may contain several transfer functions (Figure 2). Transfer functions facilitate the transfer of soil management research results from experimental sites to other locations.

The models discussed above help to explain differences in the performance of cropping systems over the years, and the extent to which these differences are caused by deficits of soil moisture, N, P and K. All models require data sets for calibration and validation. For the development of transfer functions, good-quality data sets are also required.

Assessing other key land qualities

Effects of the physical conditions of the surface soil, weed population, and pests and diseases on crop growth and yield cannot yet be assessed in a (semi-) quantitative way in tropical areas by the use of modelling techniques. These qualities and the results of interventions (e.g. tillage techniques, mulching, cover, shade manipulation by cropping pattern/rotation design, and herbicide or pesticide use) have to be assessed by careful and systematic site observations.

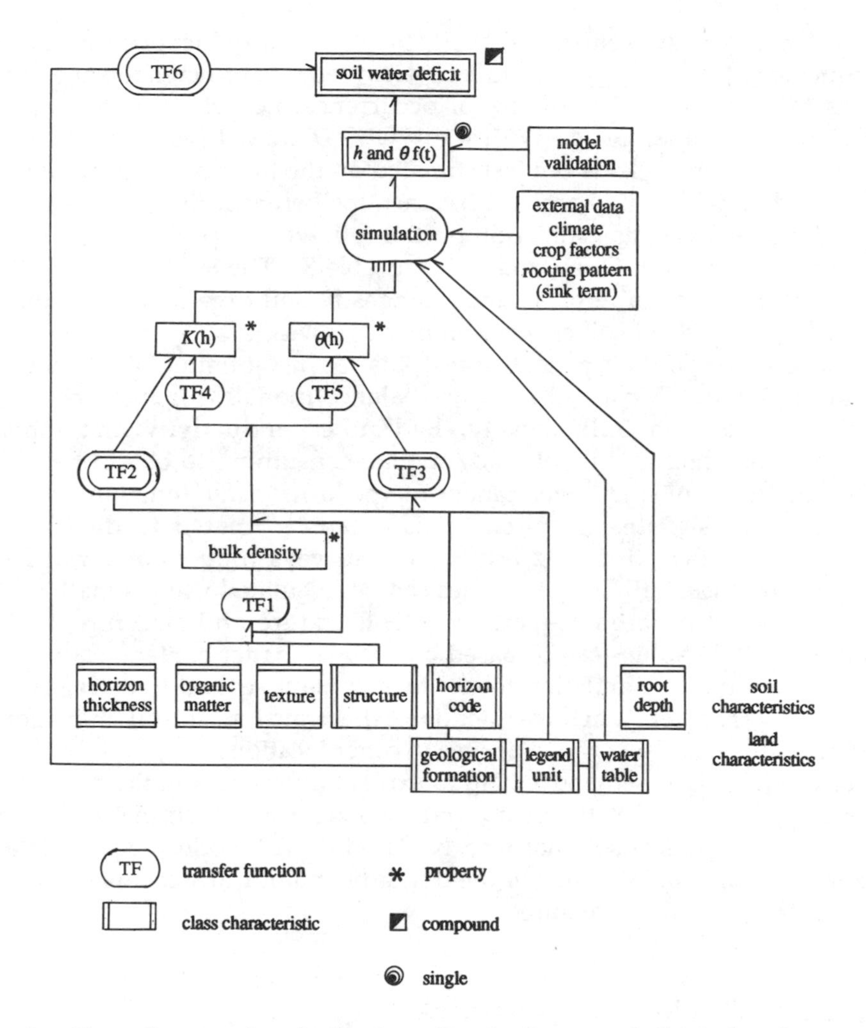

Figure 2. Flow diagram for the land quality "soil water deficit" showing transfer functions (Bouma and van Lanen, 1987).

Evaluation of sustainability cropping patterns

The maintenance of productivity and the conservation of soil resources for long periods are the main criteria used in the evaluation of the sustainability of cropping systems. The evaluation generally has to be made for three situations:

- current cropping systems,
- experiments, and
- transfer of experimental results from research sites to farmers' fields.

The evaluation of the sustainability of current cropping systems should be based on "site-specific" observations, covering landform, soil and vegetation/ crop cover, and interviews with farmers on land-use history, crop rotations/ cropping patterns, technology used, management, and yields/yield history. The interpretation of sequential coverages of airphotos or satellite images is an efficient tool for the assessment of the history and sustainability of land uses, particularly when land cover/land-use and landform information obtainable from the photos and images is combined and used for stratified sampling by means of site-specific observations in the field. Field observations should preferably include a comparison of (i) different land uses and land-use history for similar landform/soil situations, and (ii) similar land uses and land-use history for different landform/soil situations. Yield information provided by farmers is subject to the money bias. In addition, yields are influenced by fluctuations of weather and pest/disease incidence over the years. However, major differences in yield reported by farmers using comparable technologies and management, when related to major soil differences, can give important information on the sustainability of current cropping systems (Table 1).

Table 1. Yield decline of main-season maize crop in relation to soil parent material and cultivation period (Funnpheng, 1988).

Soil parent material	Cultivation period (years)	Yield (kg/ha)	Average yield decline (%)
Siltstone/shale	3-5	3125-4060	–
-ditto-	6-8	2185-3125	26
-ditto-	10-25	935-2185	57
Andesite	3-5	4375-5625	–
-ditto-	10-25	3125-4375	25

The evaluation of the sustainability of new or improved cropping systems has to be made in two stages: (i) during the design of experiments, and (ii) after the experiments have been conducted. The first evaluation should, as far as possible, use available knowledge on crop growth and soil processes to predict the expected results of the experiments. The second evaluation should in a (semi-) quantitative way predict the sustainability of promising experimental results. Simulation models and comparison with long-term climatic records play an important role in this evaluation.

Transfer functions are important when results of experiments are transferred to new locations. The development of transfer functions should be an integral part of the research.

Conclusions

- There is little quantitative data on the sustainability of cropping systems for the tropics. Evaluation of the sustainability of cropping systems should therefore be based on available knowledge on plant growth and soil processes.
- Site-specific landform/soil observations combined with farmers' interviews generally provide important information on sustainability problems of current cropping systems.
- Yields of crops and productivity are difficult to relate to individual land properties. It appears easier to assess separately the effects of a number of "key land qualities" on yields. Key land qualities in tropical highlands are: soil moisture availability, nutrient availability, physical surface soil conditions, weed population, and pests/diseases.
- Two key land qualities can be assessed with a reasonable accuracy by existing simulation models: available soil moisture and available N, P and K. Other key land qualities need careful monitoring during experiments. Paired-plot techniques (Lal, 1988) will generally be useful. Preventive applications of herbicides and pesticides may be needed to assess quantitatively the effects of weeds, pests, and diseases on crop yields.
- The development of transfer functions based on observations of different experimental sites will facilitate the transfer of promising technologies developed by research to other locations.

References

BOUMA, J. and VAN LANEN, H.A.J. 1987. Transfer functions and threshold values: from soil characteristics to land qualities. In: *Quantified land evaluation*, ed. Beek, Burough and McCormack. ITC Publication no. 6. Enschede, The Netherlands: ITC.

DOORENBOS, J. and KASSAM, A.H. 1979. *Yield response to water*. FAO Irrigation and Drainage Paper no. 33. Rome: FAO.

FEDDES, R.A., KOWALIK, P.J. and ZARADNY, H. 1978. *Simulation of field water use and crop yield*. Wageningen, the Netherlands: PUDOC.

FRESCO, L.O. 1986. Cassava in shifting cultivation. *A systems approach to agricultural technology development in Africa.* Amsterdam, The Netherlands: Royal Tropical Institute.

FUNNPHENG, P. 1988. Contribution to erosion hazard assessment in land evaluation for conservation. Ban Sila area, Pa Sak valley, Thailand. Unpublished M.Sc. thesis. ITC, Enschede, The Netherlands.

JANSSEN, B.H. GUIKING, F.C.T. VAN DER EIJK, D., SMALING, E.M.A., WOLF, J. and VAN REULER, H. 1988. A system for quantitative evaluation of the fertility of tropical soils (QUEFTS). Unpublished report. Agricultural University, Wageningen, The Netherlands.

KOOIMAN, A. 1987. Relations between land cover and land use and aspects of soil erosion. Upper Komering catchment, South Sumatra, Indonesia. Unpublished M.Sc. thesis. ITC, Enschede, The Netherlands.

LAL, R. 1988. Monitoring soil erosion's impact on crop productivity. In: *Soil erosion research methods*, ed. R. Lal. Wageningen, The Netherlands: International Society of Soil Science.

ROSSITER, D.G., TOLOMEO, M. and VAN WAMBEKE, A.R. 1988. Automated Land Evaluation System (ALES). Ithaca, NY: Cornell University.

SANCHEZ, P.A., COUTO, W. and BUOL, S.W. 1982. The fertility capability soil classification system: interpretation capability and modification. *Geoderma 27.* Amsterdam, The Netherlands: Elsevier.

VAN KEULEN, H. and WOLF, J., eds. 1986. *Modelling agricultural production: weather, soils and crops.* Wageningen, The Netherlands: PUDOC.

WOOD, S.R. and DENT, J.F. 1983. *LECS: A land evaluation computer system methodology.* Bogor, Indonesia: FAO and CSR.

YOUNG, A. 1985. *The potential of agroforestry as a practical means of sustaining soil fertility.* ICRAF Working Paper no. 34. Nairobi, Kenya: ICRAF.

YOUNG, A. 1986. The potential of agroforestry for soil conservation. Part I. *Erosion control.* ICRAF Working Paper no. 4, Nairobi, Kenya: ICRAF.

YOUNG, A. 1987. *The potential of agroforestry for soil conservation.* Part II, *Maintenance of fertility.* ICRAF Working Paper no. 43, Nairobi, Kenya: ICRAF.

YOUNG, A., CHEATLE, R.J. and MURAYA, P. 1988. *The potential of agroforestry of soil conservation.* Part III. *Soil changes under agroforestry (SCUAF): A predictive model.* ICRAF Working Paper no. 44, Nairobi, Kenya: ICRAF.

Section 4: Site selection for experiments

Elements of experimental site selection

SAMARN PANICHAPONG*

Abstract

The paper discusses IBSRAM's concepts and methodologies for site selection, characterization, and the establishment of experiments for soil management networks. The suggestions in this paper are intended as a basis for discussions to determine appropriate guidelines for the ASIALAND network on sloping lands.

Introduction

This paper outlines the major points presented and discussed in previous IBSRAM training workshops. IBSRAM places strong emphasis on site selection and characterization for field experiments for the following reasons:

- Site selection is the first step in establishing a field experiment. Any experiment which starts with poor site selection is doomed to failure - failure, that is, in the sense that the experimental results will have no practical application. Studies have shown that areas on sloping terrain with uniform soils do not exist. Pushparajah and Chan (1986) concluded that lithological and pedological factors were the major causes of inherent variation. Such variations could often be large and induce a significant differential effect, even on the growth of tree crops. Biogenetic variability

* Thai adviser, IBSRAM Headquarters, PO Box 9-109, Bangkhen, Bangkok 10900, Thailand.

due to vegetation and faunal activity are also encountered. However, the presence of the latter are often confined to only a few soils. The largest variability in chemical properties is induced by management practices, such as the burning of residues and the use of fertilizers, especially in tree crop areas. Such large variability entails more intensive sampling, and affects the outcome of most of the established field experiments. The results are only valid for the specific crop variety and pedoclimatic conditions in which they were obtained. The extension of the results to other sites is always doubtful in the absence of a proper site characterization.

- The main task of IBSRAM is to develop adaptive soil management research networks. In the broadest sense, the objective is to form a linkage between individuals or institutes with a shared purpose into some form of collaborative effort. The success of any research network requires close collaboration among cooperators and the network coordinator. They must speak the same language and base their work on the same methodology. The reason for the emphasis on site selection and characterization in every training workshop is to make sure that every coordinator applies the same standard criteria in selecting sites for the experimental network, and characterizes the site in the same way so that the results of the experiments can be compared and transferred. Wambeke (1988) considered, site characterization, including site selection, to be one of the fundamental components of soil management networks. The characterization process takes on a new dimension in the research networks. It is not always feasible to reproduce site characteristics from one site to another. The best alternative is to characterize the site as well as possible and to document the results clearly and methodically (Eswaran, 1986).

There is no single standard set of criteria for site selection. However, the following basic considerations need to be borne in mind:

- The site must conform to the research requirements stipulated by the network. The main objective of the ASIALAND network is to test sustainable soil management technologies for sloping lands in the tropical Asian region. Therefore land with some degree of slope which is commonly used by farmers is selected for the network. Although the soil type is not the main consideration, it must represent the major soil of the area.
- The site must be representative of an important agroenvironment.
- The site must be available for long-term experiments.
- Acessibility to the site must be reasonably good, allowing researchers to make regular observations and sampling.
- The site should be located near a weather station - or alternatively the researcher needs to have access to standard meteorological data-collecting equipment.

The methodological guidelines for IBSRAM's soil management networks suggest five steps for site selection and characterization for experimental trials on soil management:

- Site selection at the regional level, aimed at locating the general area in which the experiments are to be located, and examining it in enough detail to ensure that it is representative of the soils and climate desired. A reconnaisance soil map, climatological data, a land-use map, and regional socioeconomic information are used in the selection process, and the information these data provide can be supplemented by field observation. Consultation with local soil scientists can be of great help in ensuring an accurate assessment.
- Site selection at the local level within the area selected, in accordance with the requirements for a description of the regional level, as given above. The area is surveyed in more detail to facilitate the final choice of the project site itself, which usually needs to be about 2 ha. A more detailed soil map, probably at detailed reconnaisance or semidetailed level, is used, together with cropping systems and local socioeconomic conditions before a final decision is made.
- After the experimental site has been selected, a detailed characterization has to be made. This involves a very detailed soil survey on a regular grid system, a detailed contour map with contour intervals of 2 m, climatological data measurements, and a description of the socioeconomic characteristics of the site. Details of the information to be collected are explained in the guidelines. Table 1 gives a list of the soil and site information which needs to be assessed and monitored.

Site characterization is considered to be another major step in experimental site selection. Without proper site characterization, the transferable value of the results obtained from the experiment will be reduced. Davidescu and Davidescu (1972) have shown that the results of field experiments were valid strictly for species, variety (hybrid), and the pedoclimatic conditions in which they were obtained. An interpretation of the experimental data under field conditions requires a detailed soil characterization followed by a physical, chemical and biochemical analysis at various stages of crop development. Nelson (1987), has recommended a suitable approach to the arrangement for the layout of the plots once the site has been chosen and characterized.

Site characterization should (1) define soils at the lowest level of classification, such as the family name in the *Soil Taxonomy*, the soil series or soil variants in the national system, (2) include important soil management properties which are not included in *Soil Taxonomy* system (such as stoniness, rockiness and slope), (3) use the existing national soil classification system, but also refer to international systems used in the cooperating countries (Wambeke, 1988).

- The design of the agronomic experiment to be carried out. The design of the experiment is one of the factors influencing the technical success of soil management research networks, and is stressed in all IBSRAM training workshops. Some basic recommendations are made in the guidelines.

Table 1. Summary of soil and site information to be assessed and monitored.

Climate	Soil	Drainage and hydrology (field)	Topography and vegetation	Socioeconomics
Rainfall	Field arrangement of horizons	Depth to water table	Slope, erosion hazard	Land tenure
Temperature	Texture	Infiltration rate		Cropping systems
Potential evaporation	Structure	Hydraulic conductivity	Existing vegetation	Cost of production
	Stoniness	Moisture retention and available moisture capacity		Income (on farm and off-farm)
Storms	Depth of rooting zone		Position and accessibility	
Seasonal variability	Consistence in the laboratory			Marketing problems related to crop production
Length of growing season	Particle size distribution			
Solar radiation	Bulk density and porosity			
	O.M. content			
	C.E.C. and base saturation			
	Electrical conductivity			
	pH			
	Soluble and exchangeable cations (especially Na and Mg)			
	Nutrient levels (including possible toxicities)			
	Sulphate and carbonate level			

- During the course of the experiment, there needs to be careful monitoring of the growth and yield of the crop in relation to the treatments selected, of the weather, and of changes in soil morphology, soil physical properties, and soil fertility. This is crucial to the success of the experiment.

The intention of this presentation is to introduce participants to the methodological guidelines prepared by IBSRAM. The guidelines are intended for use by all cooperators. They should not, however, be thought of as a rigid set of requirements. In fact, they need need to be frequently modified to suit local conditions. This workshop will consider and modify the guidelines in order to meet the perceived requirements of the ASIALAND network.

References

DAVIDESCU, D. and DAVIDESCU, V. 1972. *Evaluation of fertility by plant and soil analysis*. Tunbridge Wells, Kent, England: Abacus Press.

ESWARAN, H. 1986. Soil and site characterization for soil-based research networks. In: *Soil management under humid conditions in Asia - ASIALAND*, 169-186. IBSRAM Proceedings no. 5. Bangkok: IBSRAM.

NELSON, L.A. 1987. Design of experiments for land development and management of acid soils in Africa. In: *AFRICALAND - Land development and management of acid soils in Africa II*, 209-216. IBSRAM Proceedings no. 7. Bangkok: IBSRAM.

PANICHAPONG, S. 1986. Site selection for agricultural research or experimentation in Thailand. In: *Soil management under humid conditions in Asia - ASIALAND*, 117-132. IBSRAM Proceedings no. 5. Bangkok: IBSRAM.

PUSHPARAJAH, E., and CHAN, H.Y. 1986. Site selection for agronomic experimentation: Malaysian experience. In: *Soil management under humid conditions in Asia - ASIALAND*, 133-154. IBSRAM Proceedings no. 5. Bangkok: IBSRAM.

WAMBEKE, A. Van. 1988. Site selection and soil variability. In: *First Training Workshop on Site Selection and Characterization*, 43-52. IBSRAM Technical Notes no.1. Bangkok: IBSRAM.

Agroclimatic characterization: Basic processes and measurements

R. H. B. EXELL*

Abstract

The important physical processes of agrometeorology are: exchanges of solar and longwave radiation at the surface, transfers of heat above and below the surface, transfers of water by rainfall, evapotranspiration and runoff, and air movement in the surface boundary layer. The most generally useful data include air temperature, precipitation, insolation, soil temperature, soil moisture content and evaporation, although measurements of some of these quantities may be unreliable. Evapotranspiration from crops is generally calculated from other data; methods used include the radiation method from temperature and insolation data, and the Penman method (in which wind and humidity data are also needed). Field agrometeorological stations may have instruments for measuring rainfall, air temperature, insolation, wind, air humidity, evaporation, soil temperature, and soil moisture content.

The physical processes of agrometeorology

The physical processes of agrometeorology are energy and mass transfer processes, mainly transfer of radiation, heat, water, momentum, carbon dioxide, and other airborne materials.

---

* Energy Technology Division, Asian Institute of Technology, PO Box 2754, Bangkok 10501, Thailand.

Radiation

The principal input is shortwave solar radiation, while longwave thermal radiation is the principal output. Apart from the predictable diurnal and seasonal motion of the sun, the effects of cloud and the slope of the terrain on the distribution of the solar input are important, together with the optical properties of vegetation and soil. In the case of thermal loss by longwave radiation, temperature, cloud cover, and air humidity are important.

Heat

Above the surface, heat transfer is by convection, which may be free as a result of surface heating, or forced through the effect of the wind. Below the surface, the temperature distribution in the soil is a result of conduction.

Water

The inputs of water are rainfall and dew formation. Dew formation is less important, except possibly in arid areas or during long periods of drought. The distribution of rainfall in time and space is highly variable and is a major question of agrometeorology. The output of water from the surface is via evapotranspiration, runoff, and percolation through the soil. Evaporation is governed principally by the solar radiation input (to provide the necessary latent heat), by air humidity, and by turbulence produced by the wind at the surface. Soil moisture is a major limiting factor in crop growth; too much water in the soil leads to anaerobic conditions, and too little leads to wilting and plant death. The distribution of soil moisture in depth and time are therefore important, while the leaching of soil nutrients and soil erosion are further problems.

Momentum

Air movement at the surface generated by the wind aloft is very complicated. It depends not only on the surface roughness, the presence of obstructions and the general lie of the terrain, but also on the thermal stability of the air in the boundary layer at the surface (normally of the order of ten to a hundred metres deep) due to variations in the heating and cooling of the surface by radiation processes. Air movement, besides having important effects on heat and water vapour transfer, may be important in damage to crops and in soil erosion. Exchange of gases in the atmosphere, such as carbon dioxide, is also influenced by air movement.

Meteorological observations

Pressure

Pressure is measured at all synoptic meteorological stations. It is not usually needed, but when it is, the pressure can be obtained from nearby synoptic stations with sufficient accuracy. Corrections for distance are normally small, but corrections for height may be important on high ground.

Air temperature

The air temperature, as normally reported, is measured by thermometers in a screen above the ground. This is not necessarily the temperature experienced by crops or the soil. It does, however, serve as a good reference standard.

Air humidity

The humidity of the air is best measured by simultaneous readings of dry and wet bulb thermometers placed in the white louvered screen. It may be expressed in several different forms, e.g. relative humidity, dew-point temperature, or vapour pressure. Dew-point and vapour pressure, which are direct indications of the absolute humidity, are fairly conservative in uniform terrain. On the other hand, relative humidity (the ratio of the actual amount of water vapour in the air to that in fully saturated air) is sensitive to changes in air temperature. This is because the saturation vapour pressure is a strong function of temperature.

Wind

Wind force and direction vary strongly with time, height and location, and are very difficult to measure in a standard uniform manner. The exposure of the instrument can make a big difference to the readings obtained. For this reason, standard wind data at synoptic stations are not often representative of a local site.

Precipitation

Rainfall measurements can be affected both by the design and by the exposure of the rain gauge. Rainfall data must therefore be treated with caution. Rainfall totals or averages in the long term can be useful in uniform terrain, but in mountain regions local variations can be large. Totals for single storms and short periods are valid only for the point of measurement.

Sunshine and solar radiation

Daily duration of sunshine measurements are often valuable, and are representative of a considerable area around the point of observation, except near a sea coast and in mountain areas. It is possible to estimate solar radiation from sunshine records with fair accuracy, but for more specialized work direct measurements of solar radiation can be made. Care must, however, be taken in accepting solar radiation measurements, since the instruments need good maintenance and regular calibration.

Evaporation

The direct measurement of evaporation is very difficult. The reason for this is that the water vapour flux upward from a surface depends on the profiles of temperature, humidity, and momentum over the surface. Moreover, measurement of the evaporation from a pan requires that observations be made of the small difference between two levels that are close together, with consequent loss of accuracy. The pan may also be a source of drinking water for birds and animals, and hidden leaks may be another difficulty.

Surface and soil temperatures

Surface temperature measurements, obtained by placing an unscreened thermomenter on the ground at night, are not reliable. On the other hand, soil temperatures at various depths are very useful as reference values, and are often uniform over large areas. At a depth of about 0.3 m, the temperatures obtained eliminate short-term variations of air temperature.

Soil moisture

When it is attempted, the measurement of soil moisture may be unreliable, and cannot be regarded as representative of anything other than the point of observation. Even with the use of modern methods, it is doubtful whether any network would obtain adequate sampling efficiency.

Calculation of evapotranspiration

Because the measurement of evapotranspiration is difficult, a number of methods of calculating it have been developed. They require information on air temperature, solar radiation, air humidity, and wind. The 'reference crop evapotranspiration' ET_o is defined as the rate of evapotranspiration from an extensive surface of 8-15 cm-tall green grass cover of uniform height, actively growing, completely shading the ground and not short of water. This quantity is then estimated from the meteorological data by means of an empirical formula. The 'crop evapotranspiration' ET_c of a specific crop is then calculated from ET_o by means of the formula:

$$ET_c = k_c . ET_o,$$

where k_c is a crop coefficient that depends on the crop characteristics, the time of planting or sowing, the stage of crop development, and the general climatic conditions.

The radiation method

If air temperature, together with cloudiness, sunshine or solar radiation data are available, but wind and humidity data are not, then the recommended relationship is:

$$ET_o = c(W.R_s),$$

where ET_o = reference crop evapotranspiration (mm/d)
R_s = solar radiation in equivalent evaporation (mm/d)
W = weighting factor, depending on temperature and altitude
c = adjustment factor, depending on humidity and wind

Direct measurements of solar radiation are used for R_s if they are available. Otherwise, R_s is estimated from sunshine or cloudiness. The weighting factor W, reflecting the effects of temperature and altitude, increases as the temperature and the altitude increase. The adjustment factor c is a weak function of mean relative humidity and daytime wind conditions, rough approximations of which are sufficient to estimate ET_o.

The Penman method

The form of the equation used in this method is:

$$ET_o = c[W.R_n + (1 - W).f(u).(e_a - e_d)].$$

Here W is the weighting factor for temperature and altitude, and R_n is the net radiation, i.e. the amount of solar radiation absorbed at the surface (incident radiation less reflected radiation) minus the net longwave output (longwave radiation emitted by the surface less the downward atmospheric radiation) in equivalent evaporation (mm/d). The factor $f(u)$ is a wind function:

$$f(u) = 0.27(1 + u/100),$$

where u is the 24-hour wind run (km) at height 2 m. This expression requires that the next factor $(e_a - e_d)$, which is the difference between the saturation water vapour pressure e_a at the mean daily air temperature and the mean actual water vapour pressure e_d, is given in millibars. The factor c is an adjustment that depends on the mean humidity and the diurnal wind variations.

Instruments of measurement

A field agrometeorological station may have instruments for measuring rainfall, air temperature, sunshine duration or solar radiation, wind speed and direction, air humidity, evaporation, soil temperature, and soil moisture. In the traditional field station the instruments are placed on a plot of land 10 m x 10 m in size surrounded by a wire fence (Figure 1). They are distributed in such a way that they do not interfere with each other. The instruments are simple, robust and easy to maintain. Readings are recorded daily by the observers. Today, however, manufacturers are producing compact automatic weather stations with electronic sensors mounted on a small mast. Electronic data loggers record the observations in computer-readable form for collection weekly or monthly by the operation personnel.

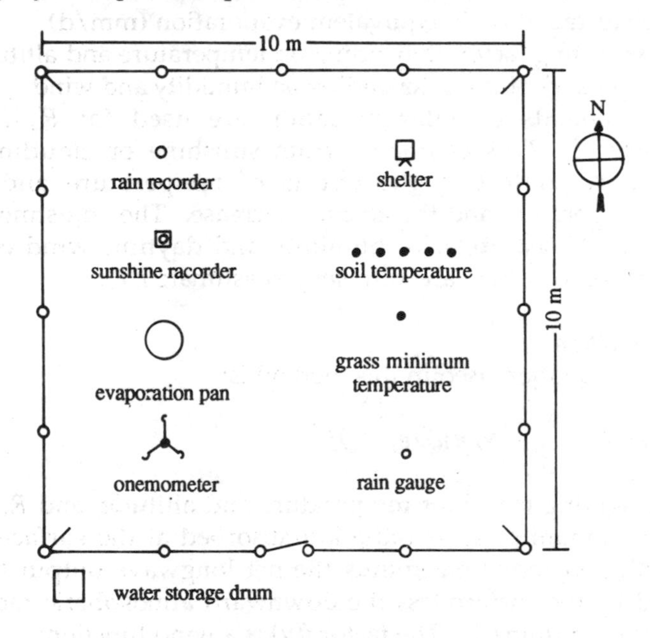

Figure 1 Layout of field agrometeorological station.

Rainfall

Measurement of rainfall is obtained from the depth to which rain has accumulated in a vessel free from obstructions, with no splash-in from the surroundings or evaporation from the vessel. A cylindrical knife-edge rim of precisely known area (200-500 cm²) is placed above a funnel leading to a receiver with a narrow neck to prevent evaporation. It is designed so that water cannot splash out (Figure 2). The rim must be horizontal at a height of 30 cm above the ground; at greater heights, overexposure to wind affects the accuracy of measurement. The amount of water collected is determined daily with the help of a graduated measuring cylinder. There should be no objects closer to the rain gauge than 2-4 times their height.

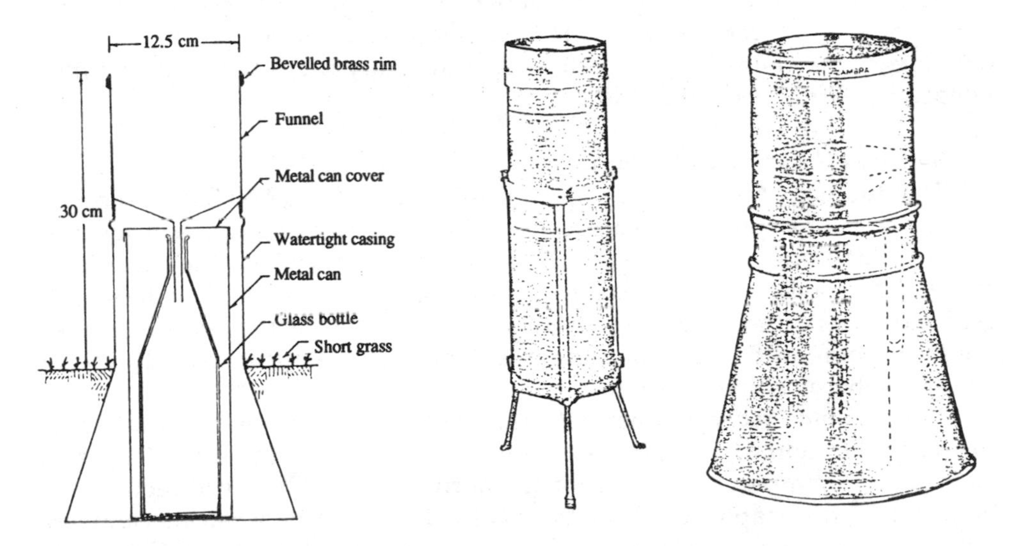

Figure 2. Rain gauges.

For automatic measurements, a tipping bucket rainfall recorder may be used. The collecting chamber is divided into two equal parts placed centrally under a collecting funnel and tilted so that water from the funnel flows into one half only. The chamber is balanced so that it tips each time 0.2 cm of rain has fallen, bringing the other half under the funnel. Each move is recorded on a chart, or is counted electrically.

Air temperature

A mercury-in-glass thermometer is used to measure air temperature. It must be shielded from solar radiation and rain, but there should be a free circulation of air around the instrument. Thermometers are therefore placed in a standard louvered screen painted glossy white and mounted 1.5-2 m above the ground. The ground cover beneath the screen should be grass.

The maximum and minimum temperatures each day may be measured with special thermometers. The maximum thermometer is mercury-in-glass, with a small constriction in the capillary near the bulb to prevent the mercury falling when the temperature falls. The minimum thermometer is alcohol-in-glass, with a small index inside the liquid in the stem. When the alcohol contracts due to cooling, the meniscus drags the index with it; when it expands, the alcohol flows past the index, leaving the position of the index unchanged. Both of these special thermometers must be reset daily.

For automatic recording, thermistors (whose electrical resistance varies with temperature) may be used. The resistance is measured electrically. The data logger may be programmed to record the daily maximum and minimum temperatures from the same sensor.

Sunshine duration and solar radiation

The duration of bright sunshine is measured by a Campbell Stokes recorder. This consists of a solid glass sphere about 10 cm in diameter, concentric with a metal collar slotted to take specially treated cards on which the records are made. The sphere acts as a lens that concentrates the rays of the sun onto the card and burns a trace when the sun is shining. The movement of the sun changes the focal point on the card, and the total length of the burnt trace during the day gives the sunshine duration.

The sunshine recorder does not measure the radiation intensity, although the daily duration of sunshine is closely correlated with the total daily solar radiation. A simple instrument for measuring solar radiation directly is the bimetallic actinograph, which measures the difference between the temperatures of a blackened bimetallic strip exposed to solar radiation and two similar bimetallic strips painted white on either side of it. A system of levers connected to the bimetallic strips causes the movement of a pen to record the temperature difference (which is proportional to the solar radiation intensity) on a paper chart. The instrument needs regular calibration and is accurate enough only for determining integrated daily totals, but it has the advantage of working without electricity.

A more accurate, but also more costly, solar radiation instrument is the solarimeter (Figure 3). This has a thermopile with a blackened set of junctions acting as the receiver, protected by a pair of concentric glass domes. The other set of junctions is in good thermal contact with a heavy metal base plate. The thermal EMF produced when solar radiation falls on the receiver is measured by a sensitive galvanometer, or recorded by a data logger. A less expensive electrical instrument that can be used with data loggers is a small photovoltaic cell, but care has to be taken that it has been properly calibrated against an accurate solarimeter, since the thermopile responds uniformly to the whole solar spectrum, whereas the photovoltaic cell does not.

Figure 3. Solarimeter.

Wind speed and direction

Wind speed is measured by a cup anemometer. A revolution counter gives the wind run between observations, which may be expressed in kilometres. Wind direction is obtained from a wind vane. In automatic weather stations the data logger can be programmed to record mean wind speed and the mean wind vector as well as the instantaneous values.

The siting of the instruments should be as free as possible from obstructions, but should be chosen with due regard to the purpose of the weather station. The height of the instruments is important. At synoptic stations it is 10 m above the ground; adjacent to the runways at airports it is 6 m; at agrometeorological stations it is 2 m, and sometimes another instrument is placed at a height 0.5 m above the water level in the evaporation pan.

Air humidity

Wet and dry bulb thermometers placed inside the thermometer screen may be used to measure air humidity. The wick that carries water from a small reservoir to the wet bulb must be kept clean, and distilled water should be used. Evaporation from the water in the wick covering the wet bulb lowers the bulb temperature by an amount that depends on the air humidity. The air humidity itself is obtained from tables and may be expressed in various ways, such as relative humidity, dew-point temperature or vapour pressure. If instead of being naturally ventilated the bulbs are ventillated by a small clockwork fan (the aspirated psychrometer), better results are obtained.

In automatic weather stations, the air humidity sensor may be a cracked chromium oxide capacitor inside the thermometer screen. Variations in the air humidity produce variations in the dielectric constant of the capacitor, which are measured electronically and recorded by the data logger.

Evaporation

The usual method of measuring evaporation for studies of crop water requirements involves the use of an evaporation pan 100 cm square and 55 cm deep sunk into the ground, with its rim projecting 5 cm above the surface (Figure 4). The water level is kept near ground level (5 cm below the rim) by adding or removing water as required, depending on the rates of evaporation and rainfall. Daily readings of the level are made by measuring the amount of water needed to raise the water surface to a reference level, or by means of a micrometer gauge, and the evaporation (in millimetres) is deduced.

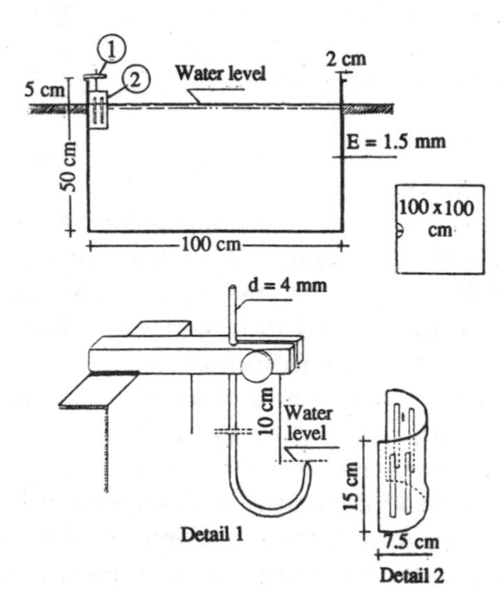

Figure 4. Colorado-type sunken evaporation pan.

Soil temperatures

Soil temperatures are normally observed at the standard depths below the surface of 5, 10, 20, 50 and 100 cm. Soil cover may be grass or bare ground, depending on requirements. Mercury thermometers may be used with the bulbs at the required depths. Down to 20 cm, the thermometer may have a right-angled bend with the graduations on the horizontal part of the stem. At the larger depths, each thermometer is suspended inside a tube let into the soil. The thermometer is embedded in wax within an outer glass tube which can be withdrawn to obtain the soil temperature reading. In automatic weather stations, thermistors may be buried in the ground at the required depths; their resistances are measured electrically and are recorded with the data logger.

Soil moisture

The traditional method of measuring soil moisture content consists of removing soil samples by augering and determining their moist and dry weights. The dry weights are found by drying the samples in an oven at 105°C. Samples should be taken from the rooting zone of the crops. The gravimetric method is a destructive one and may disturb an experimental plot sufficiently to distort the results.

An indirect nondestructive method of determining soil moisture content is to use neutron moderation. The neutron probe contains a source of fast neutrons and a detector selectively sensitive to only to slow neutrons. It is lowered into an access tube inserted vertically into the soil (Figure 5). Some of the fast neutrons are slowed down and scattered by hydrongen nuclei in the surrounding soil water, and the response of the detector provides a measure of the moisture content. A range of soil moisture regimes is needed to calibrate the instrument for a given soil type.

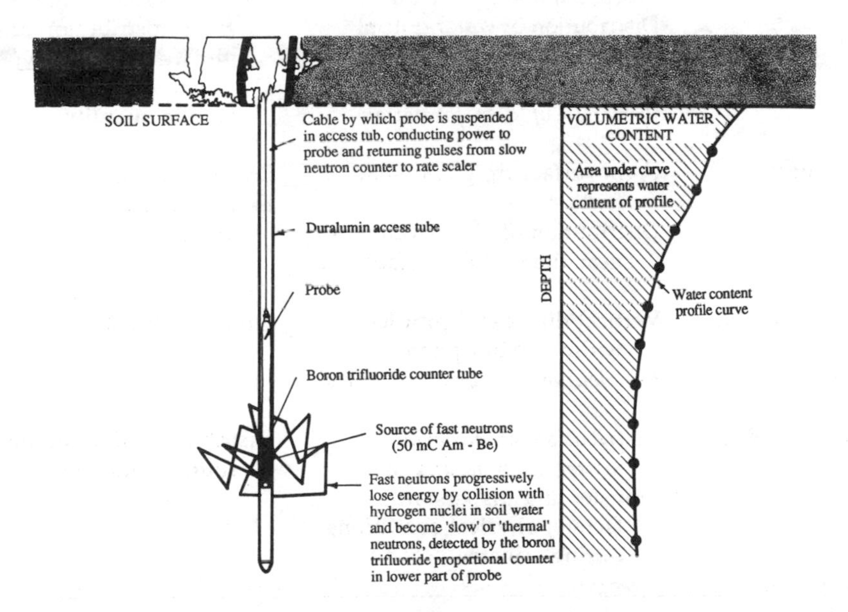

Figure 5. The probe in use in the field, showing diagrammatically the scattering of fast neutrons and the return of slow neutrons in soil around the centre of sensitivity of the probe; the graph shows an example of a soil moisture profile.

Appendix

Summary of basic processes and data

Table 1. Basic processes and related scientific topics.

Process subject	Atmospheric sciences	Soil sciences
Radiation	Distribution of radiation *Temperatures of surfaces*	Optical properties of the soil
Heat	Convectional and turbulent transfer distribution and associated processes	Temperature distribution Conduction of heat within the soil
Water	Distribution of water input (precipitation climate) Distribution of water output (evaporation climate) *Wetness of surface* *Processes of dew formation*	*Infiltration of water* Internal percolation, drainage and redistribution of soil moisture *Distribution of soil moisture,* both in quantity and quality
Momentum	Form surface drag and skin friction Distribution of air movements near the earth's surface	Erosion processes Exchange of gases between soil and air
Carbon dioxide	Vertical fluxes and profiles within the biosphere *Distribution in time and space*	Flux from the soil
Other materials	*Distribution of airborne matter in time and space* Transfer and dispersion factors; physical conditions during transfer	Distribution of pollutant sources and sinks

Table 2. A brief summary of standard data and its usefulness.

	Generally useful	Limited use	Treat with caution
Generally available	Air temperature Precipitation Sunshine Synoptic maps	Pressure Wind General weather data	Surface temperature Air humidity
Limited availability	Soil temperature Radiation	Air quality	Soil moisture Evaporation Transpiration

Further reading

SMITH, L.P. 1975. *Methods in agricultural meteorology*. Amsterdam, Oxford, New York, Tokyo: Elsevier.

DOORENBOS, J. and PRUITT, W.O. 1977. *Guidelines for predicting crop water requirements*. FAO Irrigation and Drainage Paper no. 24. Rome: FAO.

DOORENBOS, J. 1976. *Agro-meteorological field stations*. FAO Irrigation and Drainage Paper no. 27. Rome: FAO.

SINGH, P., HUDA, A.K.S. and VIRMANI, S.M. 1988. Agroclimatic monitoring and analysis. In: *First Training Workshop on Site Selection and Characterization*, 53-88. IBSRAM Technical Notes, no. 1. Bangkok: IBSRAM.

Socioeconomic site characterization for soil management experiments: guidelines and interpretation

AREE WIBOONPONGSE and SONGSAK SRIBUNJIT*

Abstract

Socioeconomic characterization is necessary for designing relevant soil management experiments. The paper discusses survey and sampling methods and the design of questionnaires. Relevant socioeconomic variables are outlined. Examples of variables and interpretations of the information analyzed are presented. The actual questionnaire has been incorporated in IBSRAM's methodological guidelines, which appears as an appendix in the proceedings

Introduction

This paper briefly discusses methods of data collection, examples of preparing master tables, and the interpretation of results in connection with socioeconomic surveys. It is not intended as a comprehensive guide to the principles or techniques, since training materials produced by the Farming Systems Support Project (FSSP, 1986) and IRRI (1986) are easily accessible.

---

* Department of Agricultural Economics and Multiple Cropping Centre, Faculty of Agriculture, Chiang Mai University, Chiang Mai, Thailand; and Department of Economics, Faculty of Social Science and Multiple Cropping Center, Chiang Mai University.

Data acquisition

The required information can be obtained either from existing data sources or by conducting a survey. The former usually provides aggregated information, which is sometimes very general. A survey, on the other hand, can be made specific to serve the purposes of researchers.

Survey methods

A survey entails asking people for information using either verbal or written questions. For research on farming systems, information on farmers' practices, input utilization, output levels, and farm sizes are included. A field survey can be conducted in two different ways - informal or formal. An informal survey, as defined by the FSSP (1986), is a field study in which researchers conduct informal interviews with farmers, and visit farms to develop an understanding of the farming system. This is particularly useful for ascertaining farmers' values, opinions, and objectives, and for understanding the reasons underlying complex management strategies. The sample included in the survey may not be representative of the populations, so statistical tests cannot be applied. The FSSP (1986) suggests that multidisciplinary teams should conduct informal surveys. Even though the method is not useful for statistical tests, it can provide quick results and an overall view, and hypotheses can be drawn for further research at a relatively low cost.

In contrast to the informal survey, a formal survey, generally called a sampling survey, is one that uses a formal sampling method and pretested questionnaires which allow statistical analysis to be applied. In many research projects, an informal survey is often followed by a formal survey, and in some circumstances the two methods are combined when a specific survey is required.

For the purposes of this paper, a survey is considered to be a means of providing sufficient information to assist in identifying the major characteristics of the selected sites so that more relevant experiments can be outlined and their priorities can be set. With this objective in mind, a decision on the survey methods to be used can be made, and the major points listed and questions for interviews formulated.

Sampling methods and sample sizes

There are two general classes of sampling techniques, called non-probability sampling and probability sampling. Accidental selection and purposive selection of samples fall into the first class; the main limitation is that conclusions about the entire population cannot be drawn. The second method includes simple, systematic, stratified, or clustered sampling techniques.

If the selected sites are small, the probability sampling survey will cost about the same as the informal survey. However, statistical tests would be an additional benefit, and the representativeness of the respondents will provide a more accurate picture of the circumstances. In the case of large sites, the additional cost to the formal survey results from the money and time spent in defining a population frame and the sampling procedure. Given the same number of respondents, other costs would remain the same for both methods. A practical simple random sampling, by randomly picking the first farm and then every other *n*th farm along the village road is a suggested approach. The value for *n* is calculated by dividing the total farm population by the sample size decided upon.

To determine the sample size the Central Limit Theorem is used, wherein the optimum sample size increases as population variability and confidence levels increase, and tolerable error decreases. The formula is:

$$N = \frac{V^2}{E^2} C^2$$

where:
- N = optimum sample size
- C = degree of confidence desired in the sample estimate
- V = population variability as measured by the standard deviation
- E = error allowable in the sample estimate

However, a rule of thumb is often used, i.e. using a large proportion for a small population:

* at least 10% for a population of 1000
* at least 5% for a population of 10 000

for an error of 5-10% (IRRI, 1986).

Questionnaire

Structured questions in written form are usually used in a formal survey. Once the objectives are set, an appropriate questionnaire can be designed. This includes taking into account types of questions (open-ended or choices), variables and their types (e.g. for the age variable asking for specific age or ranges of age), the sequence of questions, and the layout of the questionnaire. It is essential to check whether each question is relevant. Pretesting a questionnaire is necessary. This can be done quickly, and will eventually save much time and expense.

What should be asked? The major aspects of the socioeconomic setting which could affect the farmers' need for knowledge of soil and agronomic technology in the Chiang Mai highlands can be outlined as follows:

- *Social structure*
    - ethnicity
    - religious belief
    - household structure
- *Land*
    - farm size, field location
    - land ownership
    - land utilization
- *Labour*
    - labour availability
    - labour utilization patterns
    - access to hired labour
    - off-farm employment opportunities
- *Capital*
    - savings
    - access to credit
    - cash flow
- *Farming patterns and production*
    - major crops/livestock
    - production sequences
    - major inputs
    - major problems
    - yield
    - potential crops for expansion
- *Basic needs*
    - food consumption
    - education
    - others
- *Other socioeconomic factors*
    - access to inputs and markets
    - social obligations and regulations on production
    - sources of information on farming
- *Soil management by villagers: problems and practices*
    - soil fertility
    - soil erosion

- *Soil conservation*
  - existing conservation practices
  - soil conservation problems, i.e. constraints of conservation practices, for example:
    - awareness of the importance of soil conservation
    - knowledge on conservation
    - cost-benefit of conservation methods
    - tools and equipment used for conservation practices
    - monetary capital and credit for soil conservation
- *Soil improvement and development*
  - existing soil improvement and development practices
  - soil improvement and development problems

It is suggested that questions should be organized under subheadings such that omission and repitition of information can be avoided. Some information in the questionnaire can be better gathered by observation and measurement than through an interview. Information can be drawn from different subheadings to check cross-relationships. For example:

- soil fertility level and yield
- monetary capital and soil conservation practices, and
- the relationship between crops and soil management practices.

How to question? Several variables outlined in the above survey may be quantitative or qualitative when the information is obtained for a general purpose. The choice between these two types of data depends on the analytical methods to be applied. The purpose of asking each question and the reason for formulating it in a particular way will be exemplified.

A few major properties of good questions should be maintained, namely:

- Relevance: unnecessary questions should not be asked,
- Accuracy: information gathered must be reliable and valid, and
- Simplicity: questions should be easily understood and avoid bias, ambiguity, and annoyance to those to whom the question is posed.

Types of questions

The questions can be open-ended or closed-ended.

- *Open-ended questions.* These are valuable for use at the beginning of the interview to motivate respondents to cooperate. They are free-answer questions, and thus most beneficial for exploratory surveys and when ranges of responses are unknown.
- *Closed-ended or alternative questions.* In this type of question, choices for answers are provided. It is necessary to ensure that responses are

covered. However, space for unexpected answers should be provided. The major problem in developing multiple-choice questions is the framing of alternative responses, and ensuring that there are no overlapping answers. The question posed for a given set of alternatives should only deal with one dimension of an issue.

Examples and interpretation

The questionnaire developed was discussed and finalized by the participants and is attached as a guideline in this volume. Some of the questions and interpretations of the information are discussed below.

Ethnicity

For some areas, such as the highlands in northern Thailand, ethnicity is believed to be highly correlated with activeness and ambition. Mongs, Lisu, and Haws are very active and progressive farmers compared with Karens. It can be expected that the adoption rate of a new recommended technology is relatively high by the Mongs, Lisu, or Haws. Thus information on ethnicity is necessary for this study area, whereas it may not be relevant elsewhere.

Level of education of the household members

Prior information on the levels of education of the household members is relevant to show the progressiveness of the household. The results of a survey by the authors is shown in Table 1.

Table 1. The highest level of education of the household members.

Level of education	Number of households	%
Primary school	60	78.95
Secondary school	14	18.42
Vocational college	2	2.63
Total	76	100.00

It seems that this village had good access to educational facilities, and that the level of education is above average. This indicates that a moderate level of technology could easily be adopted in this village.

Religion

A new recommended technology should conform to the religious beliefs of the villagers. At Khun Jae, there are two religious groups, i.e. Christians and Buddhists. Sunday is the religious day for Christians, whilst for Buddhists it is once every two weeks. Thus the religious days can be a part of the labour availability and time-schedule constraints.

Structure of population

The classification of a population according to age, tribes, and occupation will show the availability of labour for each tribe in the village. Classification of the population by tribes can be important, at least in terms of typical occupations. For example, Haws readily adapt as traders or farmers, depending on the opportunity to earn higher profits, whereas Lisu prefer to be farmers. Therefore, if Haw people can obtain more profit from trading it would be difficult to transfer Haw labour from trading to farming. Moreover the supply of labour for farming may be short during peak periods if the proportion of Haw people in any area is high.

Income, consumption, and savings

The structure of household income, consumption, and savings indicates the capacity of the villagers to adopt the recommended soil management technology. Low savings levels will inhibit capital-intensive technology, and consequently an alternative capital saving technology should be recommended.

Table 2 shows that in the village investigated the farmers could be in a position to consider capital-intensive technology.

Table 2. Income, consumption, and savings per household in 1986 and 1987.

	1986	1987	% increase
Income (baht)*	22,440	26,295	14.66
Regular consumption (baht)	13,638	15,009	9.14
Savings (baht)	8,802	11,286	22.01
Savings as % of income	39.22	42.91	--

* 1 US$ = app. 25 baht.

Credit system

Sources of credit, interest rates, the minimum, maximum or average amounts of credit, and credit management in the village should also be identified. An assessment of the credit system, together with income, consumption, and savings in the village will indicate the capability of the farmers to adopt alternative recommended technologies. If the technology is too expensive, then such technology should not be included in the programme.

Structure of the cropping calendar

A cropping calendar will privide information on the seasonal demand for labour. This information will help extension workers to find the appropriate time to introduce the new technologies.

Benefit-cost analysis of alternative soil management methods

Other things being equal, the soil management method that provides the highest benefit-cost ratio is preferable.

Tools and equipment

An assessment of the availability of tools and equipment is a necessary part of the survey. This information can be used to evaluate the capability of the farmers to adopt the alternative soil management methods which are recommended.

Structure of land holding

Farm size is always of prime interest to all parties concerned in agricultural development. The information on farm size is also useful for a decision on the selection of soil management techniques and equipment.

A survey in one area showed that most villagers had very small farms (Table 3). Furthermore, only 30% are reported as having adequate land for farming. Under these circumstances, any technology (e.g. silt pits or alley crops) which would immediately reduce the effective land area would be unacceptable to most of the small farms.

Table 3. Structure of land holding.

Size (rai)*	% of households
Less than 1	57.14
1 - 5	26.19
6 - 10	9.53
Greater than 10	7.14
Total	100.00

* 6.25 rai = 1 ha.

Cross relationships between soil conditions and management

Some relationships can be drawn from the questionnaire in the guidelines (Tables 4 to 6). The following are examples pertaining to soil conditions and management practices.

Table 4. Classification of households according to the severity of soil erosion and levels of slope.

Steepness of agricultural land	Severity of soil erosion (%)		
	High	Moderate	Low
High	5.33	55.37	0
Moderate	0	9.92	0
Low	0	0	29.38

Table 5. Classification of households according to the degree of soil conservation and levels of slope of agricultural land.

Steepness	Degree of soil conservation (%)			
	High	Moderate	Low	None
High	0	0	0	100
Moderate	0	0	0	100
Low	0	0	0	90

Table 6. Percent of households classified by degree of soil improvement and levels of slope of agricultural land.

Steepness	Degree of soil conservation (%)			
	High	Moderate	Low	None
High	0	0	0	100.00
Moderate	0	0	0	100.00
Low	10	67.78	22.22	0

Conclusion

The socioeconomic characteristics of a selected site provide a basic guide for the initiation of relevant experiments and the development of a technology. Ethnicity, educational levels, and religious beliefs are important social factors in some regions. Economic factors, such as resource endowment and credit availability, indicate the potential for the adoption of technology. Farming patterns and cultivation practices reveal the seasonal variations in resource utilization and indicate the possibility for change. The socioeconomic information required for a network does not require an intensive survey. A small well-designed questionnaire to be used during the interviews is sufficient, and some of the information needed can be gathered by observation and actual measurements. A probabilistic random sampling is recommended, since the population of each site is rather small.

References

FSSP (Farming Systems Support Project). 1986. *Diagnosis in farming systems research and extension, FSR/E training units,* vol. 1. Institute of Food and Agricultural Sciences. Florida, USA: University of Florida.

IRRI (International Rice Research Institute). 1986. *Training materials.* Vol. 1, *Farming systems socioeconomic research training course.* Agricultural Economic Department. Philippines: IRRI.

Section 5: Site characterization

Soil variability on experimental sites

E. PUSHPARAJAH*

Abstract

The paper considers the factors that affect the variability of soil on a given site. The major and readily recognizable causes of inherent variations are lithological and pedological factors. These variations are often large and can exert considerable differential influence on the growth and productivity of plants. Biogenetic variability due in particular to vegetation is often encountered, whereas variability due to faunal activity is confined to only a few soils. Man's activities in the management of the soils, particularly in subsistence farming or in intercropping with tree crops, induces great variability in both physical and chemical properties. These activities give rise to scattered distribution of residues from burns, different depths of effective topsoil induced by tillage practices, and (in the case of plantation tree crops) differential levels of fertility.

The consequent microvariability can readily exert considerable influence on the growth of plants, often masking treatment effects. The adverse influences of such variations can be minimized by a detailed study and mapping of the experimental site, which can eventually be used to assist in planning and layout.

Introduction

Soil microvariability poses a major difficulty in setting up field experiments. Variations in soil texture, soil drainage, soil depth, and slope (which

* Programme officer, IBSRAM, PO Box 9-109, Bangkhen, Bangkok 10900, Thailand.

reflect the inherent properties of the soil) have been shown to affect the performance even of perennial tree crops. Chan *et al.* (1974) showed that such soil variations within a given site could account for up to 24% of the differences in rubber yield. The effect of soil microvariability on the growth and performance of annual crops can be more pronounced, leading to uneven stands even over much shorter distances. Such variations are particularly more marked with low levels of management, as is often the case in the tropics (Moorman and Kang, 1978).

More often than not, agronomists have ignored soil variability. The costly consequence of this omission has often been realized too late, notably in cases where statistical analysis has shown high residual error and minimal to nonsignificant treatment differences.

Lithological and pedological variability

The geological and/or parent material from which a soil is derived can vary over short distances. The influence of variable contents of coarse or fine fragments in the rooting horizons on crop productivity can be considerable. Localized outcrops of bedrock also contribute to the microvariability. Major lithological variations and geochemical gradients at short distances are also important, especially in older transported materials and sedimentary rock formations (Beckett and Webster, 1971). Lithological variability, particularly on the surface soil can further be induced or reinforced by soil losses due to surface wash, soil creep or gullying, and local deposition of erosion products, resulting in alterations within short periods of time, and hence creating an increase in soil variability.

The influence of the combined effects, both lithological and pedological, on soil microvariability over short distances in a spatial distribution of landscapes has been documented by Zainol (1984) and Chan (1985) in Malaysia, and Chauvel *et al.* (1987) who reported on findings in Guyana.

In a site-specificity study carried out in Malaysia over an area of about 30 ha, six contrasting soil units with five other distinguishable soil units differentiated by microvariations of soil depth and slope were encountered (Figure 1). The sizes of these units varied from 1 ha to 6 ha.

However, it is also possible to obtain areas with one soil series but with different phases occurring within a small area (Figure 2). The inclusion of colluvial deposits is often to be expected.

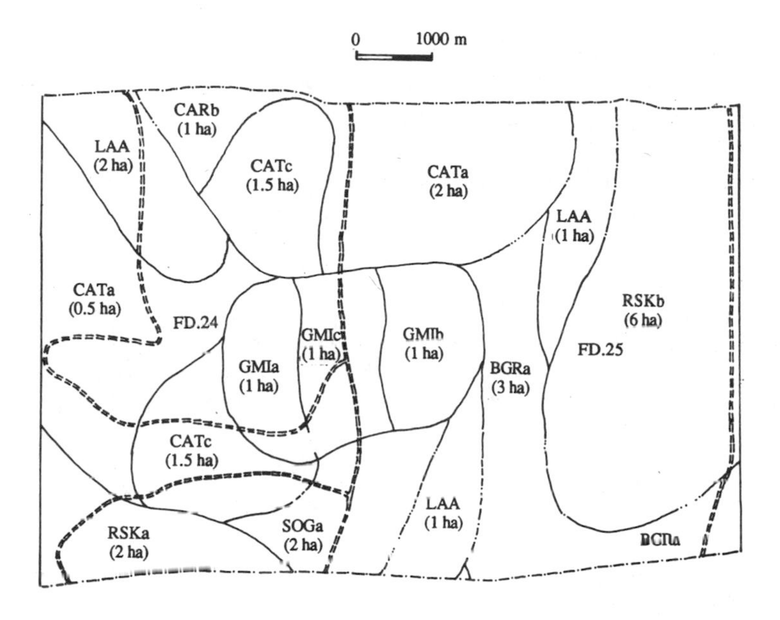

Legend: ===== Road

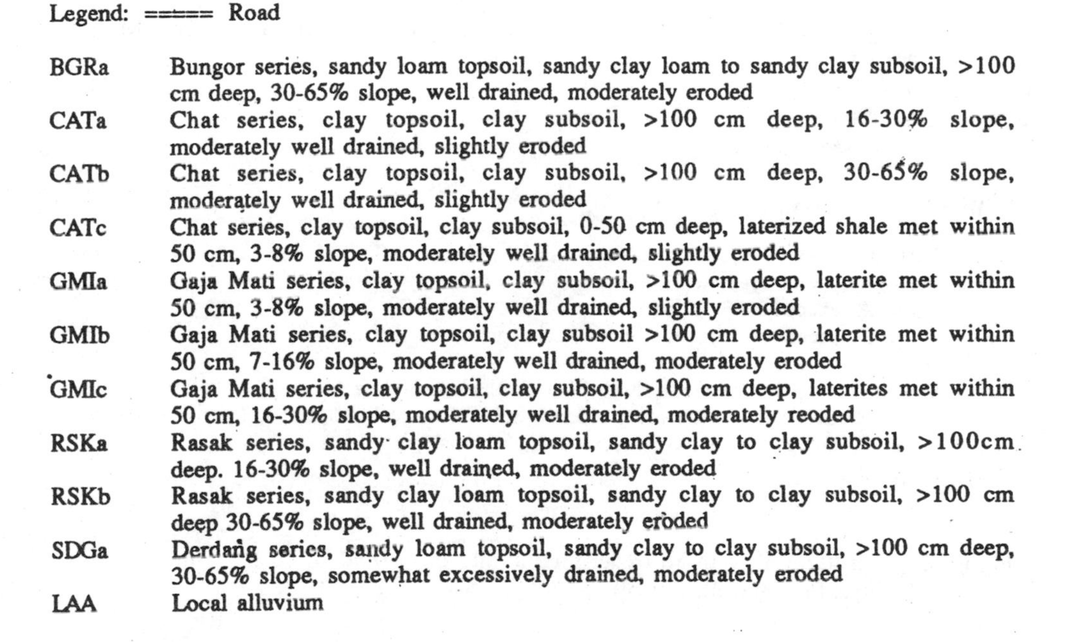

BGRa	Bungor series, sandy loam topsoil, sandy clay loam to sandy clay subsoil, >100 cm deep, 30-65% slope, well drained, moderately eroded
CATa	Chat series, clay topsoil, clay subsoil, >100 cm deep, 16-30% slope, moderately well drained, slightly eroded
CATb	Chat series, clay topsoil, clay subsoil, >100 cm deep, 30-65% slope, moderately well drained, slightly eroded
CATc	Chat series, clay topsoil, clay subsoil, 0-50 cm deep, laterized shale met within 50 cm, 3-8% slope, moderately well drained, slightly eroded
GMIa	Gaja Mati series, clay topsoil, clay subsoil, >100 cm deep, laterite met within 50 cm, 3-8% slope, moderately well drained, slightly eroded
GMIb	Gaja Mati series, clay topsoil, clay subsoil >100 cm deep, laterite met within 50 cm, 7-16% slope, moderately well drained, moderately eroded
GMIc	Gaja Mati series, clay topsoil, clay subsoil, >100 cm deep, laterites met within 50 cm, 16-30% slope, moderately well drained, moderately reoded
RSKa	Rasak series, sandy clay loam topsoil, sandy clay to clay subsoil, >100cm deep. 16-30% slope, well drained, moderately eroded
RSKb	Rasak series, sandy clay loam topsoil, sandy clay to clay subsoil, >100 cm deep 30-65% slope, well drained, moderately eroded
SDGa	Derdang series, sandy loam topsoil, sandy clay to clay subsoil, >100 cm deep, 30-65% slope, somewhat excessively drained, moderately eroded
LAA	Local alluvium

Figure 1. Detailed soil map of Serapoh Estate, Mukim of Blanja, Malaysia (Fields 24 and 25).

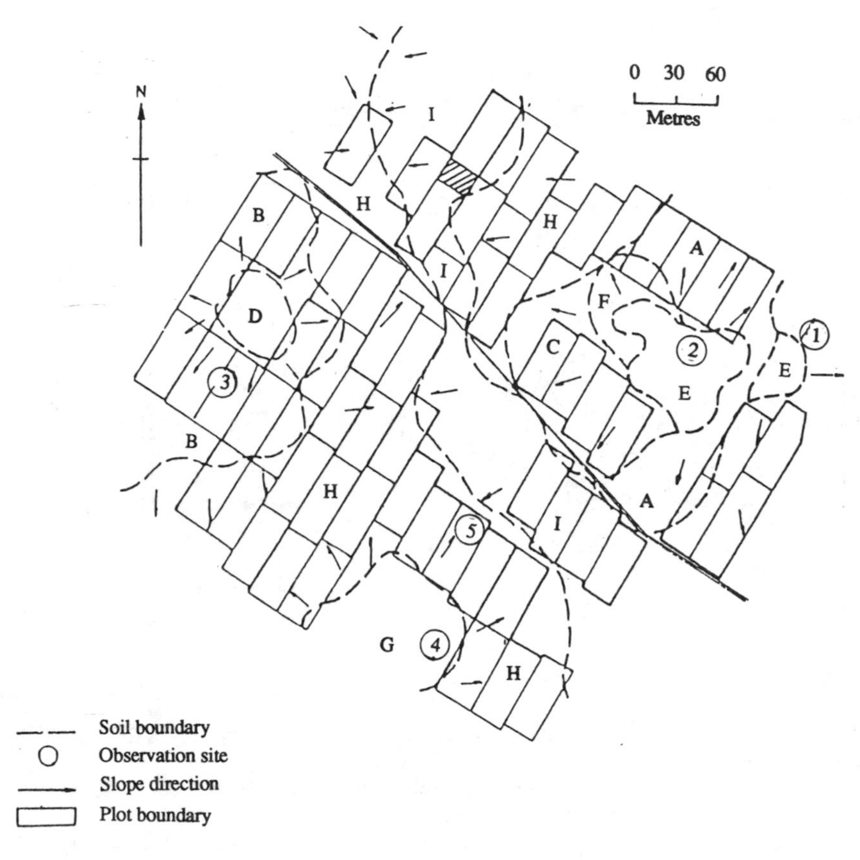

Legend

A. Durian series, clay loam topsoil, clay subsoil, 100-125 cm deep, 1-3% slope, moderately well drained
B. As in A, except 3-5% slope
C. As in B, except soil has medium laterite phase 50-125 cm
D. As in C, except 5-8% slope
E. Durian series, 8-12% slope, high laterite phase 0-50 cm
F. Durian series, sandy clay loam topsoil, subsoil clay, 1-3% slope, laterite 0-50 cm
G. As in F, except fine sandy clay subsoil, 5-8% slope
H+I Colluvium

Figure 2. Soil variability in an experimental site on a field of 10 ha (Durian series - Typic Plinthudult; Ultisol).

The effects of such a mosaic distribution of soil units scattered over small areas will lead to uneven productivity of crops. Chauvel *et al.* (1987) have clearly shown that these variations affect the performance of soybean and maize from the initial germination right through to the final yield obtained. Tillage practices, particularly the use of a tooth scarifier, reduced the differential effects on initial growth, but with time pedological properties still exert a major influence.

Similarly, Chan *et al.* (1974) showed that even with perennial tree crops microvariations such as slope, soil texture, and depth affected growth and yield considerably.

The above studies show that microvariations are likely to occur over short distances in most landscapes, and that such variations exert a tremendous influence on the performance of crops. They also indicate that lithological and pedological factors are major sources of soil microvariation, and that it is therefore difficult to obtain ideal homogenous units.

Biogenetic variability

Variability can be induced either by vegetation or by faunal activity.

Variability due to vegetation

The type of trees or covers present prior to clearing and preparation of the area is another important source of microvariability. Sanchez (1987) showed that the type of trees (albeit plantations) could have considerable effects on the surface soil (0-30 cm). Variations caused by different tree species on chemical properties are given in Table 1.

Table 1. Effect of a 10-year growth of various timber species on the topsoil.*

Type of plantation/forest	C(%)	Nutrient level (C mol/kg)		
		Ca	Mg	K
Cordia trichotonia	2.18	1.70	0.95	0.12
Caesalpinia echinata	1.40	1.75	1.00	0.13
Dalbergia nigra	1.92	0.95	0.60	0.07
Native rainforest	1.65	0.70	0.50	0.05
L.S.D. ($P<0.5$)	0.23	0.90	0.35	0.03

* Source: Silva (1983), as cited by Sanchez (1987).

The table indicates that if different tree species are present at a site there will be some variability in chemical properties. Similarly Pushparajah and Bachik (1987) showed that different ground cover conditions could exert considerable influence on soil properties (Table 2). Where there was no cover, not only was there a lower carbon content and less nutrients, but even soil physical properties were poorer. Such differences appear sufficiently large to influence the performance of crops and thus introduce variability.

Table 2. Effect of covers on the surface soil ((0-15 cm) of a Typic Paleudult.

Cover	at 4th yr after establishment				at 15th yr after establishment			
	%C	N%	Exchangeable (cmol/kg)					
			Ca	K	%C	MWD (mm)	Bulk density (Mg/m^3)	Permeability (cm/h)
Nil	1.36	0.11	0.28	0.06	1.1	0.95	1.21	8.4
Grass	1.64	0.12	0.56	0.12	1.4	3.39	1.13	13.2
Legume*	1.73	0.14	0.43	0.17	1.4	1.98	1.13	12.4
Dicot. weeds	1.55	0.13	0.62	0.20	1.3	2.20	1.16	12.0
L.S.D. ($P<0.05$)	0.25	0.02	0.23	0.05	-	0.18	0.05	3.1

* The legume covers died off after the third year. The subsequent effect is the residual effect.

Source: Pushparajah and Bachik (1987).

Variability due to faunal activity

Termite mounds exert a considerable and significant effect on soil micro-variability (Lee and Wood, 1971). these differences are readily reflected in the variable growth performance of crops. Such variable growth performance reflects the fact that not only the physical properties (eg. structure, bulk density, water-holding capacity), but also chemical properties (carbon, pH, cations, etc.) of the mounds are different from that of the surrounding soil area. Levelling of the mounds only leads to greater variability.

Variability due to man's activities

When land is cleared for planting (whether in shifting or even mechanized cultivation), trees, branches and shrubs are often assembled in one or several spots in the field and burnt. This results in a local concentration of plant nutrients, which would ultimately lead to an accumulation of ash in some sections of the field. The consequent variability on the soil, and hence on crop productivity, could be large. Although with time variability due to induced chemical changes would diminish, heterogeneity due to physical parameters would be persistent.

Soil physicochemical properties and their variations

Most tropical soils are dominated by low-activity clays, and thus in these soils plant nutrients are easily leached. The organic-matter content of the surface soil therefore plays a major role in regulating the nutrient levels in these soils, and variations in the organic matter within short distances could have a pronounced effect on the crop.

Small changes in the thickness of the surface soil may lead to very uneven growth. By cultivation such variations could be induced even in an area which previously had a uniform soil. This is best described by Figure 3. In fact Valentin (1988) has emphasized the need to monitor closely not only such features, but also the formation of crust features on the surface soil.

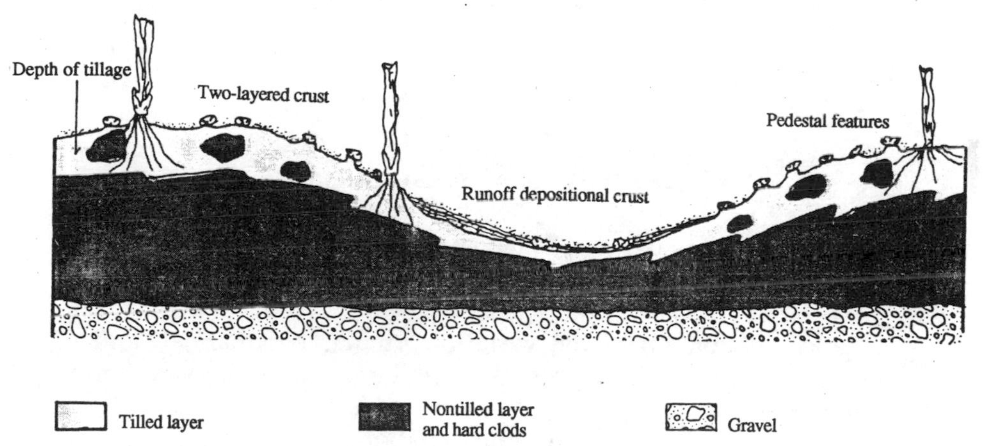

Figure 3. Example of a cultivation profile, one year after mechanical land clearing and plough ridging (Mong-Gine, 1979, as reported by Valentin, 1988).

In addition to the effect of tillage practices, the type of crop and hence fertilizer practices would also create variability. Chang *et al.* (1977) showed that a Paleudult derived from granite was chemically less variable than a Haplorthox derived from shale. However, even on the Paleudult cultivated with a tree crop, e.g. rubber, the fertilizer practice led to greater variability within the field. This was because the trees were planted in avenues 6 to 7 m apart, and the fertilizers applied on strips of one meter along the tree rows. Thus with time chemical changes were evident (Figure 4). Changes similar to those shown for phosphorus in Figure 4 also occur with respect to pH, Ca, and even on some physical parameters (Pushparajah *et al.*, 1976). Thus there is a need to appreciate the previous history of the soil.

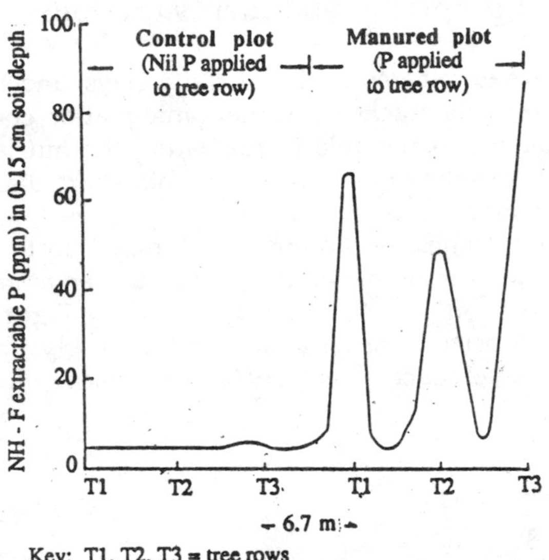

Figure 4. Influence of fertilizer on soil nutrient status (Chang *et al.*, 1977).

A detailed evaluation of the variation in pH (Pushparajah and Chan, 1987) showed that on a soil mapped as Typic Plinthudult, the variation in pH in plots of 0.16 ha was from 4.3 to 5.2 - even within a given soil unit (Figure 5). With a smaller plot size of 4 x 40 m, the variations in pH were larger, ranging from 4.45 to 6.50 (Figure 6). Such random but large variations could affect the interpretation of the final results.

The influence of such microvariability on plant growth in a 1000 m² plot, identified by a detailed soil survey as a uniform soil, was evaluated by Sudin and Chong (1987). They found that by visual observation plants could be divided into four groups, and that these groups occurred in small blocks at random. The findings are given in Table 3.

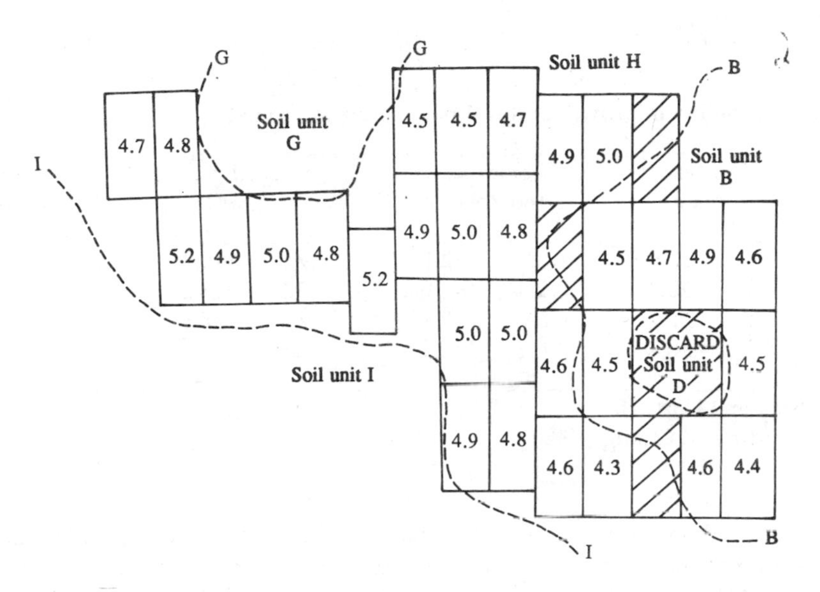

Figure 5. Variation in pH in experimental plots of 0.16 ha each (Pushparajah and Chan, 1987).

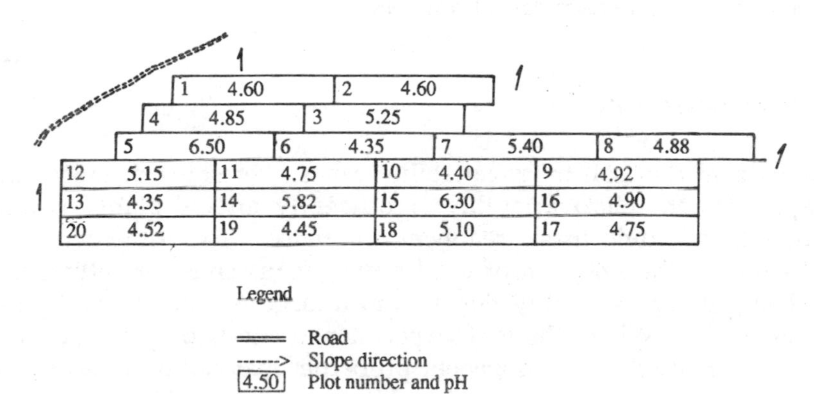

Dominant soil

Munchong series
a Typic Haplorthox

Soil characteristics

Clay loam top soil; 4-5° slope, clay subsoil; deep, gently rooling terrain; well drained; moderately eroded.

Figure 6. Variation of pH in an experimental field on an Oxisol with plot sizes of 6 x 40 m (Pushparajah and Chan, 1987).

Table 3. Variability in growth and soil parameters examined.

Plant group	Dry wt. g/plant		Nodules (no./plant)		Seed wt. (g/m^2)		NO^3-N mg/g	Avail P* mg/g
A	24.1	±2.3	6.8	±2.1	119.6	±20.7	79.8	41.1
B	11.3	±3.0	2.9	±1.0	53.2	±23.2	75.9	34.1
C	4.1	±1.0	0.9	±0.4	22.6	±8.6	46.5	27.4
D	0.7	±0.4	0.4	±0.5	4.5	±1.0	46.1	19.9

* P by Bray and Kurtz II.
Source: Sudin and Chong (1987).

A detailed evaluation showed that all the three plant parameters were related to the two soil parameters examined. The correlation coefficient between NO^3-N and the plant indices ranged from 0.88 to 0.89, while that with available P ranged from 0.94 to 0.96. This confirms the relationship between plant growth and soil chemical parameters.

Proper site selection

It is evident that soil microvariability exists over short distances in most landscapes. Such microvariability significantly affects plant growth and performance, and could mask treatment differences. Experimenters therefore face difficulty in the selection of ideal homogenous sites for setting up field trials. The type of variability depends to a large extent on the size of the experimental site. Where the total experimental site is only about 1 to 2 ha, soil microvariability due to pedogenetic properties may not be a major problem, as it is when the site is, say, 5 to 10 ha. However, microvariations in chemical and physical properties could be a major problem.

For ascertaining such microvariations, the ideal way would be to establish a uniformity trial and obtain a fertility gradient map before the trial is established. However, this is an expensive procedure and not always acceptable.

The approach suggested is to obtain initially a very detailed soil map of the area to assist in the selection of the experimental block. Attempts should be made to map variations in existing crops or any ground cover, the colour of the surface, soil erosion/depositional variations, soil depth (in particular effective depth), etc. This would give an indication of variability and help to identify contiguous sections of uniform units. Determination of some soil parameters, e.g. pH, C, and exchangeable cations within each of these units would further help in confirming the uniformity or variability of the soil.

With this knowledge, the trial should be layed out so that replicates, or even blocks within replicates, are assigned to fit into minor variations. This would enable soil variations to be accounted for as "block errors", thus minimizing interference with treatment effects.

References

BECKETT, P.H.T. and WEBSTER, R. 1971. Soil variability: a review. *Soil Fertilizer* 34(4): 1-15.

CHAN, H.Y. 1985. Classification, genesis, mapping and productivity of rubber on soils developed from sedimentary rocks in Peninsular Malaysia. Ph.D. thesis. University of Malaya. 265p.

CHAN, H.Y., WONG, C.B., SIVANADYAN, K. and PUSHPARAJAH, E. 1974. Influence of soil morphology and physiography on leaf nutrient content and performance of *Hevea*. *Proceedings of the Rubber Research Institute of Malaysia, Planters Conference*, 115-126. Kuala Lumpur, Malaysia: RRIM.

CHANG, A.K., CHAN, H.Y., PUSHPARAJAH, E. and LEONG, Y.S. 1977. Precision of field sampling intensities in nutrient surveys for two soils under *Hevea*. *Proceedings of the Conference on Chemistry and Fertility of Tropical Soils* (Kuala Lumpur, 1973), 25-37. Kuala Lumpur: Malaysian Society of Soil Science.

CHAUVEL, A., BOULER, R., GODON, P., WOROU, S. and LUCAS, Y. 1987. Environmental characterization of acid tropical soils. In: *Management of acid tropical soils for sustainable agriculture*, 217-226. IBSRAM Proceedings no. 2. Bangkok: IBSRAM.

LEE, V.E. and WOOD, T.G. 1971. *Termites and soils.* London and New York: Academic Press.

MOORMAN, F.R. and KANG, B.T. 1978. Microvariability of soils in the tropics and its agronomic implications with special reference to West Africa. In: *Diversity of soils in the tropics*, 39-43. American Society of Agronomy Special Publication no.34. Madison, USA: ASA.

PUSHPARAJAH, E., SOONG, N.K., YEW, F.K. and ZAINOL, E. 1976. Effect of fertilizers on soils under *Hevea*. In: *Proceedings of the International Rubber Conference*, (Kuala Lumpur, 1975) 37-50. Kuala Lumpur, Malaysia: Rubber Research Institute of Malaysia.

PUSHPARAJAH, E. and BACHIK, A.T. 1987. Management of acid tropical soils in Southeast Asia. In: *Management of acid tropical soils for sustainable agriculture*, 13-39. IBSRAM Proceedings no.2. Bangkok: IBSRAM.

PUSHPARAJAH, E. and CHAN, H.Y. 1987. Site selection for agronomic experimentation: Malaysian experience. In: *Soil management under humid conditions in Asia and the Pacific*, 133-152. IBSRAM Proceedings no.5. Bangkok: IBSRAM.

SANCHEZ, P.A. 1987. Management of acid soils in the humid tropics of Latin America. In: *Management of acid tropical soils for sustainable agriculture,* 63-107. IBSRAM Proceedings no.2. Bangkok: IBSRAM

SUDIN, M.N. and CHONG, K. 1987. The influence of soil heterogeneity on biological nitrogen fixation, growth and yield of soyabean. Poster presented at the Symposium on the Contribution of Biological Nitrogen Fixation to Plant Production, Cisarua, Indonesia, 3-7 August 1987.

VALENTIN, C. 1988. Degradation of the cultivation profile: surface crusts, erosion and plough pans. In: *First Training Workshop on Site Selection and Characterization,* 233-264. IBSRAM Technical Notes no.1. Bangkok: IBSRAM.

ZAINOL, E. 1984. Pedological studies of soils under rubber in Kedah (Peninsular Malaysia). Dr. Sc. thesis. State University of Ghent, Belgium. 308p.

Site characterization - physical landscape features

C.A.A. CIESIOLKA*

Abstract

In selecting experimental sites, cognizance of the following information is suggested: the available mechanical and human resources, the previous experience of personnel, the use of data to be collected, the landscape scale at which processes operate, and the efficiency/cost relating to the priority of the project.
This paper refers to landscape variables such as geology, geomorphology pedology, and anthropocentric activities that require in-depth field investigation before projects commence. Such activity aims to avoid plots with unique characteristics, such as excessive effluent flow, subsurface anomolies, masked fault lines, and sites undergoing geomorhpic processes, such as low soil creep.

Introduction

Another paper in this volume considers the properties of the soil profile which are important in site characterization[1]. However, there are a range of landscape properties which are important in selecting a site for field research and interpreting the results. These properties are considered in this paper.

---

* Paper presented by Adrian Webb, Agricultural Research Laboratories, Queensland Department of Primary Industries, Meiers Road, Indoorsopilly, Queensland, Australia.

1 See below: A.A. Webb and K.J. Coughlan, 'Physical characterization of soils'.

Use of data

Problems relating to landscape features that we have encountered are:

- Geologic features, such as sills and laterite ridges, create subsurface runoff divides and changes in soil depth. A dip of strata has facilitated drainage away from catchments on a cuesta face and influenced moisture movement down the soil catena.
- Relict landforms developed under a previous climate and stabilized under the present climate-vegetation relationships are highly susceptible to erosion when present vegetation is removed, e.g. valley alluvial fans. Recent alluvium, because of its greater interconnectedness of pore spaces, will have higher infiltration rates as compared to *in situ* derived soils on the same geology, and possibly lower shear strength. Relict erosion can be completely obliterated by modern deposition and mask the previous geomorphic history of an area, while episodic erosion may appear to be caused by man.
- Hillslope shape affects source areas of runoff and sediment production. For example, there are few rills on hillslopes that are convex both in plan and profile forms, because water naturally spreads on such slopes. More rilling occurs on rectilinear segments and effluent flow, and deposition on concave slopes generates lower sediment concentrations.

Selection of sites

It is extremely important to select study sites with considerable care, and the following basic steps should be observed:

- obtain an overview of the regional setting in relation to climate, geology, geomorphology, topography, soils, vegetation, and history of land use;
- map the pertinent variables and overlay these characteristics to ascertain spatial relationships for areas that are comparable with farming units of the locality;
- measure features and characteristics of the environment that are key variables for purposes of predicting runoff and soil erosion, e.g. landscape and soil attributes.

Given the above information, an endeavour should be made to premodel runoff and erosion so that the size and type of instruments selected can adequately handle runoff peaks and sediment loads. Such an approach may provide information concerning the need for bed-load tanks, or simply for the collection of suspended load with stage samplers.

Data collection of landscape characteristics

Firstly, there is a need to obtain an overview of the region where a particular problem is going to be investigated.

Imagery, maps, and reports

Imagery (band 1-7 satellite, SLAR, maps, survey reports) can provide information about:

- landforms - outcrops;
- approximate catchment areas;
- local relief;
- erosional status of the area;
- boundaries of geology, soil, vegetation, water bodies, and subcatchments;
- disturbances by man's activities;
- patterns of land-use and temporal changes; and
- the spatial variability of identifiable characteristics.

There is no substitute for walking over the target area and making detailed observations of soil surface characteristics, exposed rocks, vegetative growth, slope shape, length and angle, source areas of runoff and erosion, zones of deposition, variations in tree density and size, and microtopography. This necessitates observation of erosion events at the proposed site for twelve months prior to the selection of the site.

Other prior information that should be investigated at an early stage is the availability of meteorological records, including their period of record, their completeness, and their reliability and accuracy. Reports published by surveyors (field note books), lands departments, and railway and main road authorities often describe pristine conditions.

Geology

Information on the following features should be compiled:

- rock type - its lithology (dip, strike, bedding and cleavage, rills and dykes) and structure (porosity and permeability);
- stratigraphy of the surficial deposits;
- chemical composition of the parent materials;
- outcorps and concealed resistant bands of rock;
- weathering products of the parent materials;

- occurrence and location of perched water tables; and
- refilled gully lines that do not exhibit characteristics to indicate the relict processes.

Possible landscape indicators of geologic variability can be found in slope angle changes, vegetation characteristics (such as species and density), outcrops, location of stream channels as they follow the boundary between two rock types, position of the stream channel on the alluvial valley floor, and seepage points.

Topography and geomorphology

A grid survey should be carried out in order to make a detailed contour map. The aims of a topographic survey are to obtain the following information:

- An accurate description of catchment/subcatchment areas and boundaries. The scale of the survey must accommodate the size of the phenomena being observed, so that digitizing can be carried out from the constructed map.
- The drainage pattern, density, stream order, and relief.
- The lengths of channels and swales (depression lines).
- The hypsometric integral.
- The area as a percent of the first order catchments covered by each slope segment type (convex, concave, rectilinear).
- The frequency distribution of slope angles (threshold slopes) of whole catchment units.
- The slope lengths and rates of slope angle change. Where plots are located on segments of hillslopes, runoff and soil loss may be influenced by hillslope throughflow from the upslope catchment. For reliable results, a hillslope cross section should be sampled so that a representative distribution of slope angles and slope segments is being sampled.
- Surface roughness measurements need to be converted into quantitative values of depression storage. Values have been obtained from rainfall-simulator experiments, hyetograph-hydrograph analysis, and recently from advances in laser technology.
- Base-level elevations that control incisionary erosion processes. Accelerated erosion is often the result of changing the stability (shear strength) of the medium or structure that acts as the base for downcutting by channelized flow. Laboratory simulations using scale models have demonstrated that such processes eventually affect complete catchments.
- Mapping of all variables measured during the study. Three-dimensional presentations give a more realistic conception of the area under study.

From the above outline, it is obvious that a great deal of work needs to be carried out at the preparatory level. However, some of the information is often available - at least in a general form - from previous resource surveys.

Basic concepts and philosophy of *Soil Taxonomy*

HARI ESWARAN*

Abstract

Soil Taxonomy *evolved over several decades. Its early history and the principles behind it are summarized. The individual unit of soil, the pedon, is grouped on the basis of differentiating characteristics. A group of pedons forming a class is called a taxon. Groups of taxa form the six categories - from highest to lowest - of order, suborder, great group, subgroup, family and series. These categories are defined and discussed in turn.*

Introduction

Guy D. Smith was one of the prime movers behind the soil classification system described in *Soil Taxonomy* (Soil Survey Staff, 1975). Perhaps better than anyone else, he knew the details of the thinking, discussions, and arguments that shaped the development of the system.

This paper outlines the broad concepts and philosophy of *Soil Taxonomy*. In writing it I have drawn heavily from transcripts of several interviews of Dr. Smith conducted by several distinguished scientists, beginning in 1976 with Dr. Michael Leamy of New Zealand. Some of the interviews have been published in the newsletter of the New Zealand Soil Science Society and in the periodical *Soil Survey Horizons*, which is printed in the United States.

---

* Soil Conservation Service, USDA, PO Box 2890, Washington, D.C. 20013, USA.

A monograph based on the interviews is now being prepared. It will give a full account of the rationale and concepts of *Soil Taxonomy*. The monograph will be published through Soil Management Support Services, a program of international technical assistance carried out by the Soil Conservation Service of the United States Department of Agriculture, with funding from the United States Agency for International Development.

Historical background

Soil Taxonomy did not evolve overnight; it evolved over several decades. An excellent review is given by Cline (1979). Modern classification systems obtained their initial spark with the recognition by Dokuchaiev (Glinka, 1927) that soils are natural bodies; some of the concepts of Dokuchaiev are still to be found in *Soil Taxonomy*.

In the U.S., two early works on classification were written by Whitney (1909) and Marbut *et al.* (1913). There were several inconsistencies between the genetic theories of Marbut and field observations, and by 1935, when Marbut died, the stage was set for a review of the system then in use.

In 1938, a yearbook, *Soil and Men*, was published with a chapter on soil classification by Baldwin *et al.* (1938). This classification, later revised, is popularly known as the 1938 classification. It introduced to the U.S. scene, N. M. Sibertsev's concept of Zonal, Intrazonal, and Azonal soils (see Afanasiev, 1927). It also introduced the concept of great soil groups, but it did not provide a link with the lower categories. As a result, there were virtually two systems in the U.S. - one consisting of series, types, and phases, and the other of higher categories.

After the Second World War, an attempt was made to rectify this situation. The initial effort, according to Cline (1979), was mainly to "patch up the 1938 system". By then, about 5000 soil series had been identified in the U.S., and correlation of the series became an important problem.

A special issue of the journal *Soil Science* was published in 1949, devoted solely to soil classification. Soon afterwards, the Soil Conservation Service took the momentous step of deciding to develop a new classification system, and Dr. Guy D. Smith, director of soil survey investigations, was charged with leadership of the project. The system was developed through a series of seven approximations, and was finally published in 1975.

Designing the system was not an easy task. Several hundred man-years of work were required. Methods had to be developed, soils had to be analyzed, concepts had to be tested, and personal bias had to be sieved out. When we look back and evaluate all the efforts that resulted in *Soil Taxonomy*, we can see that few organizations in the world could have matched them. In addition, Dr. Smith's leadership was critical in coordinating, guiding, and encouraging the workers, and even in terminating some extreme ideas.

Logic of the system

Cline (1949) provides a succinct summary of the system's logic:

> " The purpose of any classification is to organize our knowledge so that the properties of objects may be remembered and their relationships may be understood most easily for a specific objective. The process involves formation of classes by grouping the objects on the basis of their common properties. In any system of classification, groups about which the greatest number, most precise, and most important statements can be made for the objective, serve the purpose best. As the things important for one objective are seldom important for another, a single system will rarely serve two objectives equally well."

This statement of Cline makes three important points. First, since a classification system reflects only the knowledge existing at the time, it should have the capacity to change as more information becomes available. A good example is the use of mineralogical parameters. Before 1950, soil mineralogical techniques were not well developed, and classification systems carried little or no information on soil mineralogy. Since then, we have come a long way; mineralogical family classes are present in *Soil Taxonomy*. There are still problems with these classes, mainly that of quantification. When this problem is solved, the "solution" or "quantification" will be incorporated in the system.

Second, Cline states that a classification has specific objectives. Most broadly, the objectives of a classification may be categorized as being either theoretical or practical. If a soil classification system has the sole objective of organizing genetic theories, it will be in a state of flux, and will be continuously contested as knowledge grows and concepts change. The practical objective, therefore, must be clearly determined before the system is developed. If this objective is not determined, the developed system may never be used. The objective of *Soil Taxonomy* is the making and interpretation of soil surveys used to guide land planners and managers.

Third, soil classification requires the grouping of objects on the basis of common properties. For example, Sibirtsev's Zonal, Azonal, and Intrazonal concept attempted to subdivide soils according to genetic theory. The soils within each of these genetic classes were then assigned some properties. Soon, however, it was found that some Azonal soils had Zonal properties or vice versa. Using atmospheric climate to divide soils implies that they have few similarities - the implication is that soils are different because the climate is different. Smith (1965) believed that this misconception would prevent the average mind from comparing the actual properties of the soils of two different climatic regions. Genetic theories, far from being ignored in *Soil Taxonomy,* were used for the conceptual development of the categories. But the categories and the classes within them are defined according to measurable soil properties.

The unit of classification

It is easy to describe a house, car, or fish, and identify it by placing it in a category of a classification system. These are discrete objects with well defined limits. Classifying soils is much more difficult, because soils are a continuum - their properties may change very gradually with distance over the landscape. To classify a soil, one has to identify the individual soil unit that is to be classified. In defining the individual unit, we focus on a specific part of the continuum, and in doing so we recognize that specific properties of a soil can vary within certain limits. It is only by this recognition that we can proceed to the next step, grouping the soil individuals into classes.

In *Soil Taxonomy,* the soil individual is called the pedon. A pedon is like the unit cell in a crystal. The pedon is the unit of sampling - that is, it is the smallest area that is described and sampled. It is three-dimensional, and its lateral dimensions are large enough to represent the nature of any horizons and variability that may be present.

The unit cell or pedon can theoretically be as small or as pure as one would like it to be, but such an approach would serve little practical purpose, because it would ignore the reality of soils as a continuum having variations in properties. In any given segment of a landscape, we can identify many distinct pedons.

We can also identify polypedons, which are sets of contiguous pedons that differ from other contiguous pedons in properties affecting use and management. The polypedon has shape, transitional margins, and natural boundaries. Polypedons are the real things that are classified.

Characteristics used for classification

Cline (1949) recognized three kinds of soil characteristics - differentiating, accessory, and accidental. A property that is used as the basis of grouping is a differentiating characteristic. Accessory characteristics are covarying properties about which precise statements can be made, and accidental characteristics are properties that are independent of the basis for grouping. Some properties, such as cation-exchange capacity (CEC) or the presence of gibbsite, can be considered as differentiating, accessory, or accidental characteristics, depending on the level of classification being discussed. A CEC of less than 16 m.e. per 100 g clay is a differentiating characteristic in Oxisols. A low CEC is an accessory characteristic in some Ultisols where the presence of an argillic horizon and low base saturation is used to define the order. A low CEC in the Ultisol identifies a subgroup, and is a differentiating characteristic at this level.

The presence of a small amount of gibbsite is an accidental characteristic in some Oxisols and Ultisols. If a large amount of gibbsite is present, it is a differentiating characteristic in some of the Oxisols, e.g. to define the great group of Gibbsiorthox.

Selection of the attributes becomes critical in developing the classification of a given soil. Several rules are followed in *Soil Taxonomy*. They are:

- The definition of each taxon should carry as nearly as possible the same meaning to each user. The definitions have to be precise and, whenever possible, quantitative.
- The taxonomy has to be multicategoric. The highest category, the order, has only ten classes. Succinctly, lower categories of the taxonomy contain increasing numbers of classes arranged in a logical development. One need only know a few rules to understand the relationship of the lower categories to the higher.
- The taxa should be concepts of real bodies that are known to occupy geographic areas. Hypothetical soils should not be considered in the classification. This differentiates *Soil Taxonomy* from some other hierarchical systems, which have slots for soils that theoretically might exist somewhere but have never been described. The system should be constructed, however, so that a soil can be classified if it is found and described.
- Differentiae should be soil properties that can be observed in the field, or properties that can be inferred from the observed properties or from the combined data of soil science and other disciplines. If the property cannot be measured or evaluated, it has little value in the classification. Two properties that *Soil Taxonomy* has introduced are soil moisture and temperature. At present, there are few available data on these properties because they are not routinely measured, but they can be inferred from atmospheric data.
- The taxonomy should be capable of modification to fit new knowledge with a minimum of disturbance. No individual or group can claim to have seen all the soils of the world. Soil scientists are continuously studying soils - soils they have seen before, as well as those they have not seen before. The

classification system becomes obsolete if it cannot incorporate new information that the scientists provide.

- The differentiae should allow for keeping an undisturbed soil and its cultivated (or otherwise man-modified equivalents) in the same taxon as far as possible. A classification should not be changed just because some properties of a soil have changed through use and management.
- The taxonomy must be capable of providing taxa of all soils in a landscape.
- The taxonomy should provide for all soils that are known, wherever they may be.

These last two items indicate that the system has to be complete if it is to be functional.

Taxa and categories

A group of individuals that form a class is called a taxon. A taxon has the maximum number of properties or attributes in common, and because of these similarities we can make the most numerous and most important statements about a taxon. A category is a group of taxa, defined at about the same level of abstraction and including the entire population of individuals. In *Soil Taxonomy*, there are six categories: order, suborder, great group, subgroup, family, and series. The categories serve specific purposes and the amount of detail that goes into the definition increases in the lower categories.

Order

The order are diffentiated by the presence or absence of diagnostic horizons or features. These are marks indicating differences in the degree and kinds of the dominant sets of soil-forming processes that have gone on in the past. Ten orders are described in *Soil Taxonomy* - Alfisols, Aridisols, Entisols, Histosols, Inceptisols, Mollisols, Oxisols, Spodosols, Ultisols, and Vertisols. The order category provides a conceptual grouping of the soils based on common but important properties.

Suborders

The suborders generally are defined on the basis of properties that affect the current processes of soil development. In most suborders, soil moisture and temperature regimes are the properties used. Currently, 47 suborders are recognized, and the differentiae for the suborders vary with the order. The following examples illustrate this.

Entisol	Alfisol	Aridisol	Vertisol
Aquent	Aqualf	Argid	Xerert
Arent	Boralf	Orthid	Torrert
Psamment	Ustalf		Udert
Fluvent	Xeralf		Ustert
Orthent	Udalf		

Suborders are generally used for making small scale maps - maps with a scale of 1:5 000 000 or 1:1 000 000. They are used for delineating broad geographic areas.

Great groups

In the category of great groups, a number of soil properties are identified. The whole soil is characterized, including the assemblage of horizons, and the most significant property of the whole soil is determined from the number and importance of accessory properties. Although the differentiating properties of the great group are few, there are many accessory properties.

About 200 great groups have been identified in the U.S. The great groups contain enough information to allow some general statements about the use of the soil. They are frequently used in small-scale maps.

Subgroups

The categories above the subgroup focus on the marks or causes of sets of processes that appear to dominate the course or degree of soil development. In addition, many soils have properties that are subordinate, but still are important marks of soil-forming processes.

Within a given great group, the soil may show marks of processes that are dominant in orders, suborders, or great groups, but in a given great group these processes serve only to modify other, dominant processes. Identifying these less significant processes at the subgroup level helps to show the relationship to other kinds of soils. In addition, some properties can be used to define a subgroup that are used as criteria for any taxon at a higher level.

Consequently, there are three kinds of subgroups:

- The typic subroup, which defines the central concept of the great group. The typic subgroup may not be the most extensive.
- The integrades, which define the relationship to other orders, suborders, or great groups, For example, ultic subgroups are intergrades to the order of Ultisols, orthoxic subgroups are intergrades to the suborder of Orthox, and dystropeptic subgroups are intergrades to the great group of Dystropepts.
- The extragrades, which have properties not used in higher categories. Examples are cumulic, lithic, and ruptic subgroups.

Families

In this category, the intent has been to group soils within a subgroup that have similar physical and chemical properties affecting their responses to use and management. In some instances, phases of families are used because the information contained in the family may be insufficient to make the desired kind of interpretations.

Families are defined so as to provide groupings of soils with restricted ranges in:

- particle-size distribution in horizons of major biologic activity below plow depth;
- mineralogy of the same horizons that are considered in naming particle -size classes;
- soil temperature regime;
- thickness of soil penetrable by roots; and
- a few other properties that are used to produce the homogeniety needed to define the family.

Series

The soil series is the lowest category, and in the U.S. there are about 13 000 series. Differentiating properties of a series cannot fall outside the limits established for the family. Differentiating properties of different series in the same family must meet three tests:

- the properties must be observable or inferable with reasonable assurance;
- they must fall within a unique range for each series, and this range must be significantly greater than the normal range of errors of measurment, observation, or estimation of qualified soil surveyors; and

- the properties must have some relation to horizon differentiation if horizons are present.

Polypedons, as mentioned earlier, are real things, but a soil series is conceptual. The dominant kinds of polypedons that are delineated on a map are given the names of soil series. A polypedon may be a series or, more commonly, a phase of a series.

Diagnostic horizons

One of the unique innovations in *Soil Taxonomy* is the diagnostic horizon. In developing *Soil Taxonomy,* the conventional A, B, C nomenclature of soil horizons was dropped, and in its place diagnostic surface and subsurface horizons were introduced.

The diagnostic surface horizon is called an epipedon, and it is important to note that this is not a synonym for the A horizon. An epipedon can include all or part of the B horizon if the darkening by organic matter extends from the soil surface to the B horizon. *Soil Taxonomy* identifies six diagnostic surface horizons and describes their properities.

Several diagnostic subsurface horizons are also identified. In general, these horizons are below the diagnostic surface horizon, but may include the uppermost soil horizon if the soil is truncated. *Soil Taxonomy* defines several other diagnostic soil characteristics including soil moisture and temperature regimes.

Taxonomic placement of soil

To classify a soil, it is imperative to use the keys provided for the descending hierarchy of categories in *Soil Taxonomy.* It is incorrect, for example, to have a preconceived notion of the order and suborder and go directly to great groups for examples. Even the experienced classifier must use the *Taxonomy* systematically and determine that the soil meets all the requirements of the keys for order, suborder, and so on.

Each taxon in each category also carries a definition. The classifier must check this definition before proceeding to the key of the next lower category. Recently, a few inconsistences have been found between the definition of a taxon within a category and the key of that category. Such inconsestencies are now being rectified.

Conclusion

Soil Taxonomy makes it possible to classify all soils that are known. Its structure serves equally well for organizing existing knowledge about soils for project planning and for relating test results and research to specific soil properties. It is now possible to increase our knowledge about soils in an organized manner, and this will permit application of new knowledge in planning future projects.

References

AFANASIEV, J.N. 1927. The classification problem in Russian soil science. *Russian pedological investigations*, vol. 5. Leningrad, USSR: Academy of Science.

BALDWIN, M., KELLOGG, C.E. and THORP, J. 1938. Soil classification. In: *Soils and Men*, 979-1001. Soil Conservation Service, U.S. Department of Agriculture. Year book of Agriculture. Washington DC: Government Printing Office.

CLINE, M.G. 1949. Basic principles of soil classification. *Soil Science* 7: 81-91.

CLINE, M.G. 1979. *Soil classification in the United states*, 79-12. Department of Agronomy. Agronomy Mineograms. Ithaca NY: Cornell University.

GLINKA , K. D. 1927. Dokuchaiev's ideas in the development of pedology and cognate sciences. *Russian pedological investigations*, vol. 1. Leningrad, USSR: Academy of Science. 32p.

MARBUT, C. F. 1913. *Soils of the United States*. Soil Conservation Service, U.S. Department of Agriculture. Bureau of Soils Bulletin no. 96. Washington DC: Government Printing Office.

SMITH, G. D. 1965. *Lectures on soil classification*. Pedologie, Special Bulletin no. 4. Ghent, Belgium: University of Ghent. 134p.

SOIL SURVEY STAFF. 1975. *Soil taxonomy: A basic system of soil classification for making and interpreting soil surveys*. Soil Conservation Service, U.S. Department of Agriculture. Agricultural Handbook no. 436. Washington DC: Government Printing Office. 754 p.

WHITNEY, M. 1909. *Soils of the United States*. Soil Conservation Service, U.S. Department of Agriculture. Bureau of Soils Bulletin no. 55. Washington DC: Government Printing Office.

The principles of the FAO/Unesco soil map legend

F.J. DENT*

Abstract

The legend of the FAO/Unesco soil map of the world *was elaborated in the early seventies, when the map was drawn. At that time it represented a compromise between different classification systems, with the following objectives; (i) to make an appraisal of the world's soil resources; (ii) to formulate a scientific basis for the transfer of experience; (iii) to promote a generally accepted system of classification; (iv) to provide a common framework for more detailed investigations; (v) to produce a basic document for education, research, and development; (vi) to strengthen international contacts in the field of soil science.*

The present paper gives, the principles of the legend, and discusses new developments which have been introduced since 1973. Two notable developments have been the attempt to define a third level, which will complete the first two levels, and the phase, which will allow for more detailed investigations.

Introduction

The preparation of the joint *FAO/Unesco soil map of the world* (Unesco, 1974) was commenced in 1961 in response to a recommendation of the Interna-

* FAO Regional Office for Asia and the Pacific, Maliwan Mansion, Phra Atit Road, Bangkok 10200, Thailand.

tional Society of Soil Science (ISSS) at its 7th Congress held at Madison, Wisconsin (USA) in 1960. The final two sheets of a set of 19 1:5 million scale map sheets were published in 1981. The *Soil map of the world*, therefore, took 20 years to complete and involved over 300 soil scientists from all over the world.

At the start of the project, six objectives were set for the *Soil map of the world*. The achievements of these objectives to date are discussed below.

Objective 1. To make a first appraisal of the world's soil resources

Objective achieved with the *Soil map of the world* being published, consisting of 19 map sheets and 10 volumes of explanatory text.

Objective 2. To supply a scientific basis for the transfer of experience between areas with similar environments

The *Soil map of the world* has proved of value as a basis for establishing policies of development and optimization of land use at a global level through:

- the *World map of desertification* (FAO/Unesco/WMO, 1977)
- a methodology for soil degradation assessment (FAO/UNEP/Unesco, 1979)
- a study of potential population-supporting capacity of lands in the developing world (Higgins *et al.*, 1982)
- agroecological zones studies marking suitability for the production of major agricultural commodities in different regions (FAO, 1978-81)

Objective 3. To promote the establishment of a generally accepted soil classification and nomenclature

The legend of the *Soil map of the world* is used as a common denominator for correlating different soil classification systems. The legend is often listed jointly with the national soil classification in publications and soil maps. It has also been adopted by a number of libraries and data-processing facilities as a key to stratify the soil universe.

Objective 4. To establish a common framework for more detailed investigations in developing areas

The preparation of the *Soil map of the world* has stimulated the initiation or intensification of soil surveys in a number of developing countries. The legend was designed for an inventory at 1:5 million scale, but has also proved useful for reconnaissance surveys at a larger scale. In a number of countries national soil maps have been prepared using the *Soil map of the world* legend (Indonesia and Japan in the Asia and Pacific region). A number of countries are also in the process of introducing third-level soil subunits.

Objective 5. To serve as a basic document for educational, research and development activities

Here the *Soil map of the world* has proved to be a valuable tool. It is being used for teaching, the study of soil geography, the preparation of development projects, the selection of representative sites for research, the stratification of experimental data, and ecological characterization at a regional level.

Objective 6. To strengthen international contacts in the field of soil science

Over the last 20 years, the *Soil map of the world* project has contributed significantly to strengthening international contacts in the field of soil science. However, though the different schools of thought regarding soil classification have grown closer together, the need for a generally accepted reference base for soil classification is felt by the international soils community more strongly now than ever before.

It is apparent, therefore, that the stated objectives of the *Soil map of the world* have largely been met. However, in order to retain its value, map sheets and the definition of the soil units must be updated as new knowledge develops and new surveys throw more light on the world's soil cover.

Considering the shortage of financial resources, it is unlikely that a new edition of the *Soil map of the world* could be envisaged. It is the intention, however, to continuously update the maps through digitized storage of new material. At the same time, modifications have been proposed for the legend. These are relatively few, and the general structure and rationale of the original legend are maintained. The original legend will still be used for reading and interpreting the published maps; while the revised legend will be used for updating and/or preparing new maps at a scale of 1:5 million or larger.

In the following section, the changes proposed for the revised legend are incorporated in the discussion on principles.

General principles

Lack of a generally accepted system of soil classification was a major obstacle to the preparation of the *Soil map of the world*. It was therefore necessary to establish a common denominator between different soil classification systems, and to combine into one outline the major soil units recognized in all parts of the world, both in virgin conditions and under cultivation.

The soil units adopted were selected on the basis of present knowledge of the formation, characteristics and distribution of the soils of the world, their importance as resources for production, and their significance as factors of the environment. These units do not correspond to equivalent categories in different classification systems, but they are generally comparable to the 'great group' level.

For reliable identification and correlation in areas far apart, the soil units have been defined in terms of measurable and observable properties. The system is 'natural', as differentiating criteria are essential properties of the soil itself. Key properties have been selected so as to correlate with as many other characteristics as possible, and are combined into "diagnostic horizons" used for formulating the definitions. Many key properties are relevant to soil use. As a result, the units distinguished have prediction value for the use of the soil.

The construction of the legend is based on international agreement regarding the major soils to be represented on the *Soil map of the world*. However, no consensus could be obtained on the 'weight' which each of these units should have within a classification "system". It is precisely in the concepts on which the subdivisions in categories are based - zonality, evolution, morphololgy, ecology or geography - that existing soil classification systems differ most.

Nomenclature

An attempt has been made to use as many 'traditional' names as possible, such as Chernozems, Kastanozems, Podzols, Planosols, Solonetz, Solonchaks and Regosols. Names which have recently acquired more general acceptance like Vertisols, Andosols, Gleysols and Ferralsols, have also been adopted. However, sharpening the definitions of these units by the use of precisely defined terms may have created a narrower concept than that found in the literature for units bearing the same names.

In order to avoid confusion, which has arisen from the different uses of the terms 'podzolized' and 'podzolic', the following names have been introduced for soils in which the essential characteristic is the illuvial accumulation of clay:

Luvisols - high activity clays, high base saturation
Lixisols - low activity clays, high base saturation
Acrisols - low cation-exchange capacity, low base saturation
Alisols - higher cation-exchange capacity, high total exchangeable aluminium content, low base saturation

Soils showing characteristics of both Podzols and Luvisols have been called Podzolucisols. The name Cambisols was coined as a common denominator for soils showing changes in colour, structure, and consistence resulting from weathering *in situ*. The term Phaeozems was introduced for soils occurring in the transitional belt between Chernozems or Kasatnozems and Luvisols. Hydromorphic soils containing a bleached horizon above plinthite are tentatively separated as Plinthosols. Nitosols are soils which, because of their favourable physical properties and their often higher fertility, have been separated from Ferralsols. The name Fluvisols has been introduced and redefined. Other newly introduced names are Histosols for organic soils, Anthrosols for soils resulting from human activities, Arenosols for weakly developed sandy soils, Calcisols for soils with a marked accumulation of calcium carbonate or gypsum, and Leptosols for shallow soils.

Formative elements used for naming 'major soil groupings' (level I) are given in the key to major soil groupings (see 'major soil groupings' below). Formative elements used for naming 'soil units' (level II) are as follows:

Formative elements used for naming soil units (level II)

Akric: from Gr. *akros*, ultimate; connotative of strong weathering. ≥1.5 m.e. extractable aluminium per 100 g clay in at least part of B within 125 cm of the surface.

Albic: from L. *albus*, white; connotative of strong bleaching. Albic E horizon, with minimum thickness of 125 cm from the surface.

Aric: From L. *arare*, to plough; connotative of plough layer. Showing remnants of diagnostic horizons due to deep cultivation.

Calcaric: from L. *calcium*; connotative of presence of calcium carbonate. Calcareous at least from 20 to 50 cm of the surface.

Calcic: from L. *calxis*, lime; connotative of accumulation of calcium carbonate or gypsum. Having a calcic horizon or concentrations of soft powdery lime within 125 cm of the surface.

Cambic: from late L. *cambiare*, change; connotative of change in colour, structure, or consistence. A cambic B horizon immediately below the A horizon.

Chromic: from Gr. *chromos*, colour; connotative of soils with bright colours. Hue of 7.5 YR and chroma more than 4 (rubbed soil), or a hue redder than 7.5YR.

Cumulic: from L. *cumulare*, to accumulate; connotative of accumulation of sediments. Accumulation of fine sediments more than 50 cm thick from the surface (due to irrigation or to a man-made surface layer).

Dystric: from Gr. *dys*, ill, dystrophic, infertile; connotative of low base saturation. Base saturation (NH_4OA_c) less than 50%.

Ferralic: from L. *ferrum* and *aluminium*; connotative of a high content of sesquioxides. A cambic B with oxic properties.

Ferric: from L. *ferrum,* iron; connotative of ferruginous mottling. Showing ferric properties.

Fibric: from L. *fibra,* fibre; connotative of weakly decomposed organic material. Fibre content dominant to a depth of 35 mm or more from the surface.

Fimic: from L. *fimum,* manure, slurry, mud; connotative of horizon formed by long-continued manuring. Having a fimic horizon.

Folic: from L. *folia,* leaves; connotative of undecomposed organic material. Well-drained Histosols never saturated with water for more than a few days.

Gelic: from L. *gelu,* frost; connotative of permafrost. Permafrost within 200 cm of the surface.

Gleyic: from Russian local name *gley,* mucky soil mass. Hydromorphic properties within 100 cm of the surface.

Glossic: from Gr. *glossa,* tongue; connotative of tonguing of the A horizon into the underlying layers. Tonguing of the A horizon into cambic B or C horizon.

Gypsic: from L. *gypsum;* connotative of the presence of gypsum. Having a gypsic horizon.

Haplic: from Gr. *haplos,* simple; connotative of soils with a simple, normal horizon sequence and common occurrence.

Humic: from L. *humus,* earth; rich in organic matter. Having sub-horizon of spodic B containing dispersed organic matter (does not turn redder on ignition).

Luvic: from L. *luvi,* from *luo,* to wash, 'lessiver'; connotative of illuvial accumulation of clay.

Mollic: from L. *mollis;* connotative of good surface structure. Having a mollic A horizon.

Plinthic: from Gr. *plinthos,* brick; connotative of mottled clay materials which harden irreversibly upon exposure. Having plinthite within 125 cm of the surface.

Renczic: from Polish *rzedzic,* noise; connotative of the noise made by a plough over shallow stony soil. Having a mollic A horizon containing or immediately overlying calcareous material ($CaCO_3$ equivalent >40%).

Rhodic: from Gr. *rhodon,* rose. Red to dusky red argillic or oxic B horizon.

Terric: from L. *terra,* earth, connotative of well-decomposed and humified organic materials. Very few visible plant fibres, very dark grey to black colour to a depth of 35 cm or more from the surface.

Thionic:	from Gr. *theion*, sulphur; denoting the presence of sulphidic materials. A sulphuric horizon, or sulphidic materials within 125 cm of the surface.
Umbric:	from L. *umbra*, shade; presence of an umbric A horizon.
Urbic:	from L. *urbis*, town; connotative of disposal of wastes. Accumulation of wastes, refuse; fills to a depth of more than 50 cm.
Vertic:	from L. *verto*, turn; connotative of turnover of the surface soil. Vertic properties - cracks slickensides, wedge-shaped or parallelipiped structural aggregates insufficient to qualify as Vertisols.
Vitric:	from L. *vitrum*, glass; connotative of soils rich in vitric material. Lack smeary consistence and/or have texture coarser than silt loam (weighted average) for all horizons within 100 cm of the surface.
Xanthic:	from Gr. *xanthos*, yellow. Yellow to pale yellow oxic B. Rubbed soil hues of 7.5YR or yellower; moist value 4 or more, chroma 5 or more.
Yermic:	from Spanish *yermo*, desert; connotative of soils found in arid environments. Yermic properties.

Diagnostic horizons

Soil horizons that have a set of quantitatively defined properties which are used for identifying soil units are called 'diagnostic horizons'.

The definitions and nomenclature of the diagnostic horizons used here are drawn from those adopted in *Soil Taxonomy* (Soil Survey Staff, 1975). The definitions of these horizons have been summarized and sometimes simplified in accordance with the requirements of the FAO/Unesco legend. The sombric horizon and the agric horizon of *Soil Taxonomy* have not been used as diagnostic horizons. The duripan, fragipan, petrocalcic horizon and petrogypsic horizon are used as phases.

The cation-exchange capacity (CEC), used as a criterion in the definition of diagnostic horizons, diagnostic properties, and soil units refers to the mineral component of the exchange complex. A deduction from the analytical results related to the whole soil has to be made on account of the organic-matter influence. This can be done through calculation, using an average activity of 4 m.e. per 1% of organic carbon. When the clay mineralogy is homogenous throughout the soil till at least 125 cm depth, the deduction can be made by applying the graphical method for each soil profile (Bennema and Camargo, 1979; Klamt and Sombroek, 1987).

The terminology used to describe soil morphology is the one adopted in *Guidelines for soil profile description* (FAO, 1977). Colour notations are according to the Munsell soil colour charts. Chemical and physical characteristics are expressed on the basis of *Soil survey laboratory methods and procedures for collecting soil samples* (Soil Survey Staff, 1967).

Histic H horizon

The histic H horizon is an H horizon which is more than 20 cm but less than 40 cm thick. It can be more than 40 cm but less than 60 cm thick if it consists of 75% or more, by volume, of sphagnum fibres, or has a bulk density when moist of less than 0.1.

A surface layer of organic material less than 25 cm thick also qualifies as a histic H horizon if, after having been mixed to a depth of 25 cm, it has 18% or more organic carbon and the mineral fraction contains more than 60% clay, or 12% or more organic carbon and the mineral fraction contains no clay, or has intermediate proportions of organic carbon for intermediate contents of clay. The same criterion applies to a plough layer which is 25 cm or more thick.

Mollic A horizon

The mollic A horizon is an A horizon which, after the surface 18 cm are mixed, as in ploughing, has the following properties:

- The soil structure is sufficiently strong that the horizon is not both massive and hard or very hard when dry. Very coarse prisms larger than 30 cm in diameter are included in the meaning of massive if there is no secondary structure within the prisms.
- Both broken and crushed samples have colours with a chroma of less than 3.5 when moist, and a value darker than 3.5 when moist and 5.5 when dry; the colour value is at least one unit darker than that of the C (both moist and dry). If a C horizon is not present, comparison should be made with the horizon immediately underlying the A horizon. If there is more than 40% finely divided lime, the limits of colour value dry are waived; the colour value moist should then be 5 or less.
- The base saturation is 50% or more (by NH_4OA_c).
- The organic carbon content is at least 0.6% throughout the thickness of mixed soil, as specified below. The organic carbon content is at least 2.5% if

the colour requirements are waived because of finely divided lime. The upper limit of the organic carbon content of the mollic A horizon is the lower limit of the histic H horizon.

- The thickness is 10 cm or more if resting directly on hard rock, a petrocalcic horizon, a petrogypsic horizon or a duripan; the thickness of the A must be at least 18 cm, and more than one-third of the thickness of the solum where the solum is less than 75 cm thick, and must be more than 25 cm where the solum is more than 75 cm thick. The measurement of the thickness of a mollic A horizon includes transitional horizons in which the characteristics of the A horizon are dominant - for example, AB, AE, or AC.
- The content of P_2O_5 soluble in 1% citric acid is less than 250 ppm, unless the amount of P_2O_5 soluble in citric acid increases below the A horizon, or when it contains phosphate nodules, as may be the case in highly phosphatic parent materials. This restriction is made to eliminate plough layers of very old arable soils or kitchen middens.

Fimic A horizon

The fimic A horizon is a man-made surface layer 50 cm or more thick which has been produced by long-continued manuring; otherwise it may conform to the requirements of the mollic or umbric epipedon. It includes all soils with an acid P_2O_5 content higher than 250 ppm in 1% citric acid. The fimic A horizon commonly contains artifacts, such as bits of brick and pottery, throughout its depth. The fimic horizon as defined here includes the plaggen epipedon and the anthropic epipedon of *Soil Taxonomy* (Soil Survey Staff,1975).

Umbric A horizon

The requirements of the umbric A horizon are comparable to those of the mollic A in colour, organic carbon and phosphorus content, consistency, structure and thickness. The umbric A horizon, however, has a base saturation of less than 50% (by NH_4OAc).

Horizons which have acquired the above requirements through the slow addtion of materials under cultivation are excluded from the umbric A horizon. Such horizons are fimic A horizons.

Ochric A horizon

An ochric A horizon is one that is too light in colour, has too high a chroma, too little organic carbon, or is too thin to be mollic or umbric, or is both hard and massive when dry. Finely stratified materials do not qualify as an ochric A horizon, e.g. surface layers of fresh alluvial deposits.

Argillic B horizon

An argillic B horizon is one that contains illuvial layer-lattice clays. This horizon forms below an eluvial horizon, but it may be at the surface if the soil has been partially truncated. The argillic B horizon has the following properties:

- If an eluvial horizon remains, the argillic B horizon contains more total and more fine clay than the eluvial horizon, exclusive of differences which may result from a lithological discontinuity. The increase in clay occurs within a vertical distance of 30 cm or less. Note that:
  - o if any part of the eluvial horizon has less than 15% total clay in the fine earth (less than 2 mm) fraction, the argillic B horizon must contain at least 3% more clay (for example, 13% versus 10%);
  - o if the eluvial horizon has more than 15% and less than 40% total clay in the fine earth fraction, the ratio of the clay in the argillic B horizon to that in the E horizon must be 1.2 or more;
  - o if the eluvial horizon has more than 40% total clay in the fine earth fraction, the argillic B horizon must contain at least 8% more clay (for example, 50% versus 42%).
- An argillic B horizon should be at least one-tenth the thickness of the sum of all overlying horizons, or more than 15 cm thick if the eluvial and illuvial horizons are thicker than 150 cm. If the B horizon is sand or loamy sand, it should be at least 15 cm thick; if it is loamy or clayey, it should be at least 7.5 cm thick. If the B horizon is entirely composed of lamellae, the lamallae should have a combined thickness of at least 15 cm.
- In soils with massive or single-grained structure, the argillic B horizon has oriented clay bridging the sand grains and also in some pores.
- If peds are present, an argillic B horizon either:
  - o shows clayskins on at least 1% or more of both the vertical and horizontal ped surfaces and in the pores, or shows oriented clays in 1% or more of the cross section;
  - o if the B has a broken or irregular upper boundary and meets requirements of thickness and textural differentiation as defined under the

first two properties considered above, clayskins should be present - at least in the lower part of the horizon;

- o if the B horizon is clayey with kaolinitic clay and the surface horizon has more than 40% clay, there are some clayskins on peds and in pores in the lower part of the horizon with blocky or prismatic structure; or
- o if the B horizon is clayey with 2:1 lattice clays, clayskins may be lacking, provided there is evidence of pressure caused by swelling; or if the ratio of fine to total clay in the B horizon is greater by at least one-third than the ratio in the overlying or the underlying horizon, or if it has more than 8% more fine clay; the evidence of pressure may be occasional slickensides or wavy horizon boundaries in the illuvial horizon, accompanied by uncoated sand or silt grains in the overlying horizon.

- If a soil shows a lithologic discontinuity between the eluvial horizon and the argillic B horizon, or if only a plough layer overlies the argillic B horizon, the horizon only needs to show clayskins in some parts - either in some fine pores or, if peds exist, on some vertical and horizontal ped surfaces. Thin sections should show that some part of the horizon has about 1% or more of oriented clay bodies, or the ratio of fine clay to total clay should be greater by at least one-third than the ratio in the overlying or the underlying horizon.
- The argillic B horizon lacks the set of properties which characterizes the natric B horizon.[1]

Natric B horizon

The natric B horizon has all the properties (except the last) of the argillic B horizon described above. In addition, it has:

- a columnar or prismatic structure in some part of the B horizon, or a blocky structure with tongues of an eluvial horizon in which there are uncoated silt or sand grains extending more than 2.5 cm into the horizon; and
- a saturation with exchangeable sodium of more than 15% within the upper 40 cm of the horizon; or more exchangeable magnesium plus sodium than calcium plus exchange acidity (at pH 8.2) within the upper 40 cm of the horizon, if the saturation with exchangeable sodium is more than 15% in some subhorizon within 200 cm of the surface.

---

1 This is an overall and general definition of the argillic B horizon. Several variants are described by Sombroek (1984): luvi-argillic, lixi-argillic, natri-argillic, niti-argillic and abrupto-argillic horizons.

Cambic B horizon

The cambic B horizon is an altered horizon lacking properties that meet the requirements of an argillic, natric, or spodic B horizon; lacking the dark colours, organic-matter content and structure of the histic H, or the mollic and umbric A horizons; showing no cementation, induration or brittle consistence when moist, having the following properties:

- Texture that is very fine sand, loamy very fine sand, or finer.
- Soil structure or absence of rock structure in at least half the volume of the horizon.
- Significant amounts of weatherable minerals reflected by a cation-exchange capacity (by NH_4OAc) of more than 16 m.e. per 100 g clay, or by a content of more than 3% weatherable minerals other than muscovite, or by more than 6% muscovite in the fine earth fraction.
- Evidence of alteration in one of the following forms:
  - o Gray reduction colours or artificial drainage, and one or more of the following properties: (i) a regular decrease in the amount of organic carbon with depth and a content of less than 0.2% organic carbon at a depth of 125 cm below the surface or immediately above a sandy-skeletal substratum that is at a depth of less than 125 cm, (ii) cracks that open and close in most years and are 1 cm or more wide at a depth 50 cm below the surface, (iii) permafrost at some depth, and (iv) a histic H horizon consisting of mineral soil materials or a mollic or umbric A horizon.
  - o Stronger chroma, redder hue, or higher clay content than the underlying horizon.
  - o Evidence of removal of carbonates. In particular, the cambic horizon has less carbonate than the underlying horizon of calcium carbonate accumulation. If all coarse fragments in the underlying horizon are completely coated with lime, some in the cambic horizon are partly free of coatings. If coarse fragments in the ca horizon are coated only on the underside, those in the cambic horizon should be free of coatings.
  - o If carbonates are absent in the parent material and in the dust that falls on the soil, the required evidence of alteration is satisfied by the presence of soil structure and the absence of rock structure in more than 50% of the horizon.
- Enough thickness that its base is at least 25 cm below the soil surface.

Spodic B horizon

A spodic B horizon meets one or more of the following requirements below a depth of 12.5 cm, or (when present) below an A horizon:

- A subhorizon more than 2.5 cm thick that is continuously cemented by a combination of organic matter with iron or aluminium, or with both.
- A sandy or coarse loamy texture with distinct dark pellets of coarse silt size or larger, or with sand grains covered with cracked coatings.[2]
- One or more subhorizons in which:
  - o if there is 0.1% or more extractable iron, the ratio of iron plus aluminium (elemental) extractable by pyrophosphate at pH 10 to percentage of clay is 0.2 or more, or if there is less than 0.1% extractable iron, the ratio of aluminium plus carbon to clay is 0.2 or more;
  - o the sum of pyrophosphate-extractable iron plus aluminium is half or more of the sum of dithionite-citrate extractable iron plus aluminium; and
  - o the thickness is such that the index of accumulation of amorphous material (CEC at pH 8.2 minus one half the clay percentage multiplied by the thickness in centimetres) in the horizons that meet the preceding requirements is 65 or more.

Oxic B horizon

The oxic B horizon is a horizon that is not argillic or natric[3] and that:

- is at least 30 cm thick;
- has an apparent cation-exchange capacity of the fine earth fraction of 16 m.e. or less per 100 g clay (by NH_4OAc), unless there is an appreciable content of aluminium-interlayered chlorite;
- does not have more than traces of primary aluminosilicates such as feldspars, micas, glasses, and ferromagnesian minerals;
- has the texture of sandy loam or finer in the fine earth fraction and has more than 15% clay;
- has mostly gradual or diffuse boundaries between its subhorizons; and
- has less than 5% by volume showing rock structure.

---

2 Dark pellets or cracked coatings consist of organic matter and aluminium, with or without iron.

3 The oxic B horizon is exclusive of A and E horizons, even though some of them meet the properties of an oxic B horizon.

Calcic horizon

The calcic horizon is a horizon of accumulation of calcium carbonate. The accumulation may be in the C horizon, but it may also occur in a B or in an A horizon.

The calcic horizon consists of secondary carbonate enrichment over a thickness of 15 cm or more, has a calcium carbonate equivalent content of 15% or more, and at least 5% greater than that of a deeper horizon. The latter requirement is expressed by volume if the secondary carbonates in the calcic horizon occur as pendants on pebbles, or as concretions or soft powdery forms; if such a calcic horizon rests on very calcareous materials (40% or more calcium carbonate equivalent), the percentage of carbonates need not decrease with depth.

Gypsic horizon

The gypsic horizon is a horizon of secondary calcium sulphate enrichment that is more than 15 cm thick, has at least 5% more gypsum than the underlying C horizon, and in which the produce of the thickness in centimetres and the percentage of gypsum is 150 or more. If the gypsum content is expressed in m.e. per 100 g of soil, the percentage of gypsum can be calculated from the product of the m.e. of gypsum per 100 g of soil and the m.e. weight of gypsum, which is 0.086. Gypsum may accumulate uniformly throughout the matrix or as nests of crystals; in gravelly material, gypsum may accumulate as pendants below the coarse fragments.

Sulphuric horizon

The sulphuric horizon forms as a result of artificial drainage and oxidation of mineral or organic materials which are rich in sulphides. It is at least 15 cm thick, and characterized by a pH of less than 3.5 (H_2O 1:1) and jarosite mottles with a hue of 2.5Y or more and a chroma of 6 or more.

Albic B horizon

The albic E horizon is one from which clay and free iron oxides have been removed, or in which the oxides have been segregated to the extent that the colour of the horizon is determined by the colour of the primary sand and silt particles, rather than by coatings on these particles.

An albic E horizon has a colour value moist of 4 or more, or a value dry of 5 or more, or both. If the value dry is 7 or more, or the value moist is 6 or more, the chroma is 3 or less. If the parent materials have a hue of 5YR or redder, a chroma moist of 3 is permitted in an albic E horizon where the chroma is due to the colour of uncoated silt or sand grains.

An albic E horizon may overlie a spodic B, argillic or natric B, a fragipan, or an impervious layer that produces a perched water table.

Diagnostic properties

A number of soil characteristics which are used to separate soil units cannot be considered as horizons. They are rather diagnostic features of horizons or of soils, which when used for classification purposes need to be quantitatively defined.

Abrupt textural change

An abrupt textural change is a considerable increase in clay content within a very short distance in the zone of contact between an A or E horizon and the underlying horizon. When the A or E horizon has less than 20% clay, the clay content of the underlying horizon is at least double that of the A or E horizon within a vertical distance of 8 cm or less. When the A or E horizon has 20% clay or more, the increase in clay content should be at least 20% (for example, from 30 to 50% clay) within a vertical distance of 8 cm or less, and the clay content in some part of the underlying horizon (B horizon or impervious layer) should be at least double that of the A or E horizon above.

Albic material

Albic materials are coarse-textured materials exclusive of E horizons, and have a colour value moist of 4 or more, or a value dry of 5 or more, or both. If the value dry is 7 or more, or the value moist is 6 or more, the chroma is 3 or less. If the parent materials have a hue of 5YR or redder, a chroma moist of 3 is permitted if the chroma is due to the colour of uncoated silt or sand grains.

Andic soil material

Andic soil material meets one or more of the following three requirements:

- acid oxalate extractable aluminium is 2% or more, or acid oxalate extractable aluminium plus 1/2 acid oxalate extractable iron is 2.5% or more; bulk density of the fine earth, measured in the field moist state, is less than 0.9 g/cm; and phosphate retention is more than 85%
- acid oxalate extractable aluminium is 0.4% or more, or acid oxalate extractable aluminium plus 1/2 acid oxalate extractable iron is 2.5% or more; the sand fraction is at least 30% of the fine earth and there is more than 30% of volcanic glass (or crystals coated with glass) in the sand fraction; or more than 60% by volume of the whole soil is volcanic clastic material coarser than 2 mm.
- acid oxalate extractable aluminium is between 0.4% and 2%, or acid oxalate extractable aluminium plus 1/2 acid oxalate extractable iron is between 0.5% and 2.5%; and the sand fraction is at least 30% of the fine earth; and the minimum percentage of glass required is progressively lower (from 30% to 0%) at increasing values of acid oxalate extractable aluminium (from 0.4% to 2%).

Calcareous material

The term calcareous applies to soil material which shows strong effervescence with 10% HCl in most of the fine earth, or which contains more than 2% calcium carbonate equivalent as detremined by laboratory analysis.

Continuous coherent hard rock

The term 'continuous coherent hard rock' applies to underlying material which is continuous except for a few cracks produced in place without significant displacement of the pieces, and on average 10 cm or more horizontally distant. The material is sufficiently coherent and hard when moist to make hand-digging with a spade impractical. The material considered here does not include subsurface horizons such as a duripan, a petrocalcic, a petrogypsic or a petroferric horizon, which are represented by phases.

Ferralic properties

The term 'ferralic properties' is used in connection with Cambisols and Arenosols which have a cation-exchange capacity (by NH_4OAc) of less than 16 m.e. per 100 g clay in at least some subhorizon of the cambic B horizon, or immediately underlying the A horizon.

Ferric properties

The term 'ferric properties' is used in connexion with Alisols, Lixisols and Acrisols showing one or more of the following: many coarse mottles with hues redder than 7.5YR or chromas more than 5, or both; discrete nodules, up to 2 cm in diameter, the exteriors of the nodules being enriched and weakly cemented or indurated with iron and having redder hues or stronger chromas than the interiors.

Fluvic properties

The term 'fluvic properties' is used to separate Fluvisols from other major soil groupings. The term refers to fluviatile, marine, and lacustrine sediments having one or more of the following properties:

- receiving fresh materials at regular intervals when not empoldered;
- having an organic carbon content that decreases irregularly with depth or that remains above 0.20% to a depth of 125 cm, though thin strata of sand may have less organic carbon if the finer sediments below meet the requirement, excluding buried A horizons; and
- showing stratification in at least 25% of the soil volume.

Gypsiferous material

The term gypsiferous applies to soil material which contains 5% or more gypsum.

Hydromorphic properties

The morphological characteristics which reflect waterlogging differ widely in relation to other soil properties. A distinction needs to be made between soils which are strongly influenced by groundwater in which only the lower horizons are influenced by groundwater, and those which have a seasonally perched water table. The expression 'hydromorphic properties' refers to one or more of the following features:

- saturation by groundwater, i.e. when water stands in a deep unlined bore hole at such a depth that the capillary fringe reaches the soil surface, and the water in the bore hole is stagnant and remains coloured when dye is added to it;
- occurrence of a histic H horizon;
- dominant hues that are neutral N, or bluer than 10Y; and
- saturation with water at some period of the year, or artificially drained, with evidence of reduction processes or of reduction and segregation of iron reflected by:
    - in soils having a spodic B horizon, one or more of the following:
        - o mottling in an albic E horizon or in the top of the spodic B horizon;
        - o if free iron and manganese are lacking, or if moist colour values are less than 4 in the upper part of the spodic B horizon, either:
            - no coatings of iron oxides on the individual grains of silt and sand in the materials in or immediately below the spodic horizon wherever the moist values are 4 or more, and unless an A horizon rests directly on the spodic horizon - there is a transition between the albic E and spodic B horizons at least 1 cm in thickness, or
            - fine or medium mottles of iron or manganese in the materials immediately below the spodic B horizon;
        - o a thin iron pan that rests on a fragipan or on a spodic B horizon, or occurs in an albic E horizon underlain by a spodic B horizon.
    - in soils having a mollic A horizon, If the lower part of the mollic A horizon has chromas of 1 or less, either:
        - o distinct or prominent mottles in the lower mollic A horizon; or
        - o colours immediately below the mollic A horizon or within 75 cm of the surface if a calcic horizon intervenes, with one of the following:
            - if hues are 10YR or redder and there are mottles, chromas of less than 1.5 on ped surfaces or in the matrix; if there are no mottles, chromas of less than 1 (if hues are redder than 10YR because of parent materials that remain red after citrate-dithionite extraction, the requirement for low chromas is waived);
            - if the hue is nearest to 2.5Y and there are distinct or prominent mottles, chromas of 2 or less on ped surfaces or in the matrix; if there are no mottles, chromas of 1 or less;

    - if the nearest hue is 5Y or yellower and there are distinct or prominent mottles, chromas of 3 or less on ped surfaces or in the matrix; and if there are no mottles, chromas of 2 or less;
    - hues bluer than 10Y;
    - any colour if the colour results from uncoated mineral grains;
    - colours neutral N.
  - if the lower part of the mollic A horizon has chromas or more than 1 but not exceeding 2, either;
    - o distinct or prominent mottles in the lower mollic A horizon; or
    - o basc colours immediately below the mollic A horizon that have one or more of:
      - values of 4 and chromas of 2 accompanied by some mottles with values of 4 or more and chromas of less than 2;
      - values of 4 and chromas of less than 2;
      - values of 5 or more and chromas of 2 or less accompanied by mottles with high chroma.
- in soils having an argillic B horizon immediately below the plough layer or an A horizon that has moist colour values of less than 3.5 when rubbed, one or more of the following:
  - o mottles due to segregation of iron;
  - o iron-manganese concretions larger than 2 mm, and combined with one or more of the following:
    - dominant moist chromas of 2 or less in coatings on the surface of peds accompanied by mottles within the peds, or dominant moist chromas of 2 or less in the matrix of the argillic B horizon accompanied by mottles of higher chromas (if hues are redder than 10YR because of parent materials that remain red after citrate-dithionite extraction, the requirement for low chromas is waived);
    - moist chromas of 1 or less on the surfaces of peds or in the matrix of the argillic B horizon;
    - dominant hues of 2.5Y or 5Y in the matrix of the argillic B horizon accompanied by distinct or prominent mottles.
- in soils having an oxic B horizon, and free of mottles, dominant chromas of 2 or less immediately below an A horizon that has a moist colour value of less than 3.5; or if mottled with distinct or prominent mottles having dominant chromas or 3 or less, either:
  - o in horizons with textures finer than loamy fine sand:
    - if there is mottling, chromas of 2 or less;
    - if there is no mottling and values are less than 4, chromas of less than 1; if values are 4 or more, chromas of 1 or less; or

- o in horizons with textures of loamy fine sand or coarser:
  - if hues are as red as or redder than 10YR and there is mottling, chromas of 2 or less; if there is no mottling and values are less than 4, chromas of less than 1; or if values are 4 or more, chromas of 1 or less;
  - if hues are between 10YR and 10Y and there is distinct or prominent mottling, chromas of 3 or less; if there is no mottling, chromas of 1 or less.

Interfingering

Interfingering consists of penetrations of an albic E horizon into an underlying argillic or natric B horizon along ped faces, primarily vertical faces. The penetrations are not wide enough to constitute tonguing, but form continuous skeletans (ped coatings of clean silt or sand, more than 1 mm thick on the vertical ped faces). A total thickness of more than 2 mm is required if each ped has a coating more than 1 mm thick on the vertical ped faces. A total thickness of more than 2 mm is required if each ped has a coating of more than 1 mm. Because quartz is such a common constituent of soils, the skeletans are usually white when dry, and light grey when moist, but their colour is determined by the colour of the sand or silt fraction. The skeletans constitute more than 15% of the volume of any subhorizon in which interfingering is recognized. They are also thick enough to be obvious, by their colour, even when moist. Thinner skeletans that must be dry to be seen as a whitish powdering on a ped are not included in the meaning of interfingering.

Nitic properties

The term 'nitic properties' is used for soil material having a moderate to strong angular blocky structure easily falling apart into smaller polyedric elements, showing shiny ped faces which may be thin argillans or pressure faces, and having an amorphous iron content that is at least 5% of the free iron content (Fe_2O_3 by acid oxalate extraction compared to Fe_2O_3 by dithionite-citrate extraction).

Organic soil materials

Organic soil materials are:

- saturated with water for long periods or are artificially drained and, excluding live roots, have (a) 18% or more organic carbon if the mineral fraction is 60% or more clay, (b) 12% or more organic carbon if the mineral fraction has no clay, or (c) a proportional content of organic carbon (between 12 and 18%) if the clay content of the mineral fraction is between zero and 60%; or
- never saturated with water for more than a few days and have 20% or more organic carbon.

Permafrost

Permafrost is a layer in which the temperature is perennially at or below 0°C.

Plinthite

Plinthite is an iron-rich, humus-poor mixture of clay with quartz and other dilutents. It commonly occurs as red mottles, usually in platy, polygonal or reticulate patterns, which changes irreversibly to hardpan or to irregular aggregates on exposure to repeated wetting and drying. In a moist soil, plinthite is usually firm, but it can be cut with a spade. When irreversibly hardened, the material is no longer considered plinthite but develops into a petroferric or petric phase.

Salic properties

The term 'salic properties' applies to soils which have an electric conductivity of the saturation extract of more than 15 mmhos per cm at 25°C at some time of the year within 30 cm of the surface, or of more than 4 mmhos within 30 cm of the surface if the pH (H_2O, 1:1) exceeds 8.5.

Slickensides

Slickensides are polished and grooved surfaces that are produced by one mass sliding past another. Some of them occur at the base of a slip surface where a mass of soil moves downward on a relatively steep slope. Slickensides are very common in swelling clays in which there are marked seasonal changes in moisture content.

Smeary consistence

The term 'smeary consistence' is used in connection with Andosols, and is characterized by thixotropic soil material - that is, material that changes under pressure or by rubbing from a plastic solid into a liquified stage and back to the solid condition. In the liquified stage, the material skids or smears between the fingers.

Soft powdery lime

Soft powdery lime refers to translocated authigenic lime, soft enough to be cut readily with a finger nail, precipitated in place from the soil solution rather than inherited from a soil parent material. It should be present in a significant accumulation.

To be identifiable, soft powdery lime must have some relation to the soil structure or fabric. It may disrupt the fabric to form spheroidal aggregates, or white eyes, that are soft and powdery when dry, or the lime may be present as soft coatings in pores or on structural faces. If present as coatings, it covers a significant part of the surface; commonly, it coats the whole surface to a thickness of 1 to 5 mm or more. Only part of a surface may be coated if little lime is present in the soils. The coatings should be thick enough to be visible when moist, and should cover a continuous area large enough to be more than filaments. Pseudomycelia, which come and go with changing moisture conditions, are not considered as soft powdery lime in the present definition.

Sulphidic materials

Sulphidic materials are waterlogged mineral or organic soil materials containing 0.75% or more sulphur (dry weight), mostly in the form of sulphides, and having less than three times as much carbonate ($CaCO_3$ equivalent) as sulphur. Sulphidic materials accumulate in a soil that is permanently saturated, generally with brackish water. If the soil is drained, the sulphides oxidize to form sulphuric acid; and the pH, which normally is near neutrality before drainage, drops below 3.5. Sulphidic material differs from the sulphuric horizon in that it does not show jarosite mottles with a hue of 2.5Y or more, or a chroma of 6 or more.

Takyric features

Soil with takyric features are fine textured and form a platy or massive surface crust which cracks into polygonal elements when dry.

Tonguing

As used in the definition of Podzoluvisols, the term tonguing is connotative of the penetration of an albic E horizon into an argillic B horizon along ped surfaces, if peds are present. Penetrations, to be considered tongues, must have greater depth than width, have horizontal dimensions of 5 mm or more in fine-textured argillic horizons (clay, silty clay and sandy clay), 10 mm or more in moderately fine-textured argillic horizons, and 15 mm or more in medium- or coarser-textured argillic horizons (silt loams, loams, very fine sandy loams, or coarser), and must occupy more than 15% of the mass of the upper part of the argillic horizon.

With Chernozems, the term tonguing refers to penetrations of the A horizon into an underlying cambic B horizon or into a C horizon. The penetrations must have greater depth than width, and must occupy more than 15% of the mass of the upper part of the horizon in which they occur.

Vertic properties

The term 'vertic properties' is used in connection with clayey soils which at some period in most years show one or more of the following: cracks, slickensides, wedge-shaped or parallelipiped structural aggregates, that are not in a combination, or are not sufficiently expressed, for the soils to qualify as Vertisols.

Weatherable minerals

Minerals included in the meaning of weatherable minerals are those that are unstable in a humid climate relative to other minerals, such as quartz and

1:1 lattice clays, and that, when weathering occurs, liberate plant nutrients and iron or aluminium. They include:

- Clay minerals: all 2:1 lattice clays except alluminium - interlayered chlorite. Sepiolite, talc, and glauconite are also included in the meaning of this group of weatherable clay minerals, although they are not always of clay size.
- Silt- and sand-size minerals (0.02 to 0.2 mm in diameter): feldspars, feldspathoids, ferromagnesian minerals, glasses, micas, and zeolites.

Yermic properties

The term 'yermic properties' applies to soils showing one or more of the following:

- a surface layer of gravel or stones, exposed by wind erosion of fine material, with or without the iron staining typical of desert pavements;
- surface cracks filled with in-blown sand or silt;
- sand blowing over a stable surface;
- takyric features.

Major soil groupings

In the following pages simplified definitions of the key to the 'major soil groupings' are provided, and 'soil units' are listed in order to give some idea of the structure and scope of the revised *Soil map of the world* legend. A number of rules apply, as follows:

- The soils defined have their upper boundaries at the surface, or at less than 50 cm below the surface. Horizons buried by 50 cm or more of newly deposited surface material are no longer diagnostic for classification purposes.
- Diagnostic horizons and properties are assumed to have their upper limits within 125 cm of the surface unless otherwise specified.
- The definitions are not based on differences in soil temperature and soil moisture, unless such differences are also reflected in other soil characteristics which can be preserved in samples.
- When two or more B horizons occur within 125 cm of the surface, it is the upper B horizon which is determining for the classification, if it is sufficiently developed to be diagnostic.
- The terminology "having no diagnostic horizons other than" indicates that one or more of the diagnostic horizons listed may be present.

- All definitions, with the exception of Histosols, refer to mineral soils.
- The analytical data used are based on laboratory procedures described in *Soil survey laboratory methods and procedures for collecting soil samples* (Soil Survey Staff, 1967, ISRIC, 1985).

Histosols (Gr. *hostos*, tissue; fresh/partly decomposed organic material):

Soils having an H horizon:

- of 40 cm or more either extending down from the surface or taken cumulatively within the upper 80 cm of the soil; or
- 60 cm or more if the organic material consists mainly of sphagnam or moss or has a bulk density of less than 0.1; or
- the thickness of the H horizon may be less when it rests on rocks or on fragmental material with interstices filled with organic matter.

Units:

1. Gelic Histosols (HSi)
2. Thionic Histosols (HSt)
3. Folic Histosols (HSl)
4. Fibric Histosols (HSf)
5. Terric Histosols (HSs)

Anthrosols (Gr. *anthropos*, man; soils resulting from human activities):

Other soils:

- in which human activities have resulted in profound modifications of the original soil characteristics due to:
  * removal or disturbance of surface horizons
  * cuts and fills
  * secular additions of organic materials
  * long-continued irrigation
  * other

Units:

1. Aric Anthrosols (ATa)
2. Fimic Anthrosols (ATf)
3. Cumulic Anthrosols (ATc)
4. Urbic Anthrosols (ATu)

Cumulic Anthrosols only refer to soils which have been buried to 50 cm depth or more, not just to the puddled surface layer. Several soils strongly influenced by man are excluded from Anthrosols, particularly salinized soils, drained soils, and paddy soils. This is not the case with soils puddled to below 50 cm (if they occur), which would be classified as Aric Anthrosols.

Leptosols (Gr. *leptos*, thin; weakly developed shallow soils):

Other soils:

- limited in depth by:
    * continuous coherent hard rock, or
    * highly calcareous materials, or
    * continuous cemented layer within 50 cm, or
    * soils from very stony loose material with less than 20% fine earth to 125 cm;
- with no diagnostic horizons other than the mollic, umbric, or ochric A horizon, with or without cambic B;
- with no hydromorphic properties within 50 cm;
- with no salic properties.

Units:

1. Lithic Leptosols (LPs)
2. Gelic Leptosols (LPi)
3. Yermic Leptosols (LPy)
4. Rendzic Leptosols (LPk)
5. Mollic Leptosols (LPm)
6. Umbric Leptosols (LPu)
7. Dystric Leptosols (LPd)
8. Eutric Leptosols (LPe)

Vertisols (L. *verto*, turn; turnover of surface soils):

Other soils: having

- after the upper 20 cm have been mixed, 30% or more clay in all horizons to a depth of at least 50 cm;
- cracks from the soil surface downward, which at some period in most years (unless irrigated) are at least 1 cm wide to a depth of 50 cm;
- one or more of the following:
    * gilgai
    * intersecting slickensides at some depth between 25 and 100 cm from the surface
    * wedge-shaped or parallelipiped structural aggregates at some depth between 25 and 100 cm from the surface.

Units:

1. Yermic Vertisols (VRy)
2. Gleyic Vertisols (VRg)
3. Calcic Vertisols (VRc)
4. Gypsic Vertisols (VRj)
5. Haplic Vertisols (VRh)

Fluvisols (L. *fluvius*, river; floodplains and alluvial deposits):

Other soils developed from alluvial deposits with:

- one or more of the following *fluvic* properties:

* receive fresh materials at regular intervals when empoldered;
* organic carbon content that decreases irregularly, with depth of remains above 0.20% to a depth of 125 cm (thin sand strata may have less OC if underlying finer sediments meet the requirement); buried A horizons are excluded;

- no diagnostic horizons other than:
  * ochric, mollic, umbric A horizon, or
  * histic H horizon, or
  * sulphuric horizon, or
  * sulphidic material within 125 cm of the surface.

Units:

1. Yermic Fluvisols (FLy)
2. Thionic Fluvisols (FLt)
3. Mollic Fluvisols (FLm)
4. Calcaric Fluvisols (FLc)
5. Umbric Fluvisols (FLu)
6. Dystric Fluvisols (FLd)
7. Eutric Fluvisols (FLe)

Note: Most, but not all, Fluvisols have hydromorphic properties.

Solonchaks (Russian *sol*, salt and *chak*, salty area):
Other soils with:

- the following *salic* properties:
  * EC of the saturation extract of more than 15 mmhos per cm at 25°C at some time of the year, or
  * EC of the saturation extract of more than 4 mmhos per cm at 25°C at some time of the year within 30 cm of the surface, if pH (H_2O, 1:1) exceeds 8.5;
- no diagnostic horizons other than:
  * A horizon, histic H horizon, cambic B horizon, calcic or gypsic horizon.

Units:

1. Gelic Solonchaks (SCi)
2. Gleyic Solonchaks (SCg)
3. Mollic Solonchaks (SCm)
4. Gypsic Solonchaks (SCj)
5. Calcic Solonchaks (SCk)
6. Sodic Solonchaks (SCn)
7. Haplic Solonchaks (SCh)

Gleysols (Russian *gley*, mucky soil mass; soils with an excess of water):
Other soils with:

- one or more of the following *hydromorphic* properties:

* saturation by groundwater
* occurrence of a histic H horizon
* dominant hues are neutral N, or bluer than 10Y
* saturation with water at some period yearly, or artificially drained with evidence of reduction and/or reduction and segregation of iron;

- no diagnostic horizons other than:
  * A horizon, histic H horizon, cambic B horizon, calcic or gypsic horizon;
- lack plinthite within 125 cm of the surface.

Units:

1. Gelic Gleysols (GLi)
2. Thionic Gleysols (GLt)
3. Mollic Gleysols (GLm)
4. Umbric Gleysols (GLu)
5. Calcic Gleysols (GLk)
6. Dystric Gleysols (GLd)
7. Eutric Gleysols (GLe)

Andosols (Japanese *an*, dark and *do*, soil; soil rich in volcanic glass with a dark surface):

Other soils:

- formed from *andic* materials to a depth of 30 cm or more from the surface and having:
  * a mollic or umbric A horizon possibly overlying a cambic B horizon, or
  * an ochric A horizon and a cambic B horizon
  * no other diagnostic horizons.

Units:

1. Gelic Andosols (ANi)
2. Mollic Andosols (ANm)
3. Umbric Andosols (ANu)
4. Haplic Andosols (ANh)
5. Vitric Andosols (Anz)

Arenosols (L. *arena*, sand; weakly developed coarse textured soils):

Other soils of coarse texture having:

- no diagnostic horizons other than an ochric A horizon or an albic E horizon
- no plinthite within 125 cm of the surface.

Units:

1. Gleyic Arenosols (ARg)
2. Albic Arenosols (ARa)
3. Calcaric Arenosols (ARc)
4. Luvic Arenosols (ARl)
5. Ferralic Arenosols (ARo)
6. Cambic Arenosols (ARb)
7. Haplic Arenosols (ARh)

Regosols (Gr. *rhegos,* blanket; mantle of loose material overlying hard core of earth):

Other soils having:

- no diagnostic horizons other than an ochric A horizon.

Units:

1. Gelic Regosols (RGi)
2. Gypsic Regosols (RGj)
3. Calcaric Regosols (RGc)
4. Dystric Regosols (RGd)
5. Eutric Regosols (RGe)

Podzols (Russian *pod,* under and *zola,* ash; soils with strongly bleached horizon):

Other soils having:

- a spodic B horizon

Units:

1. Gelic Podzols (PZi)
2. Gleyic Podzols (PZg)
3. Humic Podzols (PZu)
4. Ferric Podzols (PZf)
5. Cambic Podzols (PZb)
6. Haplic Podzols (PZh)

Plinthosols (Gr. *plinthos,* brick; mottled clayey materials which harden on exposure):

Other soils having:

- 25% or more plinthite by volume:
  * in a horizon which is at least 15 cm thick within 50 cm of the surface, or
  * within a depth of 200 cm when underlying an albic E horizon, or a horizon which shows hydromorphic properties within 100 cm of the surface.

Units:

1. Albic Plinthosols (PTa)
2. Umbric Plinthos (PTu)
3. Dystric Plinthosols (PTd)
4. Eutric Plinthosols (PTe)

Note: Plinthite is usually firm but can be cut with a spade. When irreversibly hardened the material is no longer considered plinthite but develops into a petroferric or petric phase.

Ferrasols (L. *ferrum,* and aluminium; a high content of sesquioxides):

Other soils having:

- an oxic B horizon;
- no argillic or natric B horizon above the oxic B horizon.

Units:

1. Yermic Ferralsols (FRy)
2. Plinthic Ferralsols (FRp)
3. Umbric Ferralsols (FRu)
4. Akric Ferralsols (FRa)
5. Rhodic Ferralsols (FRr)
6. Xanthic Ferralsols (FRx)
7. Haplic Ferralsols (FRh)

Planosols (L. *planos,* flat/level; soils developed in level or depressed topography with poor drainage):

Other soils having:

- an E horizon showing hydromorphic properties at least in part of the horizon which abruptly overlies a slowly permeable horizon within 125 cm of the surface;
- no natric or spodic E horizons.

Units:

1. Yermic Planosols (PLy)
2. Gelic Planosols (PLi)
3. Mollic Planosols (PLm)
4. Umbric Planosols (PLu)
5. Dystric Planosols (PLd)
6. Eutric Planosols (PLe)

Solonetz (Russian *sol,* salt and *etz,* strongly expressed

Other soils having:

- a natric B horizon.

Units:

1. Gleyic Solonetz (SNg)
2. Mollic Solonetz (SNm)
3. Gypsic Solonetz (SNj)
4. Calcic Solonetz (SNk)
5. Haplic Solonetz (SNh)

Gryzems (Anglo-*saxon*, grey and Russian *zemlja*, earth/land; uncoated silt and quartz grains present in layers rich in OM):

Other soils having:

- a mollic A horizon with a moist chroma of 2 or less to a depth of at least 15 cm;
- uncoated silt and quartz grains on structural ped surfaces;
- an argillic B horizon.

Units:

1. Gleyic Greysems (GRg)
2. Haplic Greyzems (GRh)

Chernozems (Russian *chern*, black and *zemlja*, earth/land):

Other soils having:

- a mollic A horizon with a moist chroma of 2 or less to a depth of at least 15 cm
- one or more of the following:
  * a calcic horizon, or
  * concentrations of soft powdery lime within 125 cm of the surface.

Units:

1. Gleyic Chernozems (CHg)
2. Luvic Chernozems (CHl)
3. Glossic Chernozems (CHw)
4. Calcic Chernozems (CHk)
5. Haplic Chernozems (CHh)

Kastanozems (L. *castaneo*, chestnut, and Russian *zemlja*, earth/land):

Other soils having:

- a mollic A horizon with a moist chroma of more than 2 to a depth of at least 125 cm;
- one or more of the following:
  * a calcic or gypsic horizon, or
  * concentrations of soft powdery lime within 125 cm of the surface.

Units:

1. Luvic Kastanozems (KSl)
2. Gypsic Kastanozems (KSj)
3. Calcic Kastanozems (KSk)
4. Haplic Kastanozems (KSh)

Phaeozems (Gr. *phaios*, dusky, and Russian *zemlja*, earth/land):

Other soils having:

- a mollic A horizon;

- a cation-exchange capacity of more than 16 m.e. per 100 g clay (by NHOAc);
- a base saturation of 50% or more throughout the upper 125 cm of the soil.

Units:

1. Gleyic Phaeozems (PHg)
2. Luvic Phaeozems (PHl)
3. Calcaric Pphaeozems (PHc)
4. Haplic Pphaeozems (PHh)

Podzoluvisols (from Podzols and Luvisols):

Other soils having:

- an argillic B horizon showing an irregular or broken upper boundary resulting from:
    * deep tonguing of the E into the B horizon, or
    * the formation of discrete nodules (commonly ranging from 2-5 cm and up to 30 cm in diameter), the exteriors of which are enriched and weakly cemented or indurated with iron and having redder hues and stronger chromas than the interiors.

Units:

1. Gelic Podzoluvisols (PDi)
2. Gleyic Podzoluvisols (PDg)
3. Dystric Podzoluvisols (PDd)
4. Eutric Podzoluvisols (PDe)

Calcisols (L. *calxis*, lime; accumulation of calcium carbonate or gypsum):

Other soils having:

- A calcic or a gypsic horizon or both within 125 cm of the surface, or
- a concentration of soft powdery lime within 125 cm of the surface, or
- redistribution of gypsum within 125 cm of the surface,
- no other diagnostic horizons than an ochric A horizon, or a cambic or argillic B horizon occuring within the calcic or gypsic horizon.

Units:

1. Arenic Calcisols (CLq)
2. Gypsic Calcisols (CLj)
3. Haplic Calcisols (CLh)

Nitosols (L. *nitidus*, shiny; shiny ped surfaces):

Other soils having:

- an argillic B horizon showing a clay distribution which does not decrease from its maximum by more than 20% within 150 cm of the surface;
- gradual to diffuse horizon boundaries between A and B horizons with a clay percentage of the B not exceeding that of the A by more than 20% within a vertical distance of 30 cm;
- nitic properties in some subhorizon within 125 cm of the surface.

Units:

1. Mollic Nitosols (NTm)
2. Umbric Nitosols (NTu)
3. Rhodic Nitosols (NTr)
4. Haplic Nitosols (NTh)

Alisols (L. *aluminium;* high aluminium content):

Other soils having:

- an argillic B horizon which has a cation exchange capacity of 16 m.e. or more per 100 g clay and a base saturation (by NH_4OAc) of less than 50% in at least in some part of the B horizon within 125 cm of the surface.

Units:

1. Plinthic Alisols (ALp)
2. Gleyic Alisols (ALg)
3. Umbric Alisols (ALu)
4. Ferric Alisols (ALf)
5. Haplic Alisols (ALh)

Acrisols (L. *acris,* very acid; low base saturation):

Other soils having:

- an argillic B horizon which has a cation-exchange capacity (by NH_4 OAc) of less than 16 m.e. per 100 g clay and a base saturation of less than 50% (by NH_4OAc) in at least some part of the B horizon within 125 cm of the surface.

Units:

1. Plinthic Acrisols (ACp)
2. Gleyic Acrisols (ACg)
3. Umbric Acrisols (ACu)
4. Ferric Acrisols (ACf)
5. Haplic Acrisols (ACh)

Luvisols (L. *luvi, luo,* to wash, 'lessiver'; illuvial accumulation of clay):

Other soils having:

- an argillic B horizon which has a cation-exchange capacity of 16 m.e. or more per 100 g clay and a base saturation (by NHOAc) of 50% or more throughout the B horizon to a depth of 125 cm.

Units:

1. Yermic Luvisols (LVy)
2. Gleyic Luvisols (LVg)
3. Albic Luvisols (LVa)
4. Vertic Luvisols (LVv)
5. Calcic Luvisols (LVk)
6. Chromic Luvisols (LVx)
7. Haplic Luvisols (LVh)

Lixisols (L. *lixivium*, washing; illuvial accumulation of clay and strongly weathered):

Other soils having:

- an argillic B horizon which has a cation-exchange capacity (by NH_4 OAc) of less than 16 m.e. per 100 g clay and a base saturation of 50% or more throughout the B horizon to a depth of 125 cm.

Units:

1. Yermic Lixisols (LXy)
2. Plinthic Lixisols (LXp)
3. Gleyic Lixisols (LXg)
4. Albic Lixisols (LXa)
5. Ferric Lixisols (LXf)
6. Haplic Lixisols (LXh)

Cambisols (L. *cambiare*, change; changes in colour, structure, and consistence):

Other soils having:

- a cambic B horizon.

Units:

1. Yermic Cambisols (CMy)
2. Gelic Cambisols (CMi)
3. Gleyic Cambisols (CMg)
4. Vertic Cambisols (CMv)
5. Calcaric Cambisols (CMc)
6. Umbric Cambisols (CMu)
7. Ferralic Cambisols (CMo)
8. Dystric Cambisols (CMd)
9. Chromic Cambisiols (CMx)
10. Eutric Cambisols (CMe)

Phases

Phases are limiting factors relating to surface or subsurface features of the land. They are not necessarily related to soil formation, and generally cut across the boundaries of different soil units. These features may form a constraint to the use of the land.

Definitions of petrocalcic and petrogypsic horizons, petroferric contact, fragipan and duripan are those formulated in the USDA *Soil Taxonomy*. In the USDA *Soil Taxonomy*, with the exception of the petroferric contact, they are diagnostic for separating different categories of soils. However, in the *Soil map of the world* they are shown as phases, as the occurrence of these horizons has not been systematically recorded in a number of countries.

Anthraquic phase

The anthroquic phase marks soils showing hydromorphic properties within 50 cm of the surface due to surface water stagnation associated with long-continued irrigation, particularly of rice.

Duripan phase

The duripan phase occurs when a subsurface horizon that is cemented by silica (accessory cements may be iron oxides and calcium carbonate) so that dry fragments do not slake during prolonged soaking in water or in hydrochloric acid. Duripans vary in appearance, but all have a very firm or extremely firm moist consistency, and they are always brittle even after prolonged wetting.

The duripan phase marks soils in which the upper level of the duripan occurs within 100 cm of the surface.

Fragipan phase

A fragipan is a loamy (uncommonly sandy) subsurface horizon having a high bulk density relative to the horizons above it. It is hard or very hard and appears cemented when dry, and weakly to moderately brittle when moist. When pressure is applied, peds or clods tend to rupture suddenly. Dry fragments slake or fracture when placed in water.

A fragipan is low in organic matter, slowly or very slowly permeable, and often shows bleached fracture planes. Clayskins may occur as patches or discontinuous streaks both on facies and in the interiors of the prisms. A fragipan commonly, but not always, underlies a B horizon. It may be from 15 to 200 cm thick, commonly with an abrupt or clear upper boundary and a gradual or diffuse lower boundary.

The fragipan phase marks soils which have the upper level of the fragipan occurring within 100 cm of the surface.

Gelundic phase

The gelundic phase marks soils showing the formation of polygons on their surface due to frost heaving.

Gilgai phase

Gilgai is the microrelief typical of cleyey soils (mainly Vertisols) that have a high coefficient of expansion with distinct seasonal changes in moisture content. This microrelief consists of either a succession of enclosed microbasins and microridges in nearly level areas, or of microridges and microvalleys that run up and down slopes. The height of the microridge ranges from a few centimetres to 100 cm. Rarely does the height attain 200 cm.

Inundic phase

An inundic phase occurs when standing or flowing water is present on the soil surface for more than 10 days during the growing period.

Lithic phase

A lithic phase occurs when continuous coherent and hard rock occurs within 50 cm of the surface. For Leptosols, the lithic phase is not shown as this is already implied in the major soil group definition.

Petric phase

A petric phase marks soils showing a layer consisting of 40% or more, by volume, of oxidic concretions or of hardened plinthite or ironstone or other coarse fragments with a thickness of at least 25 cm, the upper part of which occurs within 100 cm of the surface. The difference between this and the petroferric phase is that the concretionary layer of the petric phase is not continuously cemented.

Petrocalcic phase

A petrocalcic horizon is a continuous cemented or indurated calcic horizon, cemented by calcium carbonates, and in places by calcium and some magnesium carbonate. Accessory silica may be present.

Dry fragments do not slake in water and roots cannot enter. The petrocalcic horizon is massive or platy, extremely hard when dry (resists spade or auger) and very firm to extremely firm when moist. Noncapillary pores are filled, and hydraulic conductivity is moderately slow to very slow. It is usually thicker than 10 cm. A laminar capping is commonly present but is not required. If present, the carbonates constitute half or more of the weight of the laminar horizon.

The petrocalcic phase marks soils in which the upper part of the petrocalcic horizon occurs within 100 cm of the surface.

Petroferric phase

A petroferric horizon is a continuous layer of indurated material, in which iron is an important cement and organic matter is absent, or present only in traces. The indurated layer must either be continuous or, when it is fractured, the average lateral distance between fractures must be 10 cm or more. The petroferric layer differs from a thin iron pan and from an indurated spodic B horizon in containing little or no organic matter.

The petroferric phase marks soils in which the upper part of the petroferric horizon occurs within 100 cm of the surface.

Petrogypsic phase

The petrogypsic horizon is a gypsic horizon that is so cemented with gypsum that dry fragments do not slake in water and roots cannot enter. The gypsum content is commonly far greater than the minimum requirements for the gypsic horizon and uaually exceeds 60%.

The petrogypsic phase marks soils in which the upper part of the petrogypsic horizon occurs within 100 cm of the surface.

Phraetic phase

The pharetic phase marks soils which have a groundwater table between 3 and 5 m from the surface. At this depth the presence of groundwater is not normally reflected in the morphology of the solum; however, its presence is important for the water regime of the soil, especially in arid areas. With irrigation, special attention should be paid to effective water use and drainage to avoid salinization as a result of rising groundwater.

This phase has been used especially in the USSR. In other countries, groundwater depth is not consistently recorded during soil surveys.

Placic phase

The placic phase refers to the presence of a black to dark reddish, thin iron pan cemented by iron, by iron and manganese, or by an iron-organic-matter complex, the thickness of which ranges from 2 mm to 10 mm. In spots, it may be as thin as 1 mm or as thick as 20 to 40 mm, but this is rare. It may, but not necessarily, be associated with the stratification of parent materials. It is in the solum, roughly parallel to the soil surface. It has a pronounced wavy or convolute form. It normally occurs as a single pan; but in places it may be bifurcated. It is a barrier to water and roots.

The placid phase marks soils which have a thin iron pan within 100 cm of the surface.

Rudic phase

The rudic phase marks areas where the presence of gravel, stones, boulders or rock outcrops in the surface layer or at the surface makes the use of mechanized agricultural equipment impracticable. Fragments with a diameter of up to 7.5 cm are considered as gravel. Larger fragments are called stones or boulders.

Salic phase

The salic phase marks soils which, in some horizons within 100 cm of the surface, show EC values of the saturation extract higher than 4 mmhos/cm at 25°C. The salic phase is not shown for Solonchaks as this is implied in their definition. Soil salinity may show seasonal variations or fluctuate as a result of irrigation practices. Additionally, the effect of salinity varies greatly with the type of salts present, the permeability of the soil, climatic conditions, and the kind of crops grown.

Sodic phase

The sodic phase marks soils which have more than 6% saturation with exchangeable sodium, at least in some horizon within 100 cm of the surface. The sodic phase is not shown for Solonetz, as this is implied in their definition.

Principles for the establishment of soil subunits (level III)

In its original form, the legend and key of the *Soil map of the world* identified only two levels of classification, namely major soil groups and soil units. In recent years, there has been call for the establishment of a third level of Soil Subunits to facilitate larger-scale mapping.

To a large extent, such third-level subunits will be determined by individual country needs and their prevailing soil conditions. However, as a guide the following principles for the establishment and definitions of third-level specifiers are proposed.

Principles

1. The definitions of the third-level units should not conflict with the definitions of the first- and second-level units.
2. A first kind of third-level units is meant to accommodate intergrades between second-level units, e.g.
   Niti-Haplic Acrisols (intergrade to Nitosols)
   Gleyi-Dystric Fluvisols (intergrade to Gleysols).
   A first kind of third level units is designed to further specify important soil characteristics (preferably those influencing soil management), which are not already marked by phases.
3. Features which are represented by phases at second level should not be used as diagnostic properties at third level.
4. At third level, two specifiers or intergrades, or a combination of both, can occasionally be used by which the third-level unit is further qualified by the first adjective, e.g. Ando-Molli Eutric Cambisols and Sali-Verti Eutric Fluvisols.

Brief definitions of third-level specifiers

Albi:	soils having an albic E horizon
Ali:	soils having a saturation with aluminium of more than 50% within 100 cm of the surface
Anthraqui:	soils showing hydromorphic properties associated with surface water stagnation in long-lasting irrigation
Calcari:	soils which are calcareous within 125 cm of the surface
Calci:	soils having a calcic horizon or concentrations of soft powdery lime within 125 cm of the surface
Chromi:	soils, exclusive of Vertisols, having a strong brown to red B horizon
Chromi:	Vertisols having a moist value of more than 3 and a chroma more than 2 dominant in the soil matrix throughout the upper 30 cm
Dystri:	soils having a base saturation (NH_4OAc) of less than 50% within 125 cm of the surface
Eutri:	soils having a base saturation (NH_4OAc) of more than 50% within 125 cm of the surface
Ferri:	soils having ferric properties within 100 cm of the surface
Fluvi:	soils developed from alluvial deposits
Grumi:	Vertisols having a strongly developed fine structure in the upper 20 cm
Lepti:	soils having continuous coherent hard rock within 50 cm of the surface
Mazi:	Vertisols having a massive structure in the upper 20 cm and becoming hard when dry
Niti:	Acrisols showing nitic properties
Pelli:	Vertisols having a moist value of 3 or less and a chroma of 2 or less dominant in the soil matrix throughout the upper 30 cm
Rhodi:	soils having a red to dusky red B horizon
Sombri:	Ferralsols showing some accumulation of dark-coloured organic matter in the oxic B horizon
Stagni:	soils having hydromorphic properties related to surface water stagnation during part of the year (at least 7 days during the growing season or longer at other times of the year); lacking groundwater table within 100 cm of the surface.
Takyri:	soils, exclusive of Vertisols, having a heavy texture, cracking into polygonal elements when dry and forming a platy or massive surface crust
Yermi:	soils such as Regosols and Arenosols showing yermic properties

It is, however, difficult to devise principles that will apply worldwide to the weighting of various factors. Moreover, most agrotechnology transfer can be undertaken with 153 second-level units; for more detailed work it is necessary to go to series level at least.

References

BENNEMA, J. and CAMARGO, M.N. 1979. Some remarks on Brazilian Latosols in relation to Oxisols of Soil Taxonomy. *Proceedings of the Second International Soil Classification Workshop* (Bangkok, Thailand).

FAO (Food and Agriculture Organization of the United Nations). 1974. *FAO/Unesco soil map of the world, 1:5.000.000.* Vol. 1. Legend. Paris: Unesco.

FAO (Food and Agriculture Organization of the United Nations). 1979. *FAO/UNEP/ Unesco. A provisional methodology for soil degradation assessment.* Rome: FAO.

FAO (Food and Agriculture Organization of the United Nations). 1977. *FAO/Unesco/ WMO. World map of desertification, 1:25.000.000.* Rome. FAO.

FAO (Food and Agriculture Organization of the United Nations). 1978-1981. *Reports of the agro-ecological zones project.* World Soil Resources Reports no. 48: 1-4. Rome: FAO.

FAO (Food and Agriculture Organization of the United Nations). 1977. Guidelines for soil profile description. Rome: FAO.

HIGGINS, G.H., KASSAM, A.H., NAIKEN, L., FISCHER, G. and SHAH, M.M. 1984. *FAO/UNFPA/IIASA. Potential population supporting capacties of lands in the developing world.* Rome: FAO.

ISRIC (International Soil Reference and Information Centre). 1985. *Procedures for soil analyses,* Technical Paper no. 9. Wageningen: ISRIC.

KLAMT, E. and SOMBROEK, W.G. 1987. Contribution of organic matter to exchange properties of Oxisols. *Proceedings of the Eighth International Soil Classification Workshop* (Rio de Janeiro, Brazil).

SOIL SURVEY STAFF. 1967. *Soil survey laboratory methods and procedures for collecting soil samples.* Soil Conservation Service, U.S. Department of Agriculture. USDA Soil Survey Investigations. Washington DC: Government Printing Office.

SOIL SURVEY STAFF. 1975. *Soil taxonomy: A basic system of soil classification For making and interpreting soil surveys.* Soil Conservation Service, U.S. Department of Agriculture. Agriculture Handbook no. 436. Washington DC: Government Printing Office.

SOMBROEK, W.G. 1984. *Identification and use of subtypes of the argillic horizon.* International Soil Reference and Information Center. Working Paper and Preprint Series no. 84:1. Wageningen: ISRIC.

Soil Taxonomy and agrotechnology transfer

HARI ESWARAN*

Abstract

The common constraint to agricultural development in the LDCs is the limited information available on national resources, and a lack of experience in utilizing these resources efficiently. There is a need to develop good agrotechnology transfer to alleviate this constraint. This paper is primarily concerned with the use of Soil Taxonomy *in soil-based agrotechnologies. It points out that in the computer age of information technology, an expert system has been developed for soil surveys, and describes the research which is under way concerning the interpretation and utilization of* Soil Taxonomy *for making land-use decisions. This work is being undertaken by IBSNAT and USAID, who are curently a Decision Support System for Agrotechnology Transfer (DSSAT).*

Introduction

Many countries are increasing their investment in agricultural research and development, and many international donor agencies are also committed to this policy. As many of the less-developed countries (LDCs) have limited research capacities and resources with regard to personnel and funds, ways to increase research efficiency are being sought. Limited information on national resources and the experience to utilize these resources efficiently are common constraints

* Soil Management Support Services, PO Box 2890, Washington D.C. 20013, USA.

to agricultural development. The sustainability of agriculture which incorporates the notion of preserving resources and protecting the environment are concepts which cannot be practiced effectively in LDCs when their immediate concerns are food, fibre, and fuel production.

Although national and international agricultural research has contributed to our understanding of agriculture, the productivity of the small farmer in many LDCs, particularly in Africa, is increasing at only suboptimal levels. The current situation and the future challenges for wheat production are illustrated in Figure 1, and a similar diagram could be made for any other crop. The diagram raises the question of the impact of agricultural research in the last two or more decades in the LDCs. Although improving the quality of cultivars is a continuing challenge, and researh must be conducted to meet this, an equally important concern with respect to LDCs is resource management and optimizing the use of soil resources.

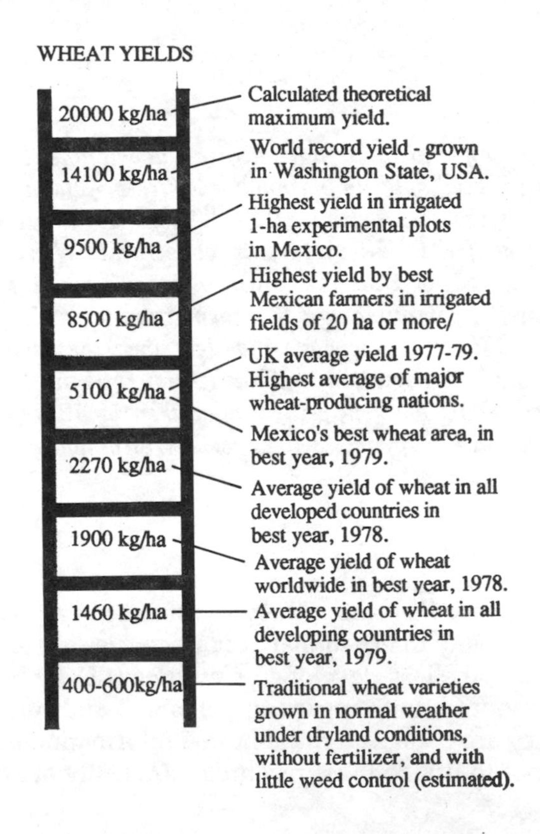

Figure 1 **Agroproductivity ladder illustrating a range of wheat yields from various environments and under various levels of management (CIMMYT, 1981).**

The gap between current productivity and the potential of the land stems from reasons which are partly socioeconomic and thus difficult to change, and partly from the fact that extension services which function to transfer research results to the farmer are poorly organized or nonexistent in many of these countries. Consequently, the millions of dollars invested in the International Agricultural Research Centers (IARCs), which are at the cutting-edge of tropical agricultural technology, have not had the impact of alleviating the livelihood of the rural poor. These commodity-based centers have done an excellent job with respect to improving genotype characteristics to enhance yield, but their thrust in resource management is only of recent origin.

In the last few decades, due to the success of hybridization and the development of high-yielding cultivars, many donors sponsoring research programs have opted to focus their attention on this type of research - regretfully at the expense of other components of resource management. It is not often realized that soil-based agricultural technology and research is site-specific to a large extent. This implies that much has to be known about individual site characteristics in order to be able to interpret the behavior of the soils, particularly with respect to response to management. Consequently, detailed knowledge about soils and their response to management is crucial to a successful exploitation of this resource. In addition, it is also necessary to know the crop or the potential uses of the soil. Matching crop requirements to soil conditions is the key not only to technology development but also to agro-technology transfer. Discriminatory use of the soil is the challenge of the next decade.

The concerns of the next decade

The current concerns which determine the action programs of the United States Agency for International Development (AID) are reflected in AID's policy on environment and natural resources, and were summarized by AID administrator, Alan Woods[1]. He indicated that the agency's central objective, which was to promote economic expansion and improve human welfare in developing countries, could only be achieved by ensuring the sustainable production of natural resources. The agency's environmental and natural resources program contains four policy objectives:

- to ensure the environmental soundness and long-term sustainability of all projects;

1 Remarks at the International Institute for Environmental Development, 15 April 1988.

- to assist developing countries in formulating policies that lead to the effective management of natural resources, and that discourage environmentally harmful activities;
- to assist developing countries in building the institutional and scientific capacity to identify and solve problems concerning their environmental and natural resources;
- to promote environmentally sound development projects funded by other donors.

AID's policies on natural resources were further elaborated by Duane Acker[2], who indicated that sustained economic development in developing countries depends heavily on the prudent and efficient use of natural resources, on the maintenance of the natural resource base and, in some instances, on the enhancement of the natural resource base.

Each country has an assemblage of problems, some unique to itself and determined by the specific physiographic setting coupled to the socioeconomic conditions, while others are shared by many. Some of the problems are man-induced, usually created out of necessity, though sometimes out of ignorance. An example of the latter is the extensive deforestation taking place in large parts of the humid tropics. The common denominator in many of these countries is that a particular country may not have the experience or technology to counteract the problem, while a neighbor or a country in another continent might already have stemmed it. The question is, how do we transfer the experience of one country to another? how can we pool information and experience so that mistakes are not repeated, potential problems are signalled, and sustainable agriculture is ensured?

The view expressed in this paper is that in this current age of information technology, we have ways of meeting the challenge of the next decade. As all forms of agriculture or other uses of the land depend on the soil, and as the goal is the use of the soil while preserving it for future generations, a knowledge of national soil resources is the first step towards sustainable agriculture.

The priciples and concepts of agrotechnology transfer are elaborated in the monograph *Soil-based agrotechnology transfer* (Silva, 1985), where it is indicated that the three requirements for successful transfer are:

- a basis for transfer,
- a common language for communicating information, and
- a source of innovations for transfer.

For soil-based agrotechnologies, soil classification terminology serves as the common language between scientists. However, there is no one inter-

---

2 Assistant to AID administrator. Testimony before the Senate Foreign Relations Subcommittee on International Economic Policy, Trade, Oceans and the Environment. 13 April 1988.

nationally recognized system. Nevertheless, *Soil Taxonomy* is now used by so many countries that it has emerged as the *de facto* soil classification system. In historical times, technology was transferred by trial and error, but with the advance of earth sciences, the concept of analogous transfer came in to vogue. The Benchmark Soils Project of the Universities of Hawaii and Puerto Rico (Silva, 1985) used the concept of transfer by analogy to test the hypothesis that the soil family categories of *Soil Taxonomy* could be the basis for such a transfer. Soil families stratify the soils of the world into relatively narrow groups, integrating the environmental information that is important for crop performance with the physicochemical characteristics of the soil that affect response to management. The Benchmark Soils Project showed the applicability of this approach, but indicated that for more refined predictions or more accurate transfers, more information than that contained in the family name is needed. This has led to the systems approach to transfer, which is more holistic and has the potential to consider all reactants. The systems approach, which requires computers, is gaining greater acceptance as advances in systems modelling, artificial intelligence and expert systems, geographic information systems, and data base management systems are developed.

As discussed later, the role of *Soil Taxonomy* has taken on a new dimension with the systems approach. Detailed soil surveys and an international soil classification system are in a greater demand today than they were in the past.

Soil Taxonomy

Soil Taxonomy is the technical language of soil science, and is designed for communication between soil scientists and as a correlation tool for soil surveys. This implies that it cannot be used in isolation. Classifying a soil at a point in the landscape, and in accordance with any system of classification, serves little purpose unless the point represents a defined segment of the landscape represented by a soil unit on a soil map. Thus although the subject of this paper is *Soil Taxonomy*, it is important to point out that the classification of the soil only becomes meaningful when it is in the context of a soil survey program.

The objective of *Soil Taxonomy* (Soil Survey Staff, 1975) is to delineate soil classes at various levels of generalization that permit us to understand, as fully as existing knowledge permits, the relationship among the soils and the factors responsible for their attributes. The system is designed for making and interpreting soil surveys, and is consequently ideally suited for general evaluations of the agricultural potential of the land. The specificity of the evaluation or interpretation made is related to the categoric level at which the soil is classified (Figure 2). This is an indication of the flexibility of the system.

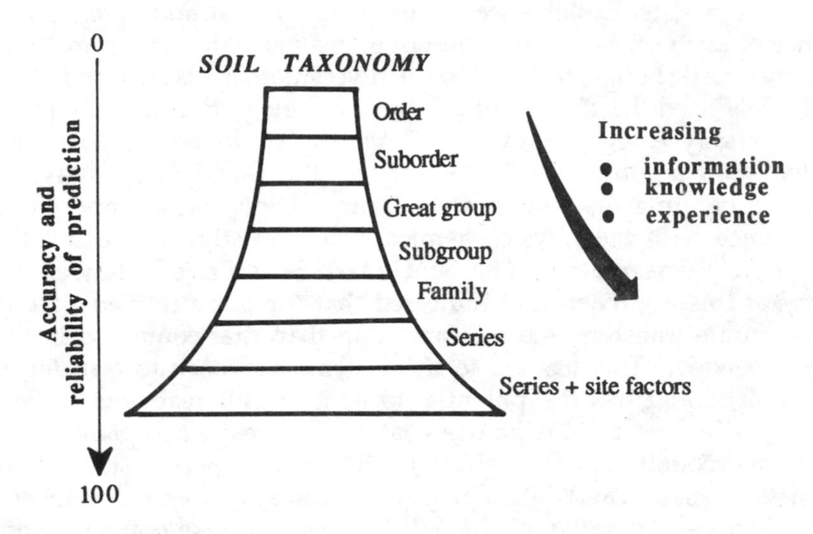

Figure 2. Categories in *Soil Taxonomy* in relation to information content and reliability of predictions of soil performance.

The greatest use of *Soil Taxonomy* is the role it plays in a soil survey program. It is the basis of soil correlation, and it is the basis for creating mapping units in a soil survey. It provides the framework for defining the mapping units. Many classification systems, particularly those which were not created with this objective in mind, do not serve this purpose. A good example is the classification system of Fitzpatrick (1980). This is an example of a theoretical classification where soils are divided into *n* boxes in accordance with certain rules. The intention is to provide for all combinations and permutations of properties or 'horizons', and consequently all soils - whether they exist or are imagined - are provided for in the system. Since the system is a combination of horizons (the formula approach), it deals with one level of observation and evaluation. The same soil horizons or properties have different significance in different soils. Even in the same soil, the position in the soil at which the horizon occurs has an important effect on its use. The implications of these variations are lost in a purely theoretical classification system.

A similar problem is encountered in the application of numerical taxonomy to soils. Although significantly differing clusters may be mathematically derived, the clusters have little meaning in the real situation with regard to the position in a landscape or the use and management of a soil. For these and other reasons, a hierarchical approach is adopted in *Soil Taxonomy,* with more and more properties incorporated at the lower categories of the system. Properties may be used at different categorical levels and reflect the significance of the property to that taxon. As illustrated in Figure 2 (and later in Tables 2 and 3), properties can be derived from soil names; not only those properties included in the definition of the taxon but, by default, properties which have been employed to exclude it from other classes.

To be useful for making and interpreting soil surveys, the categoric level must be matched with the cartographic accuracy of soil maps. It is obvious that the order level cannot be used as a mapping unit for very detailed soil maps. The categoric level implies a certain level of accuracy (Figure 3) and information content, and so does the scale of the map. There are many published maps where there is no match between the two. Although the legend of the FAO *Soil map of the world* was made specifically for a scale of 1: 5 million, it has been used for all scales including 1:250 000 or more; in these maps, there is an obvious mismatch between categoric and cartographic accuracy, and this can be deceptive.

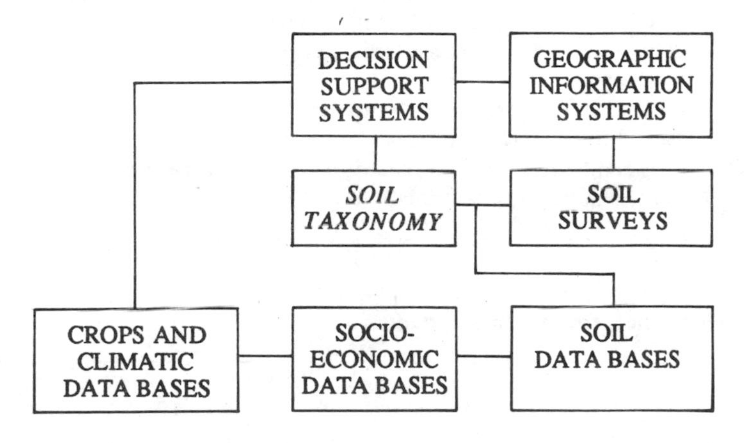

Figure 3. Systems approach to agrotechnology transfer.

In *Soil Taxonomy* there are six categories, which are, from the highest to the lowest: order, suborder, great group, subgroup, family, and series. Witty and Arnold (1987) have explained the rationale for each category, and this is summarized in Table 1. Some of the range of properties that could be derived directly from the category name for a given soil are illustrated in Table 2. The table also illustrates the increasing information content when proceeding from the order to the series category. Table 3 considers the same soils as Table 2, but shows the kinds of interpretations that could be derived or made at each categoric level.

Table 1. Basis for diferentiation of the categories.

Order: Properties resulting from the major processes of soil formation.

Suborder: Important properties which constitute major *controls*, or reflect such controls, on the current set of soil-forming processes (e.g. soil moisture regimes).

Great group: Properties that constitute subordinate or additional controls, or reflect such controls, on the current set of soil-forming processes (e.g. root restricting layers).

Subgroup: Properties resulting from either:
- a blending or overlapping of sets of processes in space or time that cause one kind of soil that has been recognized at the great group, suborder or order levels, or
- sets of processes or conditions that have not been recognized as criteria for any class at a higher level, or
- where the soil is considered to typify the central concept of the great group.

Family: Properties that reflect the potential for further change (e.g. particle-size class, mineralogy, temperature regime).

Series: Properties that reflect relatively narrow ranges of processes that transform parent materials into soils.

Table 2. Information content in soil names.

Category	1	2	3	4	5	6	7	9	10	11	12	13	14	15
Order	X	X												
Suborder	X	X	X	X										
Great group	X	X	X	X	X	X	X							
Subgroup	X	X	X	X	X	X	X	X	X					
Family	X	X	X	X	X	X	X	X	X	X	X			
Series	X	X	X	X	X	X	X	X	X	X	X	X	X	X

Table 2. (cont'd)

Soil: clayey, kaolinitic, isohyperthermic Arenic Kandiustult

1.	=	Particle-size distribution with depth	9.	=	Hydraulic conductivity
2.	=	Base saturation (pH) with depth	10.	=	Temperature constraints/ potentials
3.	=	Soil moisture regime	11.	=	Mineralogy
4.	=	Organic-matter content	12.	=	Particle size of subsoil
5.	=	Root-restricting layers	13.	=	pH of surface horizons
6.	=	Charge characteristics	14.	=	Depth to argillic horizon
7.	=	Anion fixation	15	=	Slope
8.	=	Texture of surface horizon			

Table 3. Interpretations from soil names.

Category **Derived properties**

A. Order: *Ultisols*

The soil has an argillic horizon, indicating a lighter-textured surface soil suitable for root penetration, and a heavier-textured subsoil with the capacity to store water and nutrients. The soil is generally acid, and aluminum toxicity may be anticipated. The soil is prone to erosion, particularly in monsoonal climates.

B. Suborder: *Ustults*

The prefix 'ustic' indicates a moisture deficit during some part of the year; supplemental irrigation is needed for dry-season crops. Rainfall is highly irregular, particularly during the onset of rains, and crop establishment is risky. In some Ustults, crusting is a serious problem preventing seedling emergence. As rain is in the form of high-intensity storms, and as the soil experiences alternate wet and dry conditions, erosion is very high, and appropriate soil conservation measures must be taken. Alley cropping is not recommended; contour terracing with bench terraces is required, and tree crops could be planted at the edge of the terraces.

Table 3. (cont'd).

C. Great group: *Kandiustults*

The prefix 'kandi' indicates a thick argillic horizon and the dominance of low-activity clays (LAC). The thicker the argillic, the greater the ability of the soil to store water and nutrients. If the argillic is clayey, and the dry period is not too intense, there is the possibility of a second crop using stored moisture. The LAC indicates low nutrient-holding capacity, but there are sufficient weatherable minerals to provide a slow source of nutrients. Phosphate fixation is relatively low, but the soil needs N, P, and K for good performance. Liming is essential.

D. Subgroup: *Arenic Kandiustult*

The soil has a surface layer >50 cm thick, with a sandy texture making it a very drought-prone soil. Mulching is a necessary management technique for the utilization of this soil. The soil is suitable for deep-rooted crops; without irrigation, annuals must be avoided.

E. Family: clayey, kaolinitic, isohyperthermic, *Arenic Kandiustult*

Clayey indicates that the argillic horizon has >35% clay and, as the surface is sandy, the textural contrast is an important property of this soil. Kaolinitic implies a low nutrient-holding capacity, but few adverse characteristics. Isohyperthermic implies that the soil temperature is not a constraint, but surface soil temperatures (over the top few centimeters) may rise to over 50°C during the day, a property which can be corrected by mulching.

Taxon names can be used to identify mapping units, and this is particularly useful in small scale-maps, due to the connotative nature of the names. Table 4 shows the scales of published soil maps and the categoric level most appropriate for each range in scale. The concept of categoric relevance and cartographic accuracy must be the basis for selecting the appropriate category for the map scale.

Table 4. Interpretation potential of the categories.

Order

Employed for small-scale maps (>1:6.5 million) of continents or the world. The legend is generally phases of orders. Enable an appreciation of the geographic distribution of the major landforms of the world and associated soils.
Provides the basis for the genetic differentiation of soils.

Suborders

Employed for small-scale maps (1:6.5 million to 1:1.5 million) of continents or the world. The legend is an association of phases of suborders. Enables an appreciation of major landforms, climatic regimes, and associated major soils.
Provides an outline of major agorenvironments.

Great groups

Employed for small-scale maps (1:1.5 million to 1:130 000) of continents or regions. The legend is an association of phases of great groups.
Provides an outline of the major agroecological zones of the country with the major associated soils. Map units indicate major constraints and potentials for major uses. Phases of subgroups may be used if supported by detailed field observations.

Subgroups

Employed for medium-scale maps (1:650 000 to 1:260 00). The actual categoric level is a function of detail of field observations and for reconnaissance level mapping. Phases of great groups are still adequate. Map unit names may be other than taxa names, but information of the subgroup is implied.

Family

The taxonomic name is not generally employed as a mapping unit, but the concept implies an information content of scales 1:130 000 to 1:26 000.
Maps at the detailed end of the range contain sufficient information for agrotechnology transfer.

Series

Scales of >1:1 300 to 1:26 000, with phases of series as mapping units. The ideal scale for agrotechnology transfer. Contains a very high amount of information for most uses of soils.

The systems approach

With the advent of rapid and relatively cheap computers, agricultural research and development is entering into the age of information technology. A statistical approach for determining cause and effect is being replaced by a systems approach, where it is possible to look at a number of variables simultaneously. With the emerging technology of artificial intelligence or expert systems, the ability to use heuristic knowledge becomes more promising. The key tool in the immediate future for agricultural development is the Decision Support System (DSS) which employs both soil-weather-crop modeling and expert systems. This is basically 'star wars technology' applied to agriculture.

For a DSS to be used by a scientist, agriculturist, or decision maker, it must meet two requirements. First the DSS should not be site-specific, as its function is to integrate the elements in a given landscape, catchment, watershed or any area of concern. Second, any computerized system is fueled by a data base, which must be comprehensive and reliable. The technology for both these elements exist. The geographic dimension is provided by Geographic Information Systems (GIS), which has as its base, digitized topographic or landscape information; any property or sets of properties are then added to this base to provide the geographic spread.

A GIS is an ideal tool to present soil survey information. The singular advantage compared to conventional soil maps is that a GIS has several layers of information. Any kind of information can be retrieved, and more importantly any combinations of information may be synthesized. For example, if the soil map is of great groups, one can call up any great group and request information on all the crops on a given great group, or on the yields of a given crop on this soil. A GIS is therefore a powerful planning tool.

Constraints to the development of GIS and the systems approach

Many countries are interested in or are developing some kind of GIS. The basic constraint being faced is the incompleteness or absence of a reliable data base. In LDCs, systematic soil surveys are the exception rather than the rule. Even if parts of the country have been surveyed, scales vary and methods of survey are inconsistent. Due to a lack of laboratory facilities, soil data is either unreliable or nonexistent. Consequently, though there is a desire to employ recent technology, the basic information with which to use the technology is often not available.

Soil Taxonomy in the age of information technology

The rationale of soil survey interpretations is that the soils classified in the same taxa have a common response to management practices. Smith (1965) indicated that the basic assumption is that experience with a particular kind of soil in one place can be applied to that particular kind of soil wherever it exists if consideration is taken of any climatic differences. Smith also emphasized that soil survey acts as a bridge that lets us transfer the knowledge gained by research or by the experience of cultivators from one place to all the other places in which it is applicable.

The situation in many LDCs is that there are few soil maps and even less soil information. However, soil scientists are required to make decisions on land development and other uses of the soil. An experienced soil scientist who knows *Soil Taxonomy*, can make such an assess-ment with confidence. With good data and a soil map, the assessment could be refined and improved.

Soil Taxonomy has other uses which become more apparent with the use of computers and artificial intelligence for making land-use decisions. This recent technology is referred to as expert systems and is defined as a "computing system which embodies organized knowledge concerning some specific areas of human expertise sufficient to perform as a skiful and cost-effective consultant" (Bramer, 1981). McCracken and Cate (1986) have listed some of the potential advantages and components of expert systems that may be successfully applied to soil classification as:

- the application of the judgment, experience and intuition of seasoned soil scientists through knowledge engineering;
- the marshaling of this knowledge with the use of large relational data base techniques;
- the power of computers to evaluate and establish soil relationships quickly which are not readily apparent to the human mind;
- using these techniques and technologies to check the impact of proposed changes in *Soil Taxonomy;*
- developing computers with supporting programs that can 'learn' and thereby facilitate soil classification and soil interpretations.

Soil Taxonomy, as a system of classification, already incorporates the knowledge of a spectrum of experts. To develop an expert system for soil survey interpretations utilizing *Soil Taxonomy,* a few additional steps are needed. The first is to bring together soil and crop experts to provide the basic information on matching crop requirements to soil conditions. The second is to establish a data base of both soils and crops with related information on climate. Socioeconomic considerations would help to make the interpretations more locality-specific. Finally, an expert system must be developed to manage all this knowledge and data.

This area of investigation is only in very early stages of development. There are a few schools working on the subject, and notably the University of Hawaii (Dr. Russel Yost), which is developing an expert system for acid soils. This addresses acid soils, and only taxa which imply that acid soils are included. Through interrogation of the user of the system, it determines the crops being grown or intended to be grown, the kind of soils that are present, and other pertinent information. Based on the responses, the system recommends the management practice. The main feature of this rule-based expert system is that the user can query the software at any time, and the reason for the decision (with references) is provided.

Scientists working with the International Benchmark Sites Network for Agrotechnology Transfer (IBSNAT), a project of the U. S. Agency for International Development (AID) implemented by the Universities of Hawaii and Puerto Rico, have embarked on a more ambitious project, and are in the process of developing a Decision Support System for Agrotechnology Transfer (DSSAT). The DSSAT has several components:

- soil, climate, and crop data bases;
- a geographic information system;
- a soil-weather-crop model for several crops;
- an expert system which manages the DSSAT.

When information is available for all its components, the DSSAT is a powerful and accurate tool; predictions of crop performance or management requirements are as good or better than any other means we have today. As indicated previously, the constraint for LDCs is the absence of reliable data. DSSAT is designed so that it can operate even in this environment. For example, if soil data are not available at a particular site using the extensive data base of the World Benchmark Soils Project (Eswaran *et al.*, 1987), DSSAT has the potential to develop a 'synthetic profile'. In order to do this, some kind of information must be given - the minimum being a *Soil Taxonomic* class at any category.

It was also mentioned previously that data in many countries are incomplete or unreliable. The authors of DSSAT have conceptualized a 'filtering' expert system which verifies data or fills in gaps in data. Several algorithms have been developed for the latter - and since no one algorithm will meet the needs of all soils, *Soil Taxonomy* is used to stratify the population. For example, to calculate the available water in the soil, an algorithm for Mollisols or Alfisols will not be suitable for Andisols or Oxisols. Stratifying the soil population for purposes of developing algorithms has emerged as a most important function of *Soil Taxonomy* in this area of expert systems.

References

BRAMER, M. A. 1981. Survey and critical review of expert system research. In: *Introductory readings in expert systems*, ed. D Michie, 3-22. New York: Gordon and Breach Science Publications.

CIMMYT (Centre Internacional de Majoraniento de Maiz y Trigo). 1981. *Wheat in the Third World.*

ESWARAN, H., KIMBLE, J., COOK, T. and MAUSBACH. M. 1987. World Benchmark Soils Project of SMSS. *Soil Survey and Land Evaluation* 7:111-122

FITZPATRICK, E. A. 1980. *Soils: their formation, classification and distribution.* London: Longmans.

McCRACKEN, R. J. and CATE. R.B. 1986. Artificial intelligence, cognitive science and measurement theory applied to soil classification. *Soil Science Society of America Journal* 50: 557-561.

SILVA, J. A., ed. 1985. *Soil-based agrotechnology transfer.* Benchmark Soils Project, Department of Agronomy and Soil Science. Hawaii, USA: University of Hawaii. 292p.

SMITH, G. D. 1965. *Lectures on soil classification.* Pédologie, Special Bulletin no. 4: 5-134.

SOIL SURVEY STAFF. 1975. *Soil taxonomy: A basic system of soil classification for making and interpreting soil surveys.* Soil Conservation Service, U.S. Department of Agriculture. Agricultural Handbook no. 436. Washington DC: Government Printing Office.

WITTY, J and ARNOLD, R. W. 1987. Soil taxonomy: an overview. *Outlook on Agriculture* 16:8-13.

The fertility capability soil classification system: applications and interpretations for crop production planning

IRB KHEORUENROMNE*

Abstract

The fertility capability soil classification system (FCC) is a technical system for grouping soils for agronomic management in relation to their chemicophysical properties. The concept of the system places emphasis on quantifiable parameters of both topsoil and subsoil that are directly relevant to plant growth. The structure of the system includes type (topsoil texture), substrata type (subsoil texture), and 15 basic modifiers. The system has been tested, evaluated, and revised constantly. The applications and interpretations of the system using the soil data obtained from some Alfisols, Oxisols, Ultisols, and Vertisols in Thailand indicate clearly that the system can be quite advantageous for on-site crop production planning.

Introduction

Soil classification has several objectives. Concepts for classifying natural soils have been well laid down and documented (Buol *et al.*, 1980; Cline, 1949). Each natural soil classification system generally attempts to organize all the

* Department of Soils, Faculty of Agriculture, Kasetsart University, Bangkok 10903, Thailand.

features that can be measured in soils. Such a process of recording the properties of soils can facilitate the understanding and assist in remembering the characteristics of each individual soil and its potential use. The information provided by any of the natural soil classification systems can be used for specific purposes by different users. Agronomically, it can be used to predict soil-related limitations and the soil management practices needed for agricultural production and development in each area or at each site. It can also be used as a base for extrapolation of experimental results from one place to another where comparable soil properties are found (Couto, 1988). This is called agrotechnology transfer.

Two major problems are encountered in the use of information from the natural clasdification systems for agronomic purposes. One of the problems is that there are a number of different natural soil classification systems in the world. These include *Soil Taxonomy* (Soil Survey Staff, 1975), the French System, the legend for the FAO-Unesco soil maps of the world (FAO, 1971, 1974, 1977; Buol *et al.*, 1980). The different criteria used to differentiate soil units among the different systems makes it difficult to undertake soil-related agrotechnology transfer from one place to another. Another serious problem or limitation in using a natural soil classification system for agronomic purposes has been that the users unnecessarily perceive a complicated picture of soil classification. Complicated terminology in soil classification is a major problem. With the exception of pedologists, no one seems to feel comfortable with the terms whcih are used. Besides, soil classification concerns the grouping of soils with similar properties together. Some of the properties however, may not be relevant for a particular agricultural use. Therefore, the overall picture of soil classification does not promote the effective utilization of the information on soil properties gathered by soil scientists working in soil fertility and by agronomists.

The fertility capability soil classification system FCC[1] is a technical system of soil classification developed as an attempt to bridge the gap between the subdisciplines of soil classification and soil fertility (Buol, 1972; Buol *et al.*, 1975; Buol and Couto, 1981; Sanchez *et al.*, 1982). The system seeks to identify soil fertility and management constraints which might be relevant for the use of soils for agricultural production (Couto, 1988). The purpose of this paper is to describe the structure and principles of the FCC system. It also reports on some applications and interpretations of the system, using soil data on Alfisols, Ultisols and Vertisols in Thailand for on-site crop production planning.

1 The soil Fertility Capability Classification (FCC) system.

Basic system, structure, and principles

The background, basic principles, and a detailed description of the FCC system have been published earlier (Buol *et al.*, 1975; Sanchez *et al.*, 1982; Eiumnoh, 1984; Couto, 1988). The system consists of three category levels: type (topsoil texture), substrata type (subsoil texture), and 15 basic modifiers. The term 'topsoil' refers to the plough layer (Ap) or the top 20 cm of the soil, whichever is shallower. The term 'subsoil' covers the depth interval between the topsoil and 50 cm depth (or 60 cm) (Sanchez *et al.*, 1982; Couto, 1988). The combination of class designations from the three categorical levels form an FCC unit (Sanchez *et al.*, 1982). Type and substrata type are simplified textural classes, and the modifiers are applied when a specific criterion or criteria are met. These modifiers are defined on the basis of quantifiable properties. Details of the system based on Sanchez *et al.* (1982) and Buol (1987) are shown below.

Type. Texture of plough layer or top 20 cm, whichever is shallower:

- S = sandy topsoil: loamy sands and sands (by the USDA definition);
- L = loamy topsoils: <35% clay but not loamy sand or sand;
- C = clayey topsoils: >35% clay;
- O = organic soils: >30% OM to a depth of 50 cm or more.

Substrata type. This refers to the texture of the subsoil, and is used only if there is a marked textural change from the surface, or if a hard root-restricting layer is encountered within 50 cm:

- S = sandy subsoil: texture as in type;
- L = loamy subsoil: texture as in type;
- C = clayey subsoil: texture as in type;
- R = rock or other hard root-restricting layer.

Modifiers. Where more than one criterion is listed for each modifier, only one of them needs to be met. The criterion listed first is the most desirable one, and should be used if data are available. The subsequent criteria are presented for use where data are limited:

- g' = (constant saturation): the soil is constantly saturated with no evidence of brownish or reddish mottles, except around root channels in the top 50 cm.
- g = (gley): soil or mottles <2 chroma within 60 cm of the soil surface and below all A horizons, or soil saturated with water for >60 days in most years;
- d = (dry): ustic, aridic or xeric soil moisture regimes (subsoil dry >90 cumulative days per year within 20-60 cm depth);

e = (low cation-exchange capacity): applies only to plough layer or surface 20 cm, whichever is shallower: CEC <4 m.e./100 g soil by sum of bases + KCl-extractable Al (effective CEC), or CEC <7 m.e./100 g soil by sum of cations at pH 7, or CEC <10 m.e./100 g soil by sum of cations + Al + H at pH 8.2;

a = (aluminum-toxicity): >60% Al-saturation of the effective CEC within 50 cm of the soil surface, or >67% acidity saturation of CEC by sum of cations at pH 7 within 50 cm of the soil surface, or >86% acidity saturation of CEC by sum of cations at pH 8.2 within 50 cm of the soil surface, or pH <5.0 in 1:1 H_2O within 50 cm, except in organic soils where pH must be less than 4.7;

h = (acid): 10-60% Al-saturation of the effective CEC within 50 cm of the soil surface, or pH in 1:1 H_2O between 5.0 and 6.0;

i = (high P-fixation by iron): % free Fe_2O_3/% clay >0.15 and more than 35% clay, or hues of 7.5YR or redder and granular structure. This modifier is used only in clay (C) types; it applies only to plough-layer or surface 20 cm of soil surface, whichever is shallower;

x = (X-ray amorphous): more than 1.4 oxalate extractable Al, or a pH of 10.6 in NaF in the top 20 cm, pH >10 in 1N NaF, or positive to field NaF test, or other indirect evidence of allophane dominance in the clay fraction;

v = (Vertisol): very sticky plastic clay: >35% clay and >50% of 2:1 expanding clays, or severe topsoil shrinking and swelling;

k = (low K reserves): <10% weatherable minerals in silt and sand fraction within 50 cm of the soil surface, or exchangeable K <0. 20 m.e./100 g, or K <2% of sum of bases, if bases <10 m.e./100 g;

b = (basic reaction): free $CaCO_3$ within 50 cm of soil surface (effervescence with HCl), or pH >7.3;

s = (salinity): >4 mmhos/cm of electrical conductivity of saturated extract at 25°C within 1 m of the soil surface;

n = (natric): >15% Na-saturation of CEC within 50 cm of the soil surface;

c = (cat clay): pH in 1:1 H_2O is <3.5 after drying, and jarosite mottles with hues of 2.5Y or yellower and chromas 6 or more are present within 60 cm of the soil surface (only used in Cg and Cg' substrata);

= (gravel): a prime (') denotes 15-35% gravel or coarser (>2 mm) particles by volume to any type or substrata type texture (example: S'L = gravelly sand over loamy; SL' = sandy over gravelly loam); two prime marks (") denote more than 35% gravel or coarser particles (>2 mm) by volume in any type or substrata type (example: LC" = loamy over clayey skeletal; L'C" = gravelly loam over clayey skeletal);

% = (slope): where it is desirable to show slope with the FCC, the slope range percentage can be placed in parentheses after the last condition modifier (example: Lb (3-8%) = uniformly loamy textured soil, calcareous in reaction, 3-8% slope).

The soils are classified simply by determining whether the characteristic is present or not. It should be noted here that the FCC uses the analytical data which are required for classification by *Soil Taxonomy* (Soil Survey Staff, 1975, 1987). If the raw data are available, FCC units can be derived directly from the data, otherwise inferred characteristics from the taxon are used. A similar approach can also be used with other soil classification systems provided that the same analytical methods or equivalents have been employed.

Basic interpretations of FCC nomenclature

All categorical levels of the FCC nomenclature have been designed to be interpreted for agronomic use. Basic interpretations are divided into two groups. The types and substrata types concern mainly the physical aspect of the soil characteristics such as infiltration, water-holding capacity, drainage, potential runoff, and erosion. The modifiers specify particular physico-chemical constraints within the soils. Some of the basic interpretations of types, substrata types, and modifiers based on earlier work (Sanchez *et al.*, 1982; Sanchez, 1987; Couto, 1988) are given below.

Interpretations of types and substrata types

S : high rate of infiltration, low water-holding capacity;

L : medium infiltration rate, good water-holding capacity;

C : low infiltration rates, good water-holding capacity, potential high runoff if sloping, difficult to till; when i modifier is present, these (Ci) soils are easy to till, have high infiltration rates and low water-holding capacity;

O : artificial drainage is needed and subsidence will occur; possible micronutrient deficiencies; high herbicide rates usually required;

SC, LC, SR: susceptible to severe soil degradation from erosion exposing undersirable subsoil; high priority should be given to erosion control.

Interpretations of modifiers

When only one modifier is included in the FCC unit, the following limitations or management requirements apply to the soil. Interpretations may differ when two or more modifiers are present simultaneously or when textural types are different.

g' : too wet for upland use unless expensive protection is provided;

g : denitrification frequently occurs in anaerobic subsoil; tillage operations and certain crops may be adversely affected by excess rain unless drainage is improved by tilling or other drainage procedures; good soil moisture regime for rice production;

d : moisture is limiting during the dry season unless the soil is irrigated; planting date should take into account the flush of N at onset of rains; germination problems are often experienced if first rains are sporadic;

e : low ability to retain nutrients against leaching, mainly K, Ca and Mg; heavy applications of these nutrients and of N fertilizers should be split; potential danger of overliming;

a : plants sensitive to Al-toxicity will be affected unless lime is applied; extraction of soil water below depth of lime incorporation will be restricted; lime requirements are high unless an e modifier is also indicated; this modifier is desirable for rapid dissolution of phosphate rocks;

h : low to medium soil acidity; requires liming for Al-sensitive crops, such as cotton and alfalfa, and for good latex flow in rubber; Mn-toxicity may occur on some of these soils;

i : high P-fixation capacity; requires initial application of 5-10 kg P per hectare for each percent of clay; sources and method of P fertilizer application should be considered carefully; with C texture, these soil have granular soil structure;

x : high P-fixation capacity; banding or pelletized P fertilizers are recommended; low organic N mineralization rates;

v : clayey textured topsoil with shrink and swell properties; tillage is difficult when too dry or too moist, but soils can be highly productive; P-deficiency common;

k : low ability to supply K; availability of K should be monitored and K fertilizers may be required frequently; potential K-Mg-Ca imbalances;

b : calcareous soils; rock phosphate and other non-water-soluble phosphates should be avoided; potential deficiency of certain micronutrients, principally iron and zinc;

s : presence of soluble salts; requires drainage and special management for salt-sensitive crops or the use of salt-tolerant species and cultivars;

n : high levels of sodium; requires special soil management practices for alkaline soils, including use of gypsum amendments and drainage;

c : potential acid sulphate soil; drainage is not recommended without special practices; should be managed with plants tolerant to high water table.

Applications of the FCC

Data from pedon analysis (profile description and analytical data) of a soil can be used both for taxonomic classification and fertility capability soil classification. If the pedon analysis has been conducted accurately, the FCC unit can be used as a management basis for agronomic purposes at the site. A computerized programme has been developed to assist in the utilization of the FCC (Buol *et al.*, 1988; Buol, 1987) as a draft programme for editing, and is now being used quite widely. In the programme (microcomputer 16 bits, basic language) the questions start with differentiation of types, then proceed to substrata type and modifier classifications. The printout for all steps can be done simultaneously with the classifications. On completion of the classification process, the operator can also add more details if desired. The FCC unit of the soil will be obtained, and the interpretation for the factors that constitute the FCC unit can be made. Separate interpretations to obtain agronomic recommendations for paddy rice and upland crops can also be requested through the programme. Some sample interpretations of FCC condition modifiers for rice cultivation have been reported earlier (Sanchez and Buol, 1985; Buol, 1987).

A test of FCC was carried out on six pedons of soils comprising two Alfisols, two Ultisols and two Vertisols. A working sheet (Appendix 1) was used. Data from a pedon analysis of the soils necessary for the FCC are shown in Tables 1 and 2. From these data, the FCC units obtained through the use of the computerized programme are illustrated in Table 3. From the pedon analysis data (Tables 1 and 2) it can be observed that two soils (an Oxic Haplustalf and a Ustic Palehumult) have a hyperthermic soil temperature regime. They are soils in the mountainous terrain of northern Thailand (Chiang Mai) and they also occupy a hillslope region. The Kamphaeng Saen variant is commonly found in salt-affected patches in the Mae Klong basin of the western central plain region. The Pak Chong series occupies a vast tract of karst corrosion plain in the central highlands. The Lop Buri series and Lop Buri variant are commonly found in depressions of limestone terrain in the upper central plain as well as in the central highlands.

Table 1. Physical characteristics of soils in control section for FCC.

Horizon	Depth (cm)	Particle-size distribution (USDA) Sand	Silt	Clay (------- % -------)	Texture	Colour Mottles	Structure (surface)	Slope (%)
		1. Typic Haplustalf: fine, mixed, isohyperthermic						
Ap	0-15/18	60.0	19.2	20.8	SCL	10YR 3/4	sbk	2
Bw	18-22/25	51.6	25.4	23.0	SCL	10YR 3/3	sbk	-
Bt1	25-41/48	46.2	20.3	33.5	SCL	10YR 3/3	sbk	-
Bt2	48-76/80	44.7	17.5	37.8	CL	10YR 4/4 5YR 5/6	sbk	-
Btk1	80-110	23.2	32.8	44.0	C	10YR 5/6 5YR 5/6	sbk	-
		2. Oxic Paleustalf: fine, kaolinitic, hyperthermic						
A	0-5	42.0	25.6	32.4	CL	10YR 3/4	gr	9
Bt1	5-25	32.2	24.2	43.5	C	7.5YR 4/4	sbk	-
Bt2	25-50	27.3	19.0	53.7	C	5Yr 4/4	sbk	-
Bt3	50-95	25.6	22.4	52.0	L	5YR 4/6	sbk	-
		3. Ustic Palehumult: clayey, kaolinitic, hyperthermic						
A	0-8/15	44.7	13.4	37.9	CL	7.5YR 4/4	gr	23
Bt1	8-15/20	30.7	14.8	54.5	C	5YR 4/6	sbk	-
Bt2	20-38	31.5	15.4	53.1	C	5YR 4/8	sbk	-
C1	38-75	35.8	16.5	47.7	GC	5YR 4/8 and 10YR 5/8	sbk	-
		4. Oxic Paleustult: clayey, kaolinitic, isohyperthermic						
A	0-13	10.0	12.2	77.8	C	2.5YR 3/4	sbk	3
Bt1	13-50	9.0	11.0	80.0	C	2.5YR 3/4	sbk	-
Bt2	50-105	5.7	5.3	81.0	C	2.5YR 3/6	sbk	-

Table 1. (cont'd).

Horizon	Depth (cm)	Particle-size distribution (USDA) Sand (—— %	Silt	Clay ——)	Texture	Colour Mottles	Structure (surface)	Slope (%)
		5. Typic Pellustert: very fine, montmorillonitic, isohyperthermic						
Ap	0-10	13.5	14.8	71.7	C	10YR 4/1	sbk	2
A	10-25	9.3	15.7	75.0	C	10YR 5/1	sbk	-
AB	25-50	7.1	17.9	75.0	C	10YR 5/1	sbk	-
Bw1	50-70	12.4	9.2	78.4	C	10YR 5/1	sbk	-
		6. Typic Pellustert: fine, montmorillonitic, isohyperthermic						
Ap	0-10/15	2.2	51.2	46.6	C	5YR 2.5/1	sbk	1
Bw1	15-33	3.1	50.2	46.7	C	10YR 3/1	sbk	-
Bw2	33-55	8.2	35.1	56.7	C	10YR 3/1	sbk	-
Bw3	55-74	9.4	33.9	56.7	C	10YR 4/1	sbk	-

C = clay, CL = clay loam, GC = gravelly clay, SCL = sandy clay loam, gr = granular, sbk = subangular blocky.

The factors used in the interpretation of the FCC units given in Table 3 are as follows:

Lbdns Surface crusting can occur if the soil has more than 30% of silt. The nitrogen fertilization programme should consider a flush of nitrogen at the beginning of the rainy season. Rock phosphate should not be used in this soil, and iron and zinc deficiency is common. The soil has a high soluble salt content. It should be leached with drainage. Since the soil has an alkali condition, leaching with calcium salts to prevent dispersion is recommended.

LCad Surface crusting can occur if the soil has more than 30% of silt. Conservation measures should be taken to prevent the loss of surface soil. The nitrogen fertilization programme should consider a flush of nitrogen at the beginning of the rainy season. The soil needs lime for most crops, and the root zone of aluminum-sensitive crops will be limited. Therefore the use of aluminum-tolerant varieties is recommended.

Cadi The soil has good infiltration, but conservation measures are needed to prevent surface soil loss. The nitrogen fertilization, lime

Table 2. Chemical characteristics of soils in control section for FCC.

Horizon	Depth (cm)	pH 1:1 H_2O	O.M. (%)	Avail. P (ppm)	Exchangeable bases Ca	Mg	Na	K	Sum bases	E.A.	CEC (pH 7)	ECe (dS/m)	ESP (%)	Fe_2O_3 (%)
					(m.e./100 g soil						)			
1. Typic Haplustalf: fine, mixed, isohyperthermic														
Ap	0-15/18	7.6	1.0	14.2	9.5	4.8	13.1	0.3	27.7	0.1	9.1	25.7	144.7	-
Bw	18-22/25	7.6	0.9	19.6	11.1	4.4	9.8	0.3	25.6	0.4	11.6	16.4	24.2	-
Bt1	25-41/48	8.1	0.5	3.7	9.2	6.6	12.9	0.4	29.1	0.5	16.5	11.1	78.1	-
Bt2	48-76/80	8.2	0.3	2.3	12.5	6.7	12.2	0.5	31.9	0.2	17.1	9.7	71.2	-
Btk1	80-110	8.4	0.1	1.3	25.0	10.5	9.9	0.5	45.9	0..2	18.0	3.4	55.1	-
2. Oxic Paleustalf: fine, kaolinitic, hyperthermic														
A	0-5	6.4	5.5	4.2	9.1	2.0	0.6	0.7	12.4	9.2	13.9	-	-	-
Bt1	5-25	5.2	2.1	1.5	8.4	3.3	0.4	0.8	12.9	9.2	10.7	-	-	-
Bt2	25-50	4.8	0.9	1.2	3.8	1.0	0.6	0.6	6.0	7.2	10.3	-	-	-
Bt3	50-95	4.8	0.7	0.4	4.5	0.9	0.3	0.5	6.2	6.2	10.3	-	-	-
3. Ustic Palehumult: clayey, kaolinitic, hyperthermic														
A	0-8/15	5.0	4.0	4.5	1.9	2.5	0.5	0.5	5.4	11.3	9.9	-	-	-
Bt1	15-20	4.6	2.8	4.2	5.9	2.5	1.1	0.9	10.4	13.0	13.7	-	-	-
Bt2	20-38	4.6	1.7	1.1	0.9	1.7	0.4	0.4	3.4	12.4	12.2	-	-	-
C1	38-75	4.6	0.6	0.8	0.6	0.9	0.4	0.6	2.5	12.4	12.0	-	-	-
4. Oxic Paleustult: clayey, kaolinitic, isohyperthermic														
A	0-13	5.5	3.1	2.3	6.0	1.7	0.2	0.1	8.0	8.1	15.0	-	-	6.7
Bt1	13-50	4.8	1.5	1.4	4.5	1.2	0.4	0.1	6.2	8.3	10.0	-	-	6.1
Bt2	50-105	4.3	0.7	1.5	0.8	0.7	0.3	0.1	1.9	11.2	8.6	-	-	7.1
5. Typic Pellustert: very fine, montmorillonitic, isohyperthermic														
Ap	0-10	7.8	1.4	28.3	36.4	11.4	2.9	0.2	50.9	14.3	67.4	1.1	-	-
A	10-25	7.0	1.0	7.3	29.0	28.6	2.9	0.1	60.6	16.3	76.2	0.5	-	-
AB	25-50	7.1	0.7	3.9	24.6	26.9	2.9	0.1	54.5	16.4	76.2	0.5	-	-
Bw1	50-70	7.4	0.8	3.3	28.3	32.9	2.9	0.2	64.3	14.9	75.6	0.6	-	-
6. Typic Pellustert: fine, montmorillonitic, isohyperthermic														
Ap	0-15	6.5	3.3	20.5	30.0	6.7	2.2	0.9	39.8	9.9	46.2	0.7	-	-
Bw1	15-33	7.3	1.3	5.4	32.5	7.3	2.3	0.3	42.4	9.5	47.4	0.6	-	-
Bw2	33-55	7.4	0.5	4.0	31.7	7.2	3.1	0.1	42.1	8.5	51.9	0.7	-	-
Bw3	55-74	7.7	0.4	6.4	27.8	6.4	3.3	0.1	37.6	8.9	52.6	1.2	-	-

maintenance, and aluminium toxicity are the same as those which apply to the LCad unit. The soil has a high phosphorus fixation, and an initial dose of phosphate fertilizer (% clay(app.) x 10 kg/ha P) should be applied, and subsequently an annual maintenance rate.

Cadk The soil should not be worked on when wet. The nitrogen fertilizer programme should consider a flush of nitrogen at the beginning of the rainy season. A lime maintenance programme is needed, and aluminium toxicity is present. The soil has a low capacity for supplying potassium. This should be closely monitored with soil tests. Frequent potassium application may be needed.

Cbdv Tillage is difficult and the soil should not be worked on in very wet conditions. Also, it should not be allowed to dry. The nitrogen fertilization programme can be carried out in the same way as that applicable to Lbdns, LCad, and Cadk. Rock phosphate should not be used, and iron and zinc deficiency is common for the soil.

Table 3. FCC units of some Alfisols, Ultisols, and Vertisols in Thailand.

	Taxonomic classes	FCC units	Remarks
1.	Typic Haplustalf: fine, mixed, isohyperthermic	Lbdns	Kamphaeng Saen variant
2.	Oxic Paleustalf: fine, kaolinitic, hyperthermic	LCad	Unnamed soil
3.	Ustic Palehumult: clayey, kaolinitic, hyperthermic	Cadi	Unnamed soil
4.	Oxic Paleustult: clayey, kaolinitic, isohyperthermic	Cadk	Pak Chong series
5.	Typic Pellustert: very fine, montmorillonitic, isohyperthermic	Cbdv	Lop Buri series
6.	Typic Pellustert: fine, montmorillonitic, isohyperthermic	Cbdv	Lop Buri variant

Agronomic recommendations for paddy rice and upland crops can be made in accordance with these interpretations. Some additional recommendations may also be included. For example, mulch application when the crops are planted is recommended for FCC units Cadk and Cbdv to protect emerging seedlings from high temperature and to improve germination. Using the FCC units and their interpretations as guidelines on site, the process of crop-production planning can be improved. Materials and farming techniques suggested by the guidelines can be used to obtain at least a satisfactory result in crop production.

FCC units have been tested and used for many purposes in recent years. FCC groupings have been tested as guides for N, P, and K fertilization of lowland rice (Lin, 1984, and 1985). FCC has been used to evaluate the extent of soil constraints (Sanchez and Cochrane, 1980; Trangmar *et al.*, 1984). They have also been used to group soils into smaller regions as the basis for further technology transfer and research (Paredes Arce, 1986).

It should be noted here that the FCC has included slope in the system. However, the interpretation of the slope factor is still far from being complete. Probably present research on conservation measures is not conclusive enough to indicate specific practices for a particular range of slopes. From the author's experience in studying hillslope soils, problems on the conservation of sloping land are generally more tricky than they at first appear. It will take more time to develop a clear-cut concept to tackle them.

Conclusion

The FCC can provide basic guidelines for soil maangement. The system has been designed to cover both lowland and upland soils. Nevertheless, users of the system need to understand the interactions among the factors involved if they are to use the information provided by the FCC units effectively. The system still needs certain modifications for local conditions, and it should in fact be realized that the system is not as easy to use as perhaps was originally intended. Some definitions of modifiers have been changed, and modifiers have been added to the system. Judging from the convenience in classifying the six pedons in this paper, the system has a considerable potential for on-site crop-production planning. Also, the combined use of the FCC and *Soil Taxonomy* in agrotechnology transfer, in the author's opinion, could provide a firm basis for soil management practices for crop production.

Appendix 1

FCC Work Sheet

Soil name: ______________________________

Type		Substrata type	
1. Topsoil: The shallower of Ap or top 20 cm **Yes or no (Y or N)**		2. Subsoil: Immediately under Ap, otherwise from 20 cm down to 50 cm	
1.1 Sand or loamy sand (USDA)	S	2.1 The same as type	
1.2 Loamy (<35% clay but not S)	L	2.2 S	S
1.3 Clayey (>35% clay)	C	2.3 L	L
1.4 Organic (>30% O.M. down to 50 cm+)	O	2.4 Rock or other hard root within 50 cm	R

Modifiers: mark as appropriate.

1.	Chroma <2 within 60 cm of soil surface below Ap or soil saturated >60 days/year	g	2.	Constantly saturated, with no evidence of brownish or reddish mottles except around root channels	g^1
3.	Ustic or Aridic or Xeric soil moisture regime	d (dry)	4.	Topsoil EDEC <4 m.e./100 g S or sum cations <7 m.e./100 g S or sum cations+Al+H <10 m.e./ 100 g S	S e
5.	Within 50 cm of soil surface Al-saturation of ECEC >60% or pH in 1:1 H_2O <5.0	a	6.	Within 50 cm of soil surface Al-saturation of CEC = 1-60% or pH 1:1 H_2O >5.0 and <6.0	h
7.	Topsoil in clay (C type only) % free FeO/% clay >0.15 or Hue of 7.5 or redder with granular structure	i	8.	Soil pH >10 in 1N NaF or positive field NaF test or allophane dominant in clay fraction	x
9.	Clay >35% very sticky plastic clay and 2:1 expanding clay >50% or severe topsoil shrinking and swelling or COLE >0.09	v	10.	Within 50 cm of soil surface <10% weatherable minerals in silt and sand fraction or exchangeable K <0.20 m.e./100 g S or K <2% of sum bases if sum bases <10 m.e./100 g S	k

Appendix (cont'd).

11. Within 50 cm of soil surface free $CaCO_3$ (effervescence with HCl) or pH >7.3	b	12. Within 1 m of soil surface ECe >4 dS/m	s
13. Within 50 cm of soil surface ESP >15	n	14. Within 60 cm of soil surface After drying: pH in 1:1 HO <3.5 and presence of jarosite mottles hue 2.5Y or yellower and chroma >6	c
15. With any type or substrata type 15-35% gravel or coarser 35% gravel or coarser	"	16. Record slope range if desirable _____%	%
		17. Other details	

The FCC unit of (soil name) is ____________________________

Interpretation:
Recommendation:

References

BUOL, S.W. 1972. Fertility Capability Classification system, pp. 45-50. In: *Agronomic-economic research on tropical soils,* Annual Report for 1971, Soil Science Department. Raleigh NC: North Carolina State University.

BUOL, S.W. 1987. Fertility Capability Classification system and its utilization. In: *Soil management under humid conditions in Asia (ASIALAND),* 317-331. IBSRAM Proceedings no. 5. Bangkok: IBSRAM.

BUOL, S.W., HOLE, F.D. and McCRACKEN, R.J. 1980. *Soil genesis and classification* (2d ed.). Iowa, USA: The Iowa State University Press. 404p.

BUOL, S.W., SANCHEZ, P.A. and BUOL, G.S. 1988. *FCC - Draft for editing: English, Spanish, French, Portugese. (A microcomputerized FCC programme, basic language).* Raleigh, NC: North Carolina State University.

BUOL, S.W., SANCHEZ, P.A., CATES, R.B. and GRANGER, M.A. 1975. Soil Fertility Capability Classification: a technical soil classification system for fertility management. In: *Soil management in tropical America,* ed. E. Bornemiza and A. Alvarado, 126-145. Raleigh, NC: North Carolina State University.

BUOL, S.W. and COUTO, W. 1981. Soil fertility capability assessment for use in the humid tropics. In: *Characterization and classification of some soils of the humid tropics,* 254-261. London: Wiley.

CLINE, M.G. 1949. Basic principles of soil classification. *Soil Science* 67:81-91.

COUTO, W. 1988. The Fertility Capability Classification system. In: *First training workshop on site selection and characterization*, 207-213. IBSRAM Technical Notes no. 1. Bangkok: IBSRAM.

EIUMNOH, A. 1984. Application of *Soil Taxonomy* to fertility capability classification of problem soils in the S.E. Coast of Thailand. In: *Ecology and management of problem soils in Asia*, 169-190. FFTC Book Series no. 27. Taipei, Taiwan: FFTC.

FAO (Food and Agriculture Organization of the United Nations). 1971. *Soil map of the world*. Vol. 4. *South America*, Paris: Unesco.

FAO (Food and Agriculture Organization of the United Nations). 1974. *FAO-Unesco soil map of the world*. Vol. 7. *Legend*. Paris: Unesco.

FAO (Food and Agriculture Organization of the United Nations). 1977. *Soil map of the world*, Vol. 6. *Africa*. Paris: Unesco.

LIN, C.F. 1984. Fertility capability classification as a guide to N-fertilization of lowland rice. In: *Ecology and management of problem soils in Asia*. FFTC Book Series no. 27. Taipei, Taiwan: FFTC.

LIN, C.F. 1985. Fertility Capability Classification (FCC) as a guide to P, K-fertilization of lowland rice. *Soil Taxonomy - review and use in the Asian and Pacific region*.. FFTC Book Series no. 29.

PAREDES ARCE, P.G. 1986. *Le système de classification des sols par capacité de fertilité*. Faculté des Sciences Agronomiques. Belgium: University of Gembloux.

SANCHEZ, P.A. 1987. Edaphic parameters for characterizing IBSRAM's acid tropical soils network sites. In: *Land development - management of acid soils*, 113-123. IBSRAM Proceedings no. 4. Bangkok: IBSRAM.

SANCHEZ, P.A. and COCHRANE, T.T. 1980. Soil constraints to major farming systems of tropical America. In: *Soil-related constraints to food production in the tropics*, ed. M. Drosdoff, 107-139. Los Baños, Philippines: IRRI.

SANCHEZ, P.A. and BUOL, S.W. 1985. Agronomic taxonomy for wetland soils. In: *Proceedings of the International Workshop on Wetland Soils: Characterization, Classification and Utilization* (Philippines, 1984), 207-227. Los Baños, Philippines: IRRI.

SANCHEZ, P.A., COUTO, W. and BUOL, S.W. 1982. The fertility capability soil classification system: interpretation, applicability and modification. *Geoderma* 27: 283-309.

SOIL SURVEY STAFF. 1975. *Soil taxonomy: A basic system of soil classification for making and interpreting soil surveys*. U.S. Department of Agriculture. Washington D.C: U.S. Government Printing Office.

SOIL SURVEY STAFF. 1987. *Keys to Soil Taxonomy*. SMSS Technical Monograph no. 6. Ithaca, NY, USA: Cornell University.

TRANGMAR, B.B., YOST, R.S., SUDJADI, M. SOEKARDI, M. and UEHARA. G. 1984. *Regional variation of selected topsoil properties in Sitiung, West Sumatra, Indonesia*. Research Series no. 026. Collection of Tropical Agriculture and Human Resources. Hawaii, USA: University of Hawaii.

Section 6: Physical and chemical parameters

Soil structure and rooting[1]

CHRISTIAN VALENTIN*

Abstract

The rooting of crops affects productivity. Poor roting can be a significant constraint. Rooting depends on the soil structure, taken in a broad sense to include structure, porosity, and consistency. Simple methods can be used to identify the structural units within a cultivation profile that could influence root growth: compacted volume, plough pan, pedological discontinuities, etc. Similarly, there are several approaches for a morphological characterization and quantification of rooting. The two methods are combined to assess physical constraints related to soil characteristics and tillage. A study of the cultivation profile is useful for evaluating the appropriateness of various cropping operations.

Introduction

For site characterization and monitoring experiments, samples are collected *in situ* and analyzed in the laboratory (Webb and Coughlan, 1989; Nualsri Kanchanakool, 1989). The procedure also requires a large number of field observations and measurements. Such observations include a study of the

---

1 Translated from the French by Mira Fischer.

* Institut Français de Recherche Scientifique pour le Développment en Coopération (ORSTOM), BP V-15, Abidjan, Côte d'Ivoire.

structural state of the horizons in the root zone. The definition of the soil structure is extended to porosity and consistency. The three properties which influence a large number of important rooting and crop production processes are:

- *Emergence:* Emergence is poor or delayed if the seedbed is too compacted or if the clod size does not allow sufficiently close contact between the soil and seed.
- *Water and nutrient supply:* This depends not only on soil moisture, but also on the type of root colonization, which in turn is influenced by the structural state.
- *Aeration:* At field capacity, soil pores should hold approximately 10% volume of gas, of which 10% is oxygen (Dexter, 1988). The cultivation profile should also provide favourable drainage conditions.

An earlier paper discussed the main forms of degradation of the cultivation profile (Valentin, 1988). The present paper aims to provide further information on the characterization of the soil sturcture and its relation to rooting.

Characterization of the soil structure

Structure of the tilled horizon

The description of a cultivation profile should answer two questions:

- Have the objectives of the cultivation operation under study been achieved?
- What is the response of rooting to the resulting physical conditions?

The description of a cultivation profile is linked to that of the soil profile. However, particular attention should be paid to:

- horizontal variability of the structure;
- discontinuity due to soil tillage.

In practice, it is useful to refer to an established checklist that describes various types of structure. For example, de Blic (1976a) used the following checklist in Côte d'Ivoire:

Shape of aggregates

- single grained,
- fragmentary,
- massive with blocky substructure,
- massive.

Size
- fine to medium,
- medium to coarse.

Grade
- weakly developed,
- moderately developed,
- strongly developed.

The same author distinguished four classes among tilled horizons:
- with dominant fragmentary structure,
- fragmentary to massive structure,
- dominant massive structure,
- exclusively massive structure.

The unfavourable conditions of the last three classes can be attributed to the lifting of massive and cohesive fragments from the underlying horizon, to soil compaction under the weight of tractors, or to inadequate fragmentation due to tillage under conditions which are too dry.

In France, Tardieu and Manichon (1987a) proposed the following classification for structural unit assemblage:
- fragmentary state characterized by the size of the elements (fine earth, centimeter- or decimeter-size clods);
- massive state, possibly with associated cavities and fissures.

These authors distinguished three main types of ploughed horizons. The first type is made up of fine earth and centimetre-size clods. This case is not very common and is observed in plots with little degradation. Compaction can rarely be completely avoided because of the cropping schedules. The second type is characterized by the proximity of compacted blocks and structural cavities. It is observed when the soil is highly compacted before tillage - for example, during the previous harvest. The third type presents a continuous and compact horizon. It is the structural state observed after direct planting on a previously compacted soil.

Regardless of the classification system that is adopted in relation to local conditions, the sketch of the cultivation profile should indicate, in addition to the structural units (Figure 1):
- sharpness of transition,
- the nature of discontinuities due to soil tillage (plough pan, shining surfaces, Plate 1),
- location of organic matter and biological niches,
- rooting pattern (de Blic, 1987; Fritsch and Valentin, 1987; Valentin, 1988).

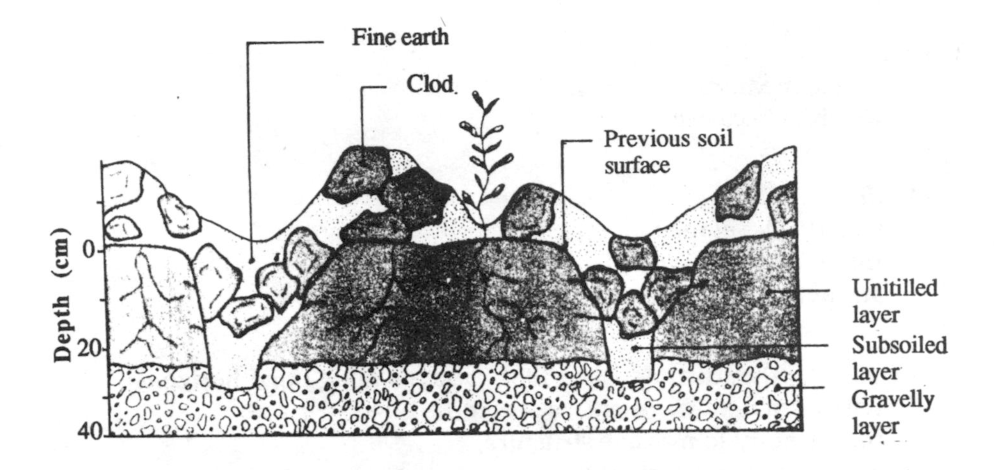

Figure 1. Cultivation profile after clearing and subsoiling on a savanna soil (adapted from de Blic and Moreau, 1977).

Plate 1. Cultivation profile in Burkina Faso. Note the plough pans (*P*) which resist root penetration (*r*) and the surface crust erosion (*E*).

Porosity

While describing the cultivation profile, inter- and intraclod porosity should be visually evaluated according to a semiquantitative scoring system. Such field observations are indispensable and are often combined with:

- field measurements, using core or excavation methods (Webb and Coughlan, 1989) to evaluate the total porosity of the different structural states identified during the description of the cultivation profile (Figure 2); and
- laboratory measurements, where bulk density is determined on clods (Lenvain *et al.*, 1987; Nualsri Kanchanakool, 1989) and the particle density of processed samples.

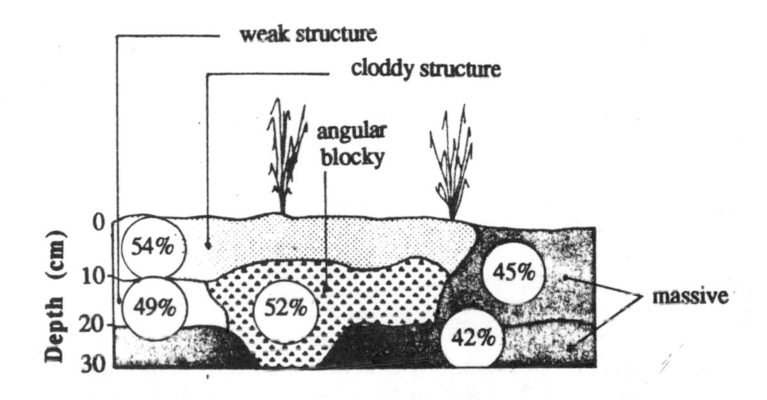

Figure 2. Porosity (%) of different structural units in the cultivation profile of an upland rice crop in the savanna - first year of cultivation after clearing and subsoiling (adapted from de Blic, 1976a).

The two series of measurements are compared to calculate inter- and intraclod porosity.

Consistency

Often, total porosity may not be a sufficient indication of the degree of soil compaction, as it only consists of the packing of particles. According to Webb and Coughlan (1989), there is a strong correlation between compaction and crop yields. Resistance to penetration is measured by various devices. Different types of penetrometers are available. A simple and inexpensive model is described in the appendix. Resistance to penetration generally decreases exponentially as soil moisture increases; each penetrometric profile should therefore be combined with the moisture profile (Figure 3).

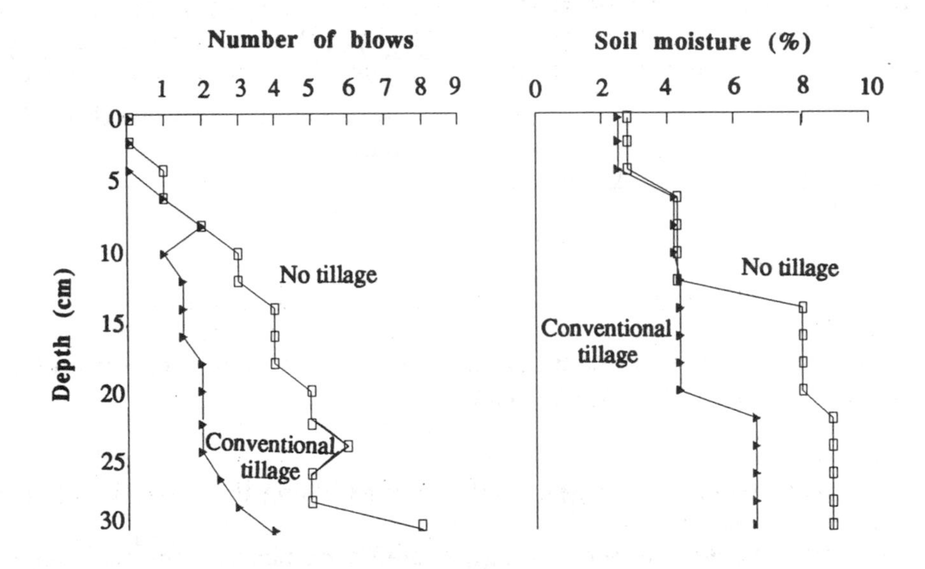

Figure 3. Comparison of penetrometric profiles (number of blows, average of 6 replicates) and soil moisture (% of mass, average of 3 replicates) for two treatments in a pineapple crop on sandy ferrallitic soils: a) without soil tillage and with mulching - harvest stubble on the surface reduces evaporation, and although soil moisture conditions are better, they are still inadequate to loosen the soil; b) with normal soil preparation (ploughing and surface operations) and incorporation of harvest residue - the soil dries more easily but soil tillage reduces resistance to penetration more than the previous treatment (adapted from Camara, 1978).

Study of rooting

Soil scientists often concentrate more on soil than on root observations. However, the establishment of the cultivation profile requires a study of the root system, as this is a vital link for understanding soil-plant relations. For this purpose, it is useful to review certain simple principles.

Characterization of the root system

Identification of root types

Rooting is characterized first by its type (fasciculate or root), and by the number and size of ramifications. The main roots develop from the seed, usually according to positive geotropism, and penetrate vertically into the soil, whereas lateral roots of the first, second, or third order may grow more horizontally.

Variables

Morphological

The area of the soil-root interface is assessed in addition to the maximum rooting depth. The soil-root contact area is calculated on the basis of root diameter. Such measurements should include variations in soil and root volume, depending on the moisture conditions.

Spatial

The main parameters to be considered in assessing the vertical and lateral distribution of roots are:

- density: This is expressed in weight/volume, number of contacts/area, root length/volume, etc; and
- the average distance separating two roots D. This variable is particularly important for studies on water supply to plants. It is generally determined from root length/unit volume L. The formulas are based on geometric models selected for root distribution. Thus:

$$D = a\,(L)^{-0.5}$$

where $a = 1.00, 1.07$, or 1.13 (authors cited by Tardieu and Manichon, 1986a).

The distribution of these parameters is generally studied according to depth and distance from the plant row.

Root growth

The preceding parameters can also be studied over time. One of the most common variables is rate of root elongation, expressed in centimetres/day, per root type (root axis, primary lateral, secondary lateral).

Methodology

Visual examination

Visual examination should precede all other measurements, as it determines the characteristics of the sampling (depth, replicates, etc.). It already indicates constraints to the root system, such as the presence of a plough pan. An examination of the roots reveals symptoms for diagnosing stress. Callot *et al.* (1982) give examples of roots that developed ferromanganic sheaths to resist pressure exerted by structural elements. The sheaths, which isolate the roots from their environment, helped evaluate differences in vigour, mainly among fruit trees.

The root profile is sketched on site and compared with the cultivation profile (McSweeney and Jansen, 1984).

Quadrat method

The quadrat method consists of placing a grill with a given mesh size (5 cm) on a vertical profile and counting the number of roots per square. The technique is frequently used for determining the total length of the roots *L* per horizon, and to calculate the average distance between roots *D*. The method was improved by Tardieu and Manichon (1986b) during a study on maize rooting:

- The grill mesh was reduced to 2 cm.
- Readings were taken:
  - o vertically, between two plants;
  - o horizontally, at five levels, within the ploughed horizon (17 cm) and below (30 cm, 60 cm, 80 cm, 100 cm).

Sampling methods

There are two main techniques for taking samples.

- In the first technique, pin boards are placed along a vertical profile that is then removed completely. The root system is carefully washed in the laboratory. If the pins are sufficiently dense - they are generally arranged 10 cm apart - the main root system remains practically intact, and the main morphological characteristics can be studied. The method requires much effort and is not suitable for quantitative studies.
- In the second technique, the samples are removed in vertical or horizontal cores (Chopart and Nicou, 1976). The cores are cut along the horizon or structural unit and washed in water. The technique is difficult to apply in gravel soils. Otherwise it is suitable for quantitative studies. It can be very usefully combined with the quadrat method. The collection of a limited number of samples helps to establish a reliable

relationship between the number of roots observed per square and the weight or length per structural unit.

Influence of physical properties on rooting

Texture

Soils with coarse texture are more easily colonized by roots than clayey soils. According to Callot *et al.* (1982), high sand rates are conducive for root ramification (through increased branching), and hence there is an increase in the quantity of roots.

Structure

Soil structure can offset the effect of texture. A very clayey soil, but one with small aggregates can be colonized as well as a sandy soil. Generally, roots in massive and compacted horizons are scanty (Figure 4), fine, and straight, whereas they are more ramified, twisted, and hairy in soils with a well-developed structure. Crumb or fine blocky structures are the most favourable to rooting (Plate 2). It is particularly useful to study lateral variations in structure within the same tilled horizon. The most massive zones, with few roots, inhibit rooting in the directly underlying layers. Heterogeneity of the surface horizon structure is thus reflected in the root distribution in the nontilled horizons (Tardieu and Manichon, 1986b).

Porosity

Small variations in porosity can bring about large differences in colonization by roots. Callot *et al.* (1982) observed that root dry weight for catstail grass (*Phleum*) increased from 1.3 g to 2.1 g/kg of soil due to a change in porosity from 40 to 45%.

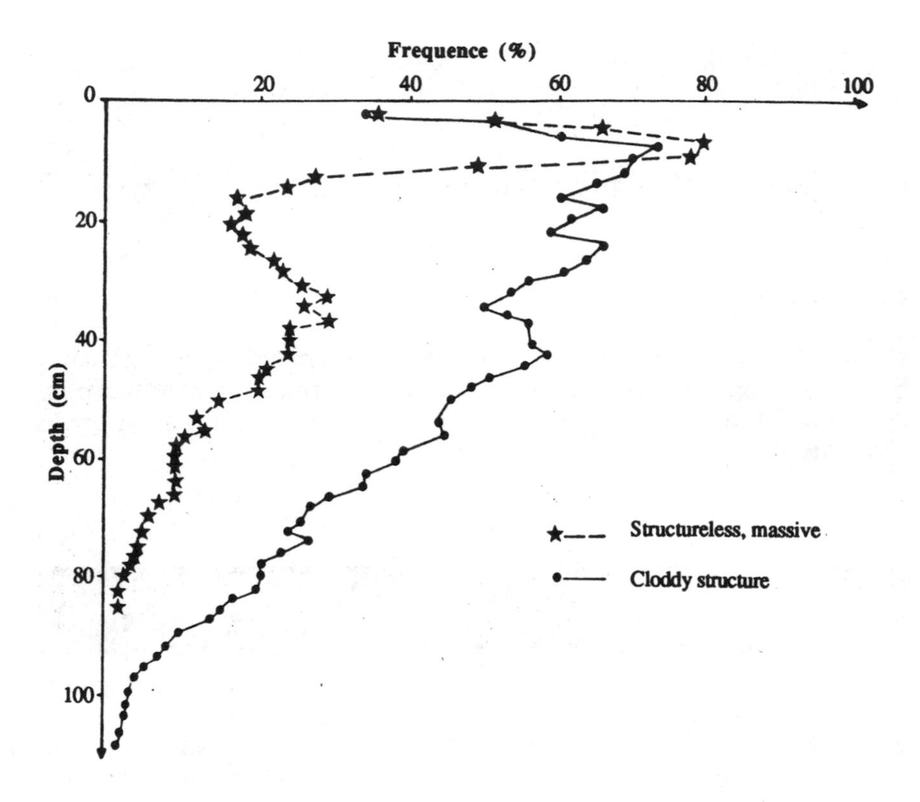

Figure 4. Examples of root profiles at the silking stage in a maize crop, using the grid method. Each point represents the average over 11 replicates of the percentage of 2 x 2 cm squares. At least one impact between the cone and one root occurred.

Consistency or compaction

Porosity provides important information on aeration and water supply to roots, but not on soil consistency. Generally, rooting is influenced more by resistance to penetration than by aeration and soil moistrue (Callot *et al.*, 1982). In order to penetrate, roots require sufficiently large and continuous pores, or have to enlarge smaller pores. Root extension is thus affected even by slight pressure (Russel, 1977). Soil compaction also reduces microporosity, with an adverse effect on drainage and gas exchange. It is therefore important to check compaction during land clearing and other mechanized operations.

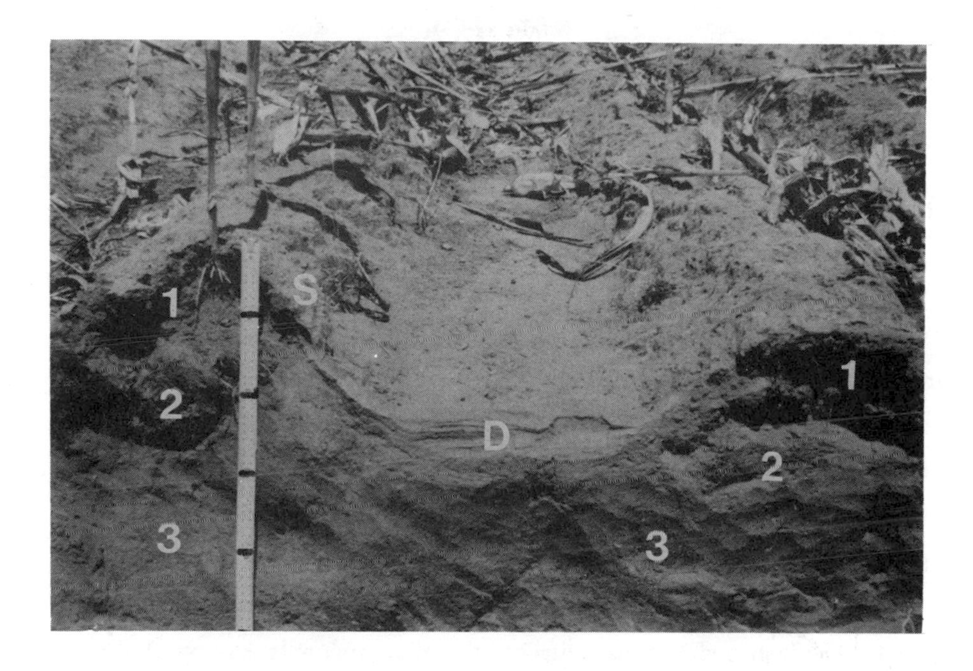

Plate 2. Cultivation profile for maize in Zambia. Note the structural units; the soil in tilled mounds is loose, and has a fine structure, which is easily penetrated by roots; under the tilled mounds, the soil is more compact, and has less root penetration; under the tilled area, the soil is compact, and there is very little root penetration. Note also the surface crust (S) and the runoff (D).

Conclusion

The study of the relationship between soil structure and rooting is a key factor in understanding soil-plant relations, and is often overlooked. The combination of characterization of the soil structure with physical field measurements is useful for analyzing the appropriateness and efficiency of soil tillage operations. Although soil tillage is usually beneficial to rooting, it can nevertheless cause localized compaction. This means that the vertical and lateral distribution of roots rather than their average density should be considered. The more regular the root colonization, the better is the uptake by plants water likely to be.

Appendix

The hammering penetrometer

Objective

To measure resistance to penetration through a vertical soil profile.

Principle

A metallic cone is driven vertically into the soil by hammering, in order to measure the resistance of the soil to penetration by the cone.

Material

Hammering penetrometers are built in accordance with the model illustrated in Figure 1. The instrument is made up of several components having the following characteristics:

- A calibrated lower rod, 1 m long and fitted with an anvil. The markings, set for example at 1-cm intervals, should be engraved rather than painted or stuck, as they last longer. The rod diameter should be between 1.2 and 1.3 of the cone diameter to minimize friction.
- An upper rod, 1 m long, along which the hammer slides. A stop at the top of the rod ensures a uniform dropping height. The stop should be removable so that the hammers can be slipped over the rod. More rods can be joined together to reach depths exceeding 1 m.
- A set of hammers of 0.2, 0.5, 1 kg or more, and up to 20 kg. The selected metal should have a high density. The hammer height should be sufficiently low to minimize friction. The hammer diameter should also be reduced to avoid double readings in case the penetrometer is not maintained in an exactly straight position.
- Cone-shaped probe, with a cross section of 2-3 cm^2 and a 90° angle. The cone should be made of very hard steel. Spare cones should be kept to replace worn cones.

In the given example, the total weight of the rods, anvil, and cone is 2 kg; it varies according to the type of cone.

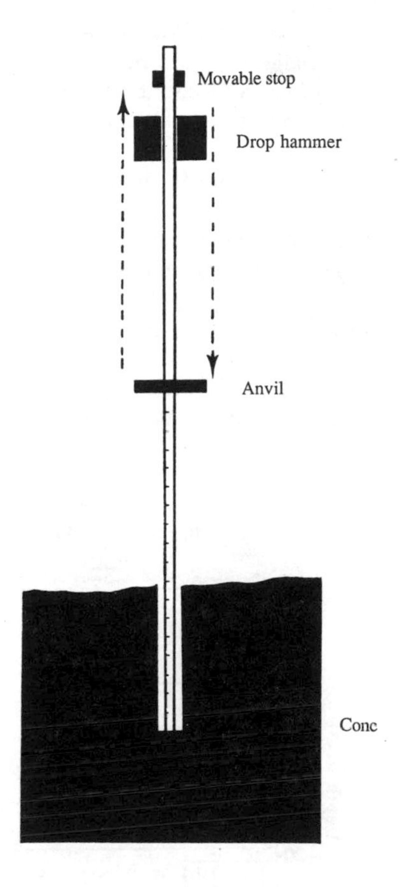

Figure 1. Diagram of a hammering penetrometer.

Procedure

Before starting the measurements, the most appropriate hammer-cone combination should be selected according to the surface horizon, to reduce the number of blows/cm, or inversely the number of centimetres/blow. The combination can be changed for deeper soils, but the penetration depth of the blow should be noted when the components are changed

Resistance can be measured in two ways. Either the penetration depth is measured after each blow, or after n blows; or the number of strokes required to reach a given depth are counted. The second method is easier.

While measuring, care should be taken that:

- penetration is vertical;
- the dropping height does not allow the hammer to rebound against the stop.

The measurements are usually carried out by two persons. One person holds the rods straight, drops the hammer, and counts the number of blows. The other checks penetration depth and notes the number of blows per given depth - for example, every 5 cm.

As resistance to penetration varies in inverse proportion to soil moisture, the soil moisture profile should also be established.

These measurements should be combined with a description of the profile in order to link them to the pedological structure and tillage-induced soil discontinuity.

Expression of the results

A preliminary bar chart of the penetrometric profile (Figure 2) should be established in the field, assuming that the same hammer-cone combination is used.

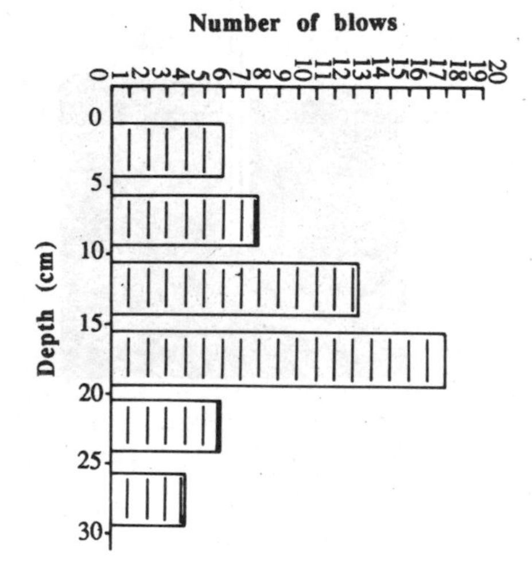

Figure 2. Example of a penetrometric profile established on site. Each bar represents a blow.

Resistance to penetration can be calculated using the following equation:

$$R = \frac{M^2.h.n}{2.(M + m).S.z}$$

where:

R : resistance to penetration (kg/cm^2)
M : weight of hammer (kg)
m : weight of rods, anvil, and cone
h : height from which hammer is dropped
S : cross section of the cone (cm^2)
n : number of blows
z : penetration depth (cm)

Accuracy

The number of replications depends on soil heterogeneity, required precision, and time-cost factor. For example, three replications are adequate to reveal the existence of a plough pan.

Generally, results of measurements are more closely linked to the resistance to penetration, which are exponential in scale, than to eventual experimental errors, which are arithmetical in scale.

Cost

The cost of the penetrometer depends on the materials used. Commercial models are relatively expensive, but the instrument can be produced locally at lower cost.

Advantages

This type of penetrometer is not very expensive if produced locally. It can be easily transported to the field and only requires two operators. Moreover, it does not disturb the environment.

It is the best way to measure rapidly variations in resistance to penetration within a profile:

- over a given area, measurements can be taken over an entire plot;
- over time, during a crop cycle for example, after a mechanical soil operation.

Disadvantages

This type of measurement can be difficult to carry out on coarse soils or those with many thick or medium-sized roots.

Very often measurements can only be carried out up to a moderate depth, as it is then difficult to extract the cone from the soil.

Conclusion

These measurements also require a visual examination or measurement of other parameters, such as moisture, root density, and bulk density.

References

AUDRY, P., COMBEAU, A., HUMBEL, F.X., ROOSE, E., and VIZIER, J.F. 1973. *Eassai sur les études de dynamique actuelle des sols.* Bulletin de groupe de travail sur la dynamique actuelle des sols, no. 2. Annexes. Paris: ORSTOM. 126p.

BILLOT, J.F. 1982. Les applications agronomiques de la pénétrométrie à l'étude de la structure des sols travaillés. *Science du Sol, Bulletin de l'Association Française de l' Etude du Sol* 3: 187-202.

BURKE, W., CABRIELS, D. and BOUMA,J. 1986. *Soil structure assessment.* Commission of the European Communities. Roterdam: Balkema. 92p.

CALLOT, G., GHAMAYOU, H., MAERTENS, C. and SALSAC, L. 1982 *Mieux comprendre les interactions sol-racine.* Incidences sur la nutrition minérale. Paris: INRA. 325p.

CAMARA, M. 1978. Caractérisation physique des profils culturaux de l'ananas. Technical Report. Abidjan: ORSTOM. 23p.

CHOPART, J.L. and NICOU, R. 1976. Influence du labour sur le développement radiculaire de différentes plantes cultivées au Sénégal. Conséquences sur leur alimentation hydrique. *Agronomie Tropicale* 31(1): 7-28.

DE BLIC, P. 1976a. Le comportement des sols ferrallitiques de Côte d'Ivoire après défrichement et mise en culture mécanisée: rôle des traits hérités du milieu naturel. *Cahiers ORSTOM, Série Pédologie* 14(2): 113-130.

DE BLIC, P. 1976b. Réorganisations pédologiques en milieu ferrallitique de Côte d'Ivoire à la suite du défrichement et de la mise en culture mécanisée. Technical Report. Abidjan: ORSTOM. 8p.

DE BLIC, P. and MOREAU, R. 1977. Structural characteristics of ferrallitic soils under mechanical cultivation in the marginal forest areas of the Ivory Coast. In: *Soil physical properties and crop production in the tropics,* ed. R. Lal and D.J. Greenland, 111-122. Chichester, UK: J. Wiley and Sons.

DE BLIC, P. 1987. Analysis of a cultivation profile under sugarcane: methodology and results. In: *Land development and management of acid soils in Africa II,* 275-285. IBSRAM Proceedings no. 7. Bangkok: IBSRAM.

DEXTER, A.R. 1988. Advances in characterization of soil structure. *Soil and Tillage Research* 11: 199-238.

FRITSCH, E. and VALENTIN, C. 1987. Characterization of a soil transect in Misamfu research station (Zambia). In: *Land development and management of acid soils in Africa II,* 287-310. IBSRAM Proceedings no. 7. Bangkok: IBSRAM.

LENVAIN, J.S., CHINENE, V.R.N., GILL, K.S. and PAUWELYN, P.L.L. 1987. Aspects of compaction, water transmissibility, and erosion on a kaolinitic clay soil. In: *Land development and management of acid soils in Africa II,* 173-185. IBSRAM Proceedings no. 7. Bangkok: IBSRAM.

McSWEENEY, K. and JANSEN, I.J. 1984. Soil structure and associated rooting behavior in minesoils. *Soil Science Society of America Journal* 48(3): 607-612.

NUALSRI KANCHANAKOOL. 1989. Chemical and physical analysis for monitoring sustainability. This volume.

RUSSEL, R.S. 1977. *Plant root systems: their function and interaction with the soil.* London: McGraw-Hill. 298p.

TARDIEU, F. and MANICHON, H. 1986a. Caractérisation en tant que capteur d'eau de l'enracinement du maïs en parcelle cultivée. I. Discussion de critères d'étude. *Agronomie* 6(4): 345-354.

TARDIEU, F. and MANICHON, H. 1986b. Caractérisation en tant que capteur d'eau de l'enracinement du maïs en parcelle cultivée. II. Une méthode d'étude de la répartition verticale et horizontale des racines. *Agronomie* 6(5): 415-425.

TARDIEU, F. and MANICHON, H. 1987a. Etat structural, système racinaire et alimentation hydrique du maïs. I. Modélisation d'états structuraux types de la couche labourée. *Agronomie* 7(2): 123-131.

TARDIEU, F.and MANICHON, H. 1987b. Etat structural, système racinaire et alimentation hydrique du maïs. II. Croissance et disposition spatiale du système racinaire. *Agronomie* 7(3): 201-211.

VALENTIN, C. 1988. Degradation of the cultivation profile: surface crusts, erosion and plough pans. In: *First Training Workshop on Site Selection and Characterization,* 233-264. IBSRAM Technical Notes no. 1. Bangkok: IBSRAM.

WEBB, A.A. and COUGHLAN, K.J. 1989. Physical characterization of soils. This volume.

Chemical parameters for site characterization and fertility evaluation

E. PUSHPARAJAH*

Abstract

The paper reviews the minimum chemical properties to be measured in order to characterize the site and monitor fertility changes in soil management experiments. The analytical methods to be used are indicated. The need for a critical evaluation of the properties to be used is also emphasized.

Introduction

For any field experiments, the need for detailed characterization of the experimental site and for regular and continuous monitoring cannot be overemphasized. An assessment of chemical parameters is required for both soil taxonomic purposes and for the evaluation of edaphic or fertility changes which relate to crop performance. The determination of the chemical parameters is only one component (albeit an important one), amongst various other physical and biological measurements.

The analysis of chemical parameters is not only time-consuming, but can also be costly. Thus it is important to realize that properties should not be measured just because methods for analysis exist and the organization has a well-equipped central laboratory. Further, when there is a massive amount of

* IBSRAM Programme Officer, PO Box 9-109, Bangkhen, Bangkok 10900, Thailand.

data available, a researcher is often at a loss to know how to use the information, so the parameters to be assessed must be confined to those of direct relevance and immediate interest. On the other hand, if a researcher feels that some additional parameters could become useful at a later date, a possible solution would be to store the soil samples for such an eventuality.

While emphasis is given to the need to curtail the analysis to those parameters considered crucial to proper evaluation of the results, it is equally important to realize that critical analysis should not be sacrificed to save time and money.

This paper therefore briefly discusses the minimum data set considered as appropriate, especially with regard to the Asian network on sloping lands, without going into detials of laboratory procedures.

Chemical parameters for soil characterization and taxonomy

General

The standard parameters and laboratory methods for *Soil Taxonomy* are documented in the Soil Survey Investigation Report no. 1 (USDA, 1984), which describes the automated procedures. However, less automated procedures are equally applicable, and those for common parameters are generally well understood. It is therefore unnecessary to discuss details of laboratory procedures,nor do all the analyses need to be made on every sample.

Sampling

The soil samples have to be obtained from the face of freshly dug pits. Each soil sample is usually obtained on a 30 to 50 cm vertical strip covering the cross section of a horizon.

A general rule is that the sample should not be collected by going deep into the face of the pedon; sampling should be confined to that portion which can be seen. This means that for shallow horizons there is a need to increase the lateral distance of sampling.

Sampling should preferably commence at the bottom of the pit. However, if the soil is light-textured, sampling has to commence from the top horizon.

Parameters and brief description of analytical methods

pH

pH is measured by a pH meter with a soil:water or soil:1N KCl solution ratio of 1:1 (i.e. one part by weight of soil to one part by weight of water or solution).

Organic carbon

Organic carbon expressed as a percentage of weight percentage is determined by the Walkey-Black wet digestion method. The sample is treated with a potassium dichromate (a strong oxidizing agent) and digested with sulphuric acid.

Nitrogen

Nitrogen is determined by kjeldahl digestion, with ammonium being steam-distilled into boric acid and titrated with HCl.

Extractable bases:

The contents of major exchangeable bases - calcium, magnesium and potassium - are expressed as m.e./100 g soil. These bases are extracted by displacing them from the cation-exchange complex by another cation, ammonium (ammonium acetate buffered at pH 7).

Extractable acidity

This is a measure (m.e./100 g soil) of acidity released from the soil by a barium chloridetriethanolamine solution buffered at pH 8.2. It includes all the acidity generated by replacement of the hydrogen and aluminium from permanent and pH-dependent exchange sites.

Extractable aluminium

This measures the exchangeable aluminium, which is a major constituent of strongly acid soils. The aluminium is extracted with 1N KCl solution.

Cation-exchange capacity (CEC)

The ammonium acetate (in neutral pH) determined value is an analytical value, and is reported as m.e./100 g soil. However, ECEC (effective CEC) derived by summation of all the extractable bases and aluminium (extracted by KCl) is more applicable for acid soils, as it closely represents CEC at the soil pH.

Parameters for soil fertility

In experiments on soil management, especially for the evaluation of sustainability, there is a need to monitor physical and chemical parameters. This paper is confined to considerations of the latter parameters, and notably those of direct relevance - pH, organic carbon, nitrogen, exchangeable bases (Ca, K and Mg), and available phosphorus.

The analytical methods used for all the nutrients, except the method for analyzing phosphorus, have been briefly discussed earlier. For acid soils in Asia, most countries have established a relationship between available P (as determined by Bray II extraction, NHF + HCl) and plant indices, and thus use this as the preferred method. This continues to be used as the main analytical method for P. However, as Olsen's extraction is used in other networks, it would be useful to include this extraction in order to allow a comparison and exchange of results between networks.

General remarks

The above are general guidelines on the parameters to be assessed. However, there are variations on what is considered a "minimum data set" in each of the networks.

In the analysis there will be some degree of variation in methods and automation procedures, which will lead to variations. It is therefore imperative that each of the procedures should be documented.

Reference

USDA (U.S. Department of Agriculture). 1984. *Soil survey laboratory methods and procedures for collecting soil samples*. Soil Conservation Services, U.S. Department of Agriculture. Washington, DC: US Government Printing Office.

Chemical and physical analysis for monitoring sustainability

NUALSRI KANCHANAKOOL*

Abstract

The paper outlines various factors which affect the analysis of the chemical and physical parameters of soils. Procedures in sampling, handling, and preparation of samples are among the more important factors influencing the accuracy of results. The need for calibration of the analytical methods by relating them to field experiments, as well as the use of standardized procedures are also emphasized. Quality control by using a number of standard samples, running duplicate analyses, and conducting interlaboratory cross-checks are procedures which are recommended to assure the reliability of the analytical results produced in a laboratory.

Introduction

Soil analyses constitute a means of monitoring fertility, and enable proper decisions to be made on appropriate management and fertilizer inputs. However, many agronomists and soil specialists often express dissatisfaction with soil analyses, probably because they expect too much from the analyses. It should therefore be kept in mind from the beginning that soil analysis by itself has a limited usage, and needs to be correlated with other factors such as climate, type of plants, cultivation practices, and nutritional status. These are

* Division of Soil Analysis, Department of Land Development, Bangkhen, Bangkok 10900, Thailand.

factors which influence production. Only with such relationships can one obtain valid recommendations.

It may be thought that it is a simple matter to make a soil test, but in fact it can be a complicated procedure if useful information is to be extracted from the analytical data.. Various factors must be taken into consideration in order to obtain reliable data.

Purpose of the analysis

Whenever there is a need to analyze soil samples for their chemical and physical properties, the first question that any soil scientist should ask himself is what the data be used for. Generally, soil samples are analyzed for the following purposes:

- for soil classification in accordance with the categories established in *Soil Taxonomy* (Soil Survey Staff, 1975);
- for evaluating the fertility status of the soil, to allow appropriate decisions on fertilizer use and soil management to be made; and
- for research.

Before initiating any work and sending samples to the laboratory, soil scientists and agronomists should consider carefully what they expect from the laboratory. They should be able to give the information and to explain their needs to the analyst. A clear understanding of these needs by the analyst will result in more reliable and usable data. One should keep in mind that the analyst is not the same person who collects the samples or works in the field. He is often carrying out research of his own, and qenerally he does not know the field conditions or the problems in the field. If he has to make a decision on his own on what is to be analyzed, it is certain that he will do so in accordance with the information he has to hand. For example, if soil samples for classification are sent to a laboratory without any specific order or detailed information, the analyst will start by carrying out a general routine analysis of chemical and physical properties needed for classification purposes. Although certain parameters are analysed routinely, many of the properties need to be analyzed only for certain soils and for different reasons, e.g. the amount of iron, and soluble salts, and the SAR ratio.

In order to understand the reasons for the analysis clearly, good contacts should exist between the soil surveyor and the analyst. A soil surveyor ought to know what information is needed to classify each soil. Further, the analyst, for his part, ought to understand or have some background of soil survey operations - sufficient at least to help make decisions on what other properties need to be looked for in addition to routine parameters so that the data produced will be of greater use.

The same also applies to the field of agronomy. In many developed countries, a research agronomist normally carries out the whole project on his own. He usually studies the problems himself, lists the purposes, decides on the methods, runs the experiments, collects the samples, carries out the analyses, and interprets the data. This work is conducted by one agronomist, or under the supervision of one agronomist. In this case, there is no problem for the analyst because the agronomist is his own analyst.

However, in many developing countries - in Thailand for instance - the approach is different. When an agronomist initiates a research project, he selects his topic, outlines the purposes and methods, forms a team, runs the experiment, and (when required) collects the samples and sends the samples to a laboratory. Problems may arise at this stage, as the analyst is often not aware what the purpose of the study was. To avoid this, there should be consultation between the analyst and agronomist on the objective of the research even before the study commences.

Sample collection

The quality of the soil samples collected will influence the value of the analytical results obtained. Therefore they should be carefully collected, with a view to providing a good representation of the whole field. The errors due to improper sampling are larger than those encountered in the chemical analysis as such.

In order to ensure that the results of the soil chemical analysis will lead to the most accurate conclusions, it is necessary to carry out preliminary studies on the soil texture, bulk density, and porosity, and also on the general cultivation conditions. The method and intensity of sample collection for different purposes differ. The different purposes include sample collection for classification, for fertility studies, or for making fertilizer recommendations. Therefore samples are always collected according to a pre-established plan.

For fertility studies, a collection of soil samples should include the following steps:

- set up a sampling plan (after the site has been studied and selected) e.g. divide the studied area into analytical units;
- collect subsamples from each unit;
- obtain a mean sample from the subsamples collected (by quartering - a sampling method); and

- put the samples obtained in containers, e.g. plastic bags) and label the samples.

The number of samples collected depends on the purpose of the analysis and also on the conditions of the area studied (e.g. slope, shape of the field, soil uniformity, and utilization of the soil). For general purposes, 15-20 subsamples from one analytical unit is considered satisfactory. For a detailed scientific study, the number of mean samples collected can be increased to 30-50.

The experimental station in Illinois recommends that complex samples should be collected. Eleven samples are taken from an analytical unit of 30 ha, each of them composed of 5 partial samples. It appears that this system is satisfactory enough for field cultures (Davidescu and Davidscu, 1972).

Sample preparation

One factor which has an influence on the analysis is sample preparation. Samples which are collected properly, if not carefully prepared, may not give reliable data. Samples should be air-dried in a clean, well-aerated place, and away from traffic (both living and nonliving). The temperature of the drying room should be kept constant. Vessels for drying samples must be clean. Large materials like big roots, stone, etc. have to be removed. Larger clods are broken with clean hands. For determinations of inorganic forms of nitrogen, for example, soil samples must be analyzed immediately after sampling, because these N forms are easily transformed.

It is difficult to avoid some delay in analysis, because the samples usually have to be transported to the laboratory, and sometimes they may have to be stored for some time before the analysis can be initiated. This causes many problems for laboratory chemists. It is known that in the course of the drying, modifications due to microbial activities or physicochemical tranformations occur which may lead to changes in some agrochemical properties. Hence various reagents (toluene or chloroform, for example) have been used to prevent microbial activities during transportation and storage.

Many workers have used deep-freeze storage to preserve samples instead of using the air-dried process. However, according to a report Allen and Grimshaw (1962), low-temperature storage is not a satisfactory method for preserving soil samples for the determination of inorganic forms of nitrogen. Some other workers have recommended that if analyses for inorganic nitrogen cannot be carried out as soon as the soil samples reach the laboratory, the samples should be dried very quickly in an oven at 55°C. This procedure, which involves drying samples at a lower temperature, is widely used, although many workers have found significant changes in the amount of ammonium and nitrate nitrogen in soils prepared by this method.

For the determination of physical properties, a knowledge of the purpose of the analysis will help in the preparation of the soil samples. To determine the bulk density of soils, for example, some laboratories use the clod method. In this method, soil clods have to be selected out of the whole range of soil samples before preparation.

The precautions noted above are very important to all the subsequent procedures.

Methods of analysis

Large numbers of analytical methods for soil chemical and physical properties have been introduced. However, much uncertainly still persists, and it is nearly impossible to find a unique method of analysis which would permit characterizing fertility and establishing fertilizer requirements under all conditions because of the numerous variable factors. In deciding a suitable form of analysis for any particular laboratory, the following considerations should be borne in mind:

- methods should be as simple and accurate as possible, depending on the availability of chemicals, utilities, efficient instruments, and maintenance services, but at the same time must satisfy the needs of the project;
- the country concerned;
- the consistency of the results should be regularly checked by means of reference methods or by participating in sample exchange programmes such as the Labex programme;
- the correlation between soil analytical results and tested plants should be studied systematically; and
- costs should be kept as low as possible.

The minimum number of chemical determinations for soil fertility required depends on local conditions. In general, an investigation of soil pH, contents of phosphorus, potassium, and soil organic matter should be undertaken. In areas where deficiencies of nutrients or toxicities are suspected, or lime requirements are evident, the analyses should include buffer pH, exchangeable acidity, contests of magnesium, calcium, and micronutrients. In some soils where salt deposits are suspected, salinity and soluble salts should also be determined.

Variation of the results

In soil analysis there is always the problem of the variation in results of the same soil samples. The variation may be due to different methods of analysis, different analysts, or different instruments.

Due to methods

As mentioned earlier, a large number of methods have been introduced. The differences either concern differences in the weight of the soil samples used, the ratio of soil solution, the type and concentration of the extractant, the pH of the system, the shaking time, or the leaching process. All of these have an influence on the analysis. For example, in measuring the bulk density of soils using core and clod methods, the DLD laboratory obtained the following results (Table 1).

Table 1. Variations in bulk density due to methods of determination.

Sample code no.	Depth (cm)	Bulk density (by core)	Bulk density (by clods)	Moisture content (% by weight)
1A/1	0-15	1.47	1.71	11.72
1B/1	15-30	1.57	1.82	10.68
1C/1	30-60	1.60	1.73	10.29
3A/1	0-15	1.76	1.88	11.55
3B/1	15-30	1.62	1.78	11.09
3C/1	30-60	1.58	1.74	11.46
4A/1	0-15	1.57	1.72	12.21
4B/1	15-30	1.65	1.89	11.97
4C/1	30-60	1.70	1.84	11.42
9A/1	0-15	1.76	1.77	11.50
9B/1	15-30	1.81	1.89	10.75
9C/1	30-60	1.94	2.06	6.8

Source: DLD laboratory.

Another example is the determination of phosphorus, for which many methods have been proposed. Each method has different advantages. For example, NH_4-F extraction (Bray 1 or 2 is appropriate for soils dominant in Al-

P; $NaHCO_3$ (pH 8.5) of Olsen promotes the hydrolysis of Al and Fe phosphates , and also activates the release of P in Ca-phosphates - so this method specifically releases P from these three forms. A double-acid solution of Mehlich dissolves some of the Al and Fe-P, and large amounts of Ca-P (Fassbender, 1977). Methods which are widely used are Bray and Kurtz (either Bray 1 or 2), Olsen, and modified Olsen. It is certain that different methods give different results (Table 2).

Table 2. Effect of extraction methods on P extracted.

Soil name	Particle-size class	pH 1:1 H_2O	ppm-P extracted by			
			Bray 1	Bray 2	Olsen	Mehlich
Sattahip (Typic Quartzisomment)	Sandy	6.3	25.71	36.31	17.35	23.91
Lopburi (Typic Pellusterts)	Fine clayey	7.6	6.63	116.90	49.6	2.87

Source: DLD-laboratory.

The analyst cannot make a decision on which method should be used unless he or she is told about the soil and crop, and an appropriate method is indicated by the person requesting the analysis.

For example, extraction of P by the Olsen method is satisfactory for soils with a pH above 7.5, while Bray's extraction is suitable for soils with lower pH's, and Mehlich 3 (Melich, 1984) for use on all soils (Kimble and Holzhey, 1988). A careful study in selecting methods which are most fitted to the condition of each situation is necessary.

Further, to ensure that the proper methods are selected, the calibration of the analytical methods for soil and plants by field experiments is widely recommended as a prerequisite. However, if under similar ecological and management conditions such calibration has already been done, the methods could perhaps be used without any further calibration. After all, whatever method is used, the procedure or any modification in the procedure will always have to be stated.

Due to analysts

Different analysts may produce different results for the same samples, supposedly analyzed using identical methods. Their techniques, carefulness, attention, and experience are all involved. From the time samples are weighed, they are usually kept in boxes or plastic bottles. It would be preferable if the samples were to be subsampled once again or at least well mixed before the weighing for analysis starts, because they are usually mixed unevenly when left in the containers.

Titration is another step where errors can occur. In some laboratories where good apparatus is available man-made errors can be easily eliminated. However, in many laboratories, the accuracy of the titration depends largely on the analyst's own experience and the sensitivity of his eyes to the final colour at the end point.

In measuring the pH of a soil, a different result may be obtained, depending on whether the electrodes are dipped in the suspension, in the clear solution, or in the soil after settlement.

Due to instruments

Another factor which may have an influence on the results of the analysis is the instrument itself. Where there are a large number of samples to be analyzed routinely, sophisticated equipment (e.g. atomic absorption apparatus, a spectrophotometer, and an autoanalyzer) are required for more accurate and time-saving work. However, the damp atmosphere and heat of the humid tropics can easily cause damage to these instruments unless the operation rooms are air-conditioned and dehumidified. Since these instruments are so expensive, many laboratories in developing countries do not have an adequate budget to replace them. Therefore no matter how old or inappropriate the instruments are, the analyst must do the best be can with them. The variation of results from faulty instruments can be wide. In a measurement of the pH of soil by the DLD laboratory using four different-pH-meters, the results obtained were as shown in Table 3.

Table 3. Variation in soil pH due to different instruments.

Brand of pH-meter used	pH of soil samles (average from 10 soils)		
	Measurement in suspension	Measurement in solution	Measurement in soil after settlement
1. A Fisher (18 yrs old)	5.745[a]	5.775[a]	5.750[a]
2. A Corning (5 yrs old)	5.827[a]	5.707[a]	5.702[a]
3. A Beckman (12 yrs old)	5.707[a]	5.732[a]	5.742[a]
4. A Beckman (5 yrs old)	5.450[b]	5.399[b]	5.485[b]
lsd 0.01	0.251	0.289	0.222
lsd 0.05	0.186	0.214	0.164
F-values	6.479**	4.797*	4.885*

Source: DLD-laboratory.

Laboratory quality control and background knowledge of the analysts.

A quality control programme is important for establishing and maintaining a fair level of accuracy and precision within the laboratory. The equipment, methods, and techniques used should be standardized. Duplication (where necessary), or check samples for a set of 10 samples are widely used. Cross-checking by nearby laboratories, or participation in a laboratory exchange programme, will be very useful.

The analyst should have a background knowledge in soil science and plants, and be aware of new developments. Of course, he will probably gain some in-service training as part of his daily routine work; but a short period of training specifically in his field will help him acquire a better understanding of the work, and encourage his personal interest. Additionally, the analyst could visit the experimental site to see the field conditions at the time of sampling, which would help him to appreciate the results of the analysis within the overall context.

After all, if variations in results do occur in the analysis, what is the use of proper collection and preparation of soil samples? At the same time, as was stated earlier, errors due to improper sampling are larger than those encountered in the chemical analysis.

References

ALLEN, S.E. and GRIMSHAW, H.M. 1962. Effect of low temperature storage on the extractable nutrient ions in soils. *Journal of the Science of Food and Agriculture* 13: 525-529.

DAVIDESCU, D. and DAVIDESCU, V. 1972. *Evaluation of fertility by plant and soil analysis*. England, UK: Abacus Press.

FASSBENDER, H.W. 1977. *Problems connected with soil testing for phosphorus and potassium*. FAO Soils Bulletin no. 38/1. Rome: FAO.

KIMBLE, J.M. and HOLZHEY, C.S. 1988. Chemical properties to be measured and monitored for soil management experiments. *First Training Workshop on Site Selection and Characterization*, 181-193. IBSRAM Technical Notes no. 1. Bangkok: IBSRAM.

MEHLICH, A. 1984. Mehlich 3 - soil test extractant: a modification of Mehlich 2 extractant. *Soil Science and Plant Analysis* 15(12):1409-1416.

SOIL SURVEY STAFF. 1975. *Soil taxonomy: A basic system of soil classification for making and interpreting soil surveys*. Soil Consevation Service, U.S. Department of Agriculture. Agriculture Handbook no. 236. Washington DC: Government Printing Office.

Surface crusting, runoff, and erosion on steeplands and coarse material[1]

CHRISTIAN VALENTIN*

Abstract

Land clearing in mountainous regions raises serious problems of water and soil conservation. There is little information available on runoff and erosion on steep soils with a high proportion of coarse material. The rare studies undertaken on the subject indicate that beyond slopes of 35-40% other processes are involved, so that methods applicable on lowland cannot be easily used on steep soils. The universal soil loss equation overestimates erosion risk on steep soils. Runoff is often stabilized and may even be reduced as the slope increases, since there is a lower occurrence of crusting. Surface coarse material may check erosion on low slopes, but tends to increase it on steep slopes. The influence of coarse material on runoff depends on its position in relation to the surface crust. Due to the interaction of a large number of parameters, models established for almost flat soils without coarse material should not be extrapolated for mountain conditions. Observations and measurements should be carried out on

---

1 Translated from the French by Mira Fischer.

* Institut Français de Recherche Scientifique pour le Développement en Coopération (ORSTOM), BP V. 51, Abidjan, Côte d'Ivoire.

site and should monitor changes in the surface crust, which is a key factor for understanding the processes.

Introduction

Increasing food requirements in several tropical countries have led to the clearing of marginal lands. Mountain regions are particularly exposed to rapid soil degradation (Latham, 1988). As they are situated above catchment basins, they produce runoff that can be disastrous to downstream fields and settlements (Kunkle and Harcharik, 1977). The traditional technique of terracing is an effective solution, but it is not always practiced by new settlers. Such expertise is in fact gradually disappearing as Noni *et al.*. (1986) observed in the Andes in Ecuador. This disquieting situation calls for an immediate implementation of a soil management approach that is suited to these parti-cular conditions. This is the objective of IBSRAM's first network operating in Asia.

This paper aims:

- to present the principles governing runoff and erosion on mountain soils that are characterized by steep slopes and coarse material on the surface;
- to show the influence of surface crusting on runoff occurrence in such an environment;
- to indicate different techniques for studying the crusting tendency of a soil and for measuring infiltrability.

Influence of slope gradient on soil and water loss

Effect of slope gradient on erosion

Physical factors

Two mechanisms can cause the removal of soil particles:

- *Splashing*: If i is the slope gradient, then the impact of the drops varies according to cos (i), with a maximum on horizontal soils. But splashing does not have a substantial effect if $i = 45°$ (or 100%), when the decrease due to cos (i) is only 29%.
- *Runoff*: Its kinetic energy increases with the square of its velocity, depending on the vertical acceleration component g, and therefore on sine^2 (i). Kinetic energy of runoff increases with sine^2 (i).

For a given intensity of runoff, the thickness of the film of water decreases as the gradient increases, soil roughness being the same and splashing intensity decreases when the thickness of the film of water increases (Palmer, 1965; Ghadiri and Payne, 1977; Poesen and Savat, 1981). An increase in the gradient thus has an ambivalent effect on splashing. It is therefore difficult to construct a model exclusively on the effect of the slope gradient on erosion. Empirical equations have been proposed to overcome this difficulty.

Empirical equations

The influence of the gradient was integrated in the universal soil loss equation (Wischmeier and Smith, 1960) by using a slope factor S. It is the proportion of soil loss on any slope in relation to that on a reference slope of 9%, and is based on the slope gradient, p, expressed as a percentage:

$$S = \frac{0.43 + 0.30\,p + 0.043\,p^2}{6.613}$$

Although the chart established from this equation may appear interesting, it should be used with caution as it can lead to error for two reasons. One of the reasons is that the equation was established on the basis of slopes equal to or less than 18%. The chart was drawn for slopes up to 50%, and any extrapolation beyond this is statistically risky.

The main reason, however, is that it is based on the hypothesis that the effect of the slope gradient is independent of other factors. But a certain number of results have contradicted this hypothesis. Most authors who have studied the influence of slope gradient on soil loss E have obtained the following type of equation:

$$E = f\,(p^m)$$

where m has values close to 1.4 (Duley and Hayes, 1932; Zingg, 1940). However, m can vary considerably according to several factors:

- *Slope* (Horvath and Erodi, 1962):

  $m = 1.6$ for $0\% < p < 5\%$

  $m = 0.7$ for $5\% < p < 11\%$

  $m = 0.4$ for $p > 11\%$

  Beyond 35-40%, erosion becomes constant (Singer and Blackard, 1982) and $m = 0$.

  These results are useful for interpreting those obtained by authors. In Senegal, Fournier (1967) showed that soil loss doubled for an increase of p from 1.0 to 1.5%. Using results from 14 tilled sandy plots distributed throughout West Africa, Collinet (1988) reported an exponential increase on slopes of between 0.9 and 6.5%. Gussak (1937), cited by Zachar (1982), working on four steeper slopes (between 9 and 58%) in the USSR obtained

m = 0.4, which is a lower increase rate. From these studies, it can be concluded that:

- the erosion increase rate is lower as the slope gets steeper.
- the use of the universal soil loss equation for steep slopes leads to overestimation of erosion (Hart, 1984; De Ploey and Bryan, 1986).

Soil. The value of m is higher as soil texture becomes coarser. In the laboratory, Gabriels *et al.* (1975) showed that m varied from 0.6 to 1.7 when the particle size increased from 0.05 mm to 1 mm. Similarly, Singer and Blackard (1982) observed a more marked influence of the slope for a soil with 40% sand compared to another with only 3% sand (Figure 1). The authors attribute the difference to the better shear strength of more clayey soils. During field measurements it is very difficult to carry out experiments on soils that only differ in terms of slope. Since slope may be a major differentiating factor in a toposequence, under natural conditions the topographic factor LS cannot be considered independently of the soil factor K.

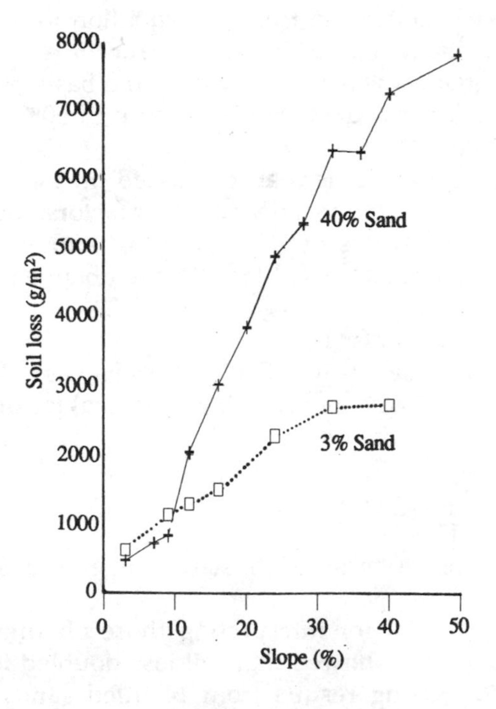

Figure 1. The influence of soil texture on the effect of the slope on soil loss, in accordance with data from Singer and Blackard (1982).

- *Climatic conditions.* As rainfall is more intense in the tropics, Hudson (1973) suggests using $m = 2$. The *LS* factor is thus also linked to the climatic factor *R*.
- *Types of land use.* In Nigeria, Lal (1985) demonstrated that on ferrallitic soils on moderate slopes, slope gradient had no influence on soil loss ($m = 0$) when the soil is mulched. On steep slopes in Trinidad, Gumbs *et al.* (1986) obtained similar results for maize and cowpea crops (Figure 2). The topographic factor *LS* is thus also linked to the crop management factor *C*.

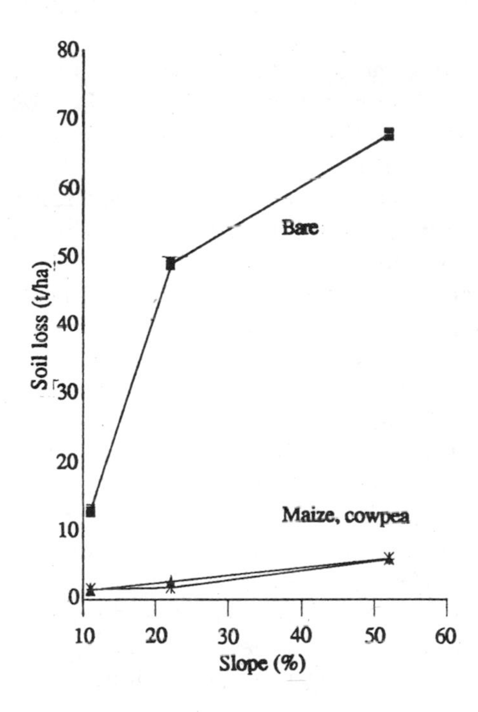

Figure 2. The influence of the soil cover on the effect of the slope on soil loss, in accordance with data from Gumbs *et al.* (1986).

Influence of slope on water loss

Apart from the plant cover, two important factors determine the time of runoff and its intensity: slope gradient and soil infiltrability. The second variable is mainly determined by the existence of a crust or seal on the soil surface. Crusting intensity tends to decrease runoff beyond a certain threshold gradient, which in turn depends on the soil type. There is interaction between several phenomena:

- In the case of heavy vertical rainfall, rainfall intensity and its impact force decrease as the gradient increases. For the same intensity and area, the number of drops reaching the soil decreases, and their impact decreases as a cosine of the gradient. It has been indicated earlier that there is no substantial effect.
- On steep slopes, runoff velocity is sufficient to carry away the particles as they detach from the surface, thus reducing the possibility of crust formation.
- The surface roughness, mainly formed by linear incisions, increases with the gradient so that runoff is reduced. This is because the infiltration area increases and the bottom and walls of the rills have better infiltrability than the interrill area.

These theoretical considerations have been confirmed by experiments. In Côte d'Ivoire, runoff decreased from 35 to 24% when the gradient increased from 4.5 to 23% on 10 m-long bare plots under natural rainfall (Roose, 1973). This reduction in runoff beyond a certain threshold was also reported in Trinidad (Gumbs *et al.*, 1986). In catchment areas of 10 km^2 or more in savanna, runoff increased with the gradient up to about 2.5%, and stabilized beyond this gradient limit (Rodier and Riebstein, in press).

The gradient factor can be studied exclusively in the laboratory under controlled conditions with rainfall simulation. Very slight modifications in runoff volume were indicated for gradients ranging from 3 to 50% (Singer and Blackard, 1982) (Figure 1). On material prone to crusting and gradients up to 60%, the influence of the gradient on the time to runoff closely depends on the resistance of the surface crust to erosion (De Ploey and Bryan, 1986). Similarly, in laboratory experiments, reduced crusting decreased runoff when the gradient was increased from 2 to 15% (Poesen, 1986a). These results demonstrate that due to the decreased crusting risk, runoff does not systematically increase with the gradient.

Effect of surface coarse material on runoff and erosion

Effect on erosion

Mountain soils are often not well developed and the surface frequently carries coarse fragments. Studies on the subject demonstrate that coarse material on the soil surface provides effective protection. In Israel, on terraces with cross slopes between 5 and 17%, erosion and runoff under simulated rainfall ceased when coarse fragments on the soil surface exceeded 60% (Seginer, *et al.*, 1962). Moreover, studies on the correlation between erodibility K, or the coefficient C in the universal soil loss equation and the percentage of coarse material on the soil surface G revealed exponentially decreasing functions:

- In Tunisia, using results from 30 plots with simulated rainfall, Dumas (1965) obtained:

$$K = 0.3855\ 10^{-0.0294\,G},\ r^2 = 0.69$$

- In West Africa, 17 1-m^2 plots, dispersed across the area from the Sahara to the tropical forest zones were studied under simulated rainfall, and the following equation was established (Collinet and Valentin, 1985):

$$C = 0.73 \exp(-0.06\,G), r^2 = 0.96$$

- A similar relation was found by Collinet (1988) in the same climatic zone on 13 longer plots (10 m):

$$C = 0.78 \exp(-0.06\,G), r^2 = 0.84$$

- An exponential type of curve established in the USA showed that the decrease is less marked than for arid or tropical soils (Box and Meyer, 1984). Soil loss with 60% coarse material on the surface is nil in Israel, 0.7% (compared with a bare soil) in Tunisia, and 2.1% in West Africa, whereas it can be as high as 22% in the USA. This demonstrates once again that the different variables of the universal soil loss equation are related. The protective effect of coarse material is stronger as rainfall intensity increases.

  This effect varies according to the gradient (De Ploey, 1981):

  o On less steep gradients (2°), marbles used to simulate coarse fragments thoroughly protected the soil against the kinetic energy of rainfall.

  o On steep slopes (16°), the marbles on the soil surface had the opposite effect of increasing soil loss. Under such conditions, the more rapid water flow tends to become more turbulent below each obstacle, causing cavity scouring. As a result, the same percentage of coarse material on the same slope can have a different effect on erosion, depending on the gradient.

Effect on runoff

Considering the physical processes, the effect of coarse material on runoff can also be ambivalent.

As coarse material is almost impermeable, it instantly causes runoff which is concentrated on bare soil (De Ploey, 1981). Runoff can therefore occur on stony soils even after a few millimetres of rain. This was observed in Israel (Yair and Klein, 1973) and in the West African arid and subarid zones (Casenave and Valentin, 1988).

At the same time, coarse fragments encourage infiltration by forming obstacles that reduce runoff velocity and thus increase water depth. This effect was observed on ferrallitic gravelly soils (Collinet and Valentin, 1979; Collinet, 1988) (Plate 1).

Plate 1. Mixed cropping with maize and yam on ridges in nothern Côte d'Ivoire. The abundance of gravelly soil in the tilled horizon reduces runoff and erosion.

The effect of gravel on runoff cannot be completely predicted by simply considering the percentage of surface gravel. The position within the surface structure is an important factor that cannot be ignored. In laboratory experiments, runoff on soils prone to crusting was greater when the gravel was half-embedded in the soil than when it was only laid on top (Poesen, 1986a). These observations were confirmed by results obtained in arid and semiarid zones in West Africa. Using results obtained from field experiments on plots under simulated rainfall, Casenave and Valentin (1988) distinguished five types of soil surfaces with more than 40% coarse material (Table 1). In the most common type, the coarse fragments were embedded in a thick crust made up of three microhorizons (Figure 3). The three following variations appeared in plots close to more humid regions:

- surface with intermittent crusting;
- surface with almost nonexistent crust and with coarse material usually laid on top;
- surface without crust, but with gravel that may form several layers (Collinet, 1988).

Table 1. The influence of the position of coarse fragments in relation to the soil surface on the main parameters of runoff and infiltration.

Parameters	TYPE OF SURFACE FEATURES				
	1	2	3	4	5
K_i (%), dry	5-20	20-35	50-60	70-80	90-100
K_i (%), wet	3-15	15-25	35-50	60-75	85-95
P_i (mm), dry	*1.5-5*	*5-10*	*10-15*	*15-25*	*10-20*
P_i (mm), wet	0-3	0-5	3-7	10-20	5-15
I_1 (mm/h)	0-2	3-5	7-12	10-20	20-30]

K_i : infilltration ratio.

P_i : preponding rainfall - the ranges of these two parameters are given for the initial dry, humid conditions.

I_1 : the minimum infiltration rate under near-saturated conditions. The different states of the soil surface are defined in the text.

In addition, there is also a particular type of soil surface that is found on hills developed on basal rocks. In this case the coarse material is spread on a well-structured and highly porous soil.

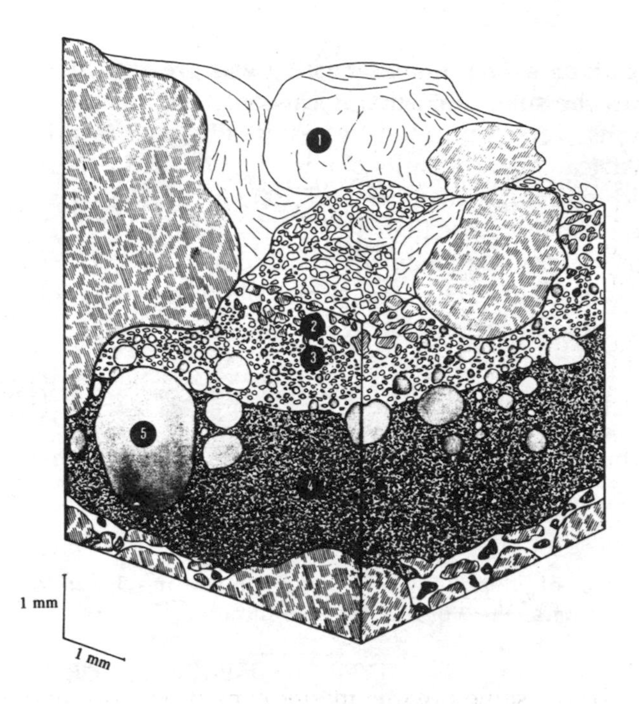

Figure 3. The most frequent type of crust with coarse material in dry areas. The coarse fragments, *1*, are present in the crust in three microhorizons: (a) as coarse, loose sand, *2*; (b) as fine, cemented sand, *3*; as plasma. The cavities tend to proup between microhorizons 2 and 3.

Study of crusting

In the earlier section, crusting was shown to be an important factor governing runoff and erosion on mountain soils. However, the phenomenon of crusting has not been sufficiently examined and taught. A preliminary report (Valentin, 1988) on a certain number of approaches to this problem was prepared to make up for this deficiency. In this paper, the influence of certain intrinsic factors on structural stability will be discussed before describing techniques for field studies.

Intrinsic factors

While studying surface crusting, the first question to be asked is whether the occurrence of crusting can be assessed from analytical data. A structural instability index (Henin and Monnier, 1958b) is used to demonstrate how crusting risk varies over time and across locations.

Variation over time: the role of organic matter

In the Central African Republic, structural instability is lowest during the dry season; it is higher in cropped soils compared with savanna and fallow soils (Combeau and Quantin, 1963) (Figure 4). In Cote d'Ivoire, on ferrallitic soils of the humid savanna, structural instability increases during the cropping season and decreases during the fallow period; it returns to the natural savanna level after about 10 years (Figure 5). Seasonal variations can be attributed to water repellency by organic matter, which increases after a long period of drought (Sebillotte, 1968; Boiffin, 1976). The type of organic matter, rather than the percentage, influences changes induced by soil management practices (Monnier, 1985). Organic matter has a favourable influence only beyond a certain value, equivalent to 7 of the following ratio:

$$\frac{\%\ \text{organic matter} \times 100}{\%\ \text{clay}}$$

Stability increase for the same organic-matter content is more marked in sandy soils than in clayey soils.

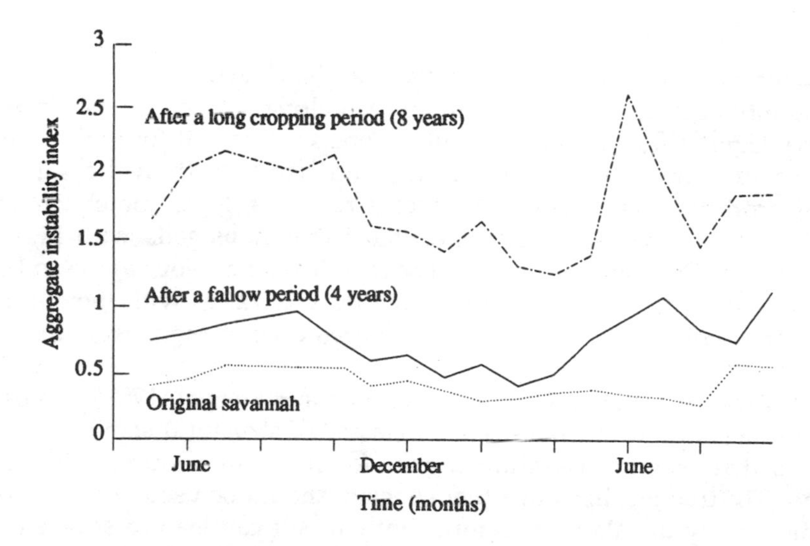

Figure 4. Seasonal variations in structural instability (Central African Republic). Note the decrease in structural instability during the dry season (Conbeau et Quantin, 1963).

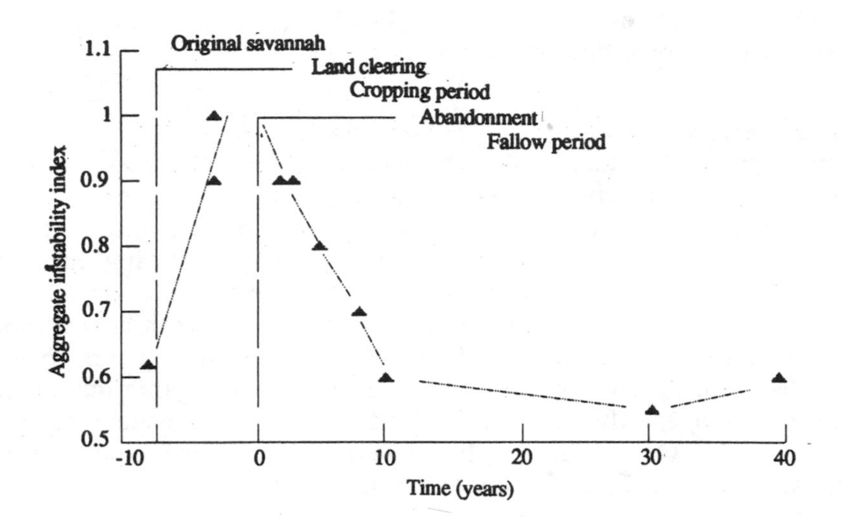

Figure 5. The development of structural instability at the commencement of cultivation and after the beginning of the fallow in a humid savanna region of Côte d'Ivoire.
Note the reccurrence of structural stability after ten years of fallow (Valentin, unpubished data).

Variation across locations: the influence of soil texture

The influence of soil texture can be seen along a toposequence in a humid savanna region (Valentin and Janeau, in press). The soil formation processes have led to a gradual leaching of clay and hematite from the soil surface, starting from the top of the slope (well-structured, ferrallitic clayey soils) to the edge of the lowlands (hydromorphic ferruginous soils with high coarse sand content). The differentiation can be seen from the colour of the soil surface, which is related to the presence of hematite and varies with the soil texture. It is a useful criterion for classifying soils according to their structural stability (Figure 6).

Using results obtained in France, Monnier and Stengel (1982) propose a soil texture triangle that integrates several classes of structural stability (Figure 7). It shows that crusting risks are higher for medium textures with high silt content. The triangle, like any other concept, should be used with caution. For example, highly unbalanced textures without silt can lead to severe crusting, which is not predicted in the diagram (Figure 7). In another instance, marked crusting mixes made up of 90% sand and 10% silt show the greatest infiltrability decreases under simulated rainfall during laboratory experiments (Poesen, 1986b).

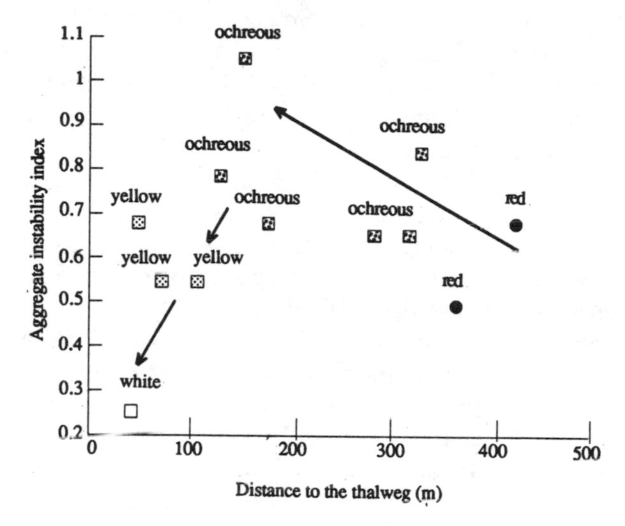

Figure 6. Variation of structural stability along a toposequence in a humid savanna area of Côte d'Ivoire.
Note that in this case the colour is a relevant indicatior of the physical behaviour of soils (Valentin et Janeau, in press).

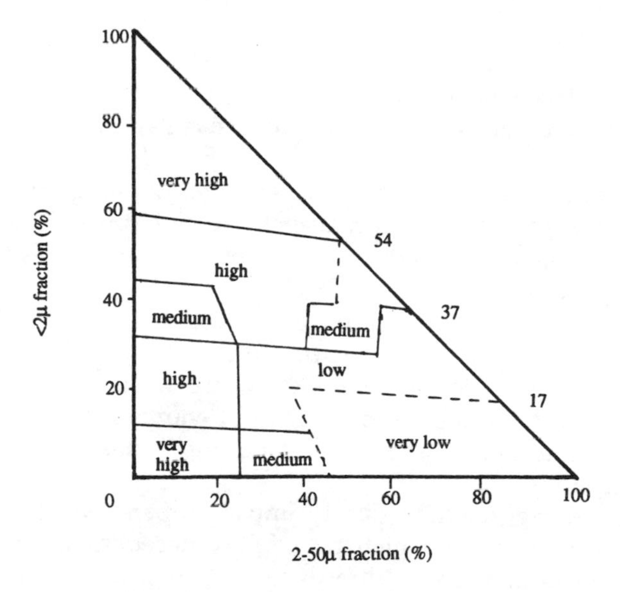

Figure 7. Structural stability classes in relation to texture (Monnier et Stengel, 1982).

Intrinsic factors other than coarse material and gradient that influence structural stability are:

- *Mineralogic composition of clays.* Smectites are more prone to liquefaction and to cracking, that lead to crusting, compared with less active clays such as kaolinite. Moreover, smectite particles have the lowest volume when dry (Tessier, 1984; Bruand, 1985), so that the crust has the highest resistance to penetration (Lemos and Lutz, 1957).
- *Chemical composition and CEC of the clay phase.* Iron (Farres, 1987) and aluminium (Deshpande, *et al.*, 1964) are two elements that ensure soil aggregate stability. Calcium in the form of carbonates is less effective than gypsum (Ben-Hur *et al.*, 1985). Magnesium has an adverse effect once it exceeds 50% of the total bases (Collinet, 1988).

Exchangeable sodium contents exceeding 15-20% facilitate clay dispersion and thus encourage crust formation. This level is lowered to 3-5% under natural high-intensity rainfall for aggregates with weak initial cohesion (Shainberg, 1985). Resistance to penetration is higher in crusts of sodium-rich soils (Alperovitch and Dan, 1973).

Similarly, the presence of soluble salts in certain alluvial soils encourages crust formation (Radwansky, 1968; Mougenot, 1983). Evaporation causes precipitation of salt on the soil surface, producing saline crusts.

Structural stability

Disaggregation mechanisms

Several phenomena can cause soil agregates to collapse. These are due to:

- Wetting
  - *Shattering.* When a dry aggregate is suddenly plunged into water, the liquid phase is subjected to a centripetal suction force. Shattering occurs when atmospheric pressure exceeds the mechanical resistance of the aggregate (Yoder, 1936; Henin, 1938).
  - *Dispersion.* Water causes aggregates to lose their structure by reducing the cohesive forces between the particles.
  - *Cracking.* Differences in swelling during wetting can lead to crack formation and fragmentation. This phenomenon occurs even in a vacuum, and is therefore independent of trapped air.
- Impact of raindrops
  - *Splitting.* Disaggregation due to impact depends on the size of raindrops, the velocity at which they fall, antecedent moisture, and the depth of water on the soil surface.

- o *Splashing*. Raindrops rebound from the soil surface and throw soil praticles over a more or less long distance.
- o *Compacting*. Splitting and splashing absorb only part of the kinetic energy of rainfall. The rest is dissipated through compaction of the soil surface, which makes the soil impermeable and thus causes runoff.

Laboratory tests

For more than 50 years, several structural stability tests have been developed to predict crusting risks from routine laboratory measurements. This approach is now felt to be unsatisfactory (Webb and Coughlan, 1989). Several criticisms have been put forward, the most serious are (Le Bissonnais, 1988):

- Classifications of soil according to these tests should also consider the initial moisture of the aggregates. Dry conditions induce shattering, making it impossible to distinguish the soils clearly. Under wet conditions, due to the mechanical disaggregation (especially if the sample is shaken) classification is more discriminatory, but it differs from the first type (Figure 8).
- As a corollary to the preceding criticism, disaggreation risks are mainly based on the interaction between the kinetic energy of rainfall and the antecedent moisture. Two soils with the same sturctural stability index can behave very differently if one is constantly under wet conditions and the other under very dry conditions.

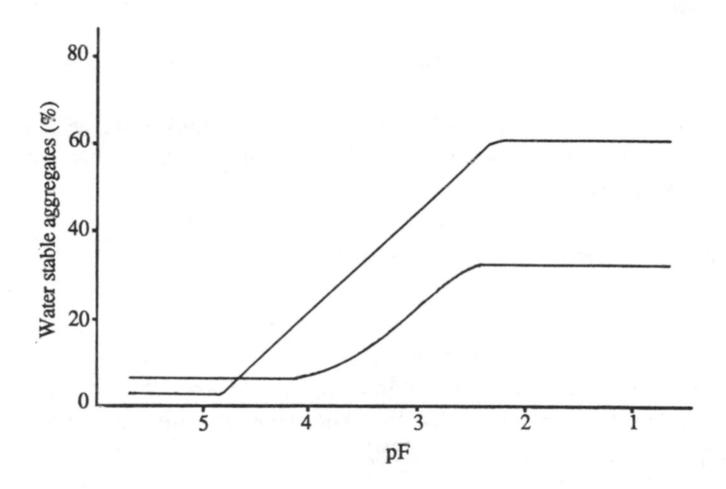

Figure 8. The influence of the initial moisture contents of aggregates before immersion in water when testing for structural instability.
Note that the aggregates are more widely dispersed when they are wet, and that they change their classification (adapted from Le Bissonnais, 1988).

The Henin and Monnier (1956) test, which induces a sudden wetting of predried aggregates, is not suited to temperate conditions where rain often falls on wet soils. It is more suited to tropical conditions where the soil usually dries before the next shower. It was therefore found to be the best of the 14 indices tested to predict erodibility in Nigeria (De Vleeschauwer *et al.*, 1978).

Crusting risks depend on the cohesion of aggregates. The Atterberg consistency tests, or the indices derived from them, can be used as crusting indicators (De Ploey and Mucher, 1981). For example, the liquidity limit was sufficient to remove more than three-fourths of preponding rainfall variance and more than half of the runoff coefficient (both measured under simulated rainfall variance), and more than half of the runoff coefficient (both measured under simulated rainfall), on 20 plots in the wet savanna region (Valentin and Janeau, in press). The indices should perhaps be adapted to climatic conditions (Figure 9).

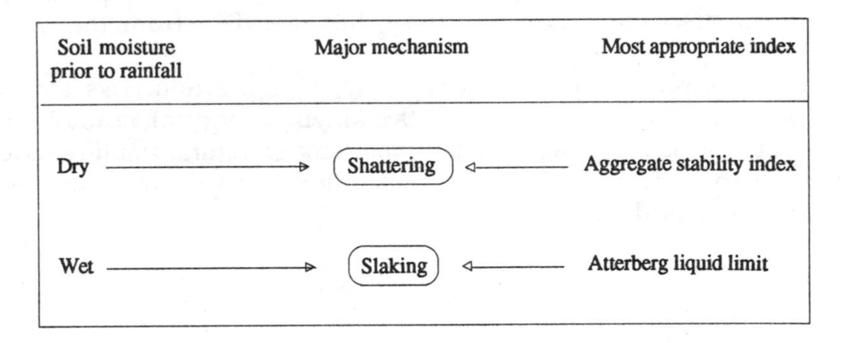

Figure 9. A suggestion for adapting structural behaviour tests with regard to their initial moisture contents - and consequently the most frequent climatic conditions.

Observations and field measurements

Unlike laboratory methods, which are not entirely satisfactory, field measurements can be used to identify crust types and to study their mechanical and hydrodynamic properties. These methods are only briefly indicated in this paper; for further information the reader should refer to an earlier IBSRAM publication (Valentin, 1988).

Crust typology

A crust typology was proposed for the West African arid and semiarid regions (Casenave and Valentin, 1988). It can be applied to other climatic zones, particularly for cultivated soils (Valentin and Camara, 1988). Seven major types of crusts were identified for cultivated soils, based on the nature and number of constituent microhorizons:

- *Drying crust.* This type of crust is characterized by the outcropping of a single sandy microhorizon, with a weak and very fragile massive structure which ranges from a few millimetres to ten or even more millimetres in thickness.
- *Structural crust.* Three types with decreasing infiltrability are distinguished:
  - o With one microhorizon. This crust forms soon during the first rainfall. It is quite thick (10 mm or more) and very rough due to the presence of partially slaked aggregates.
  - o With two microhorizons. The outcropping sandy horizon, often having an unbroken massive structure, covers a plasmic seal made up of fine elements.
  - o With three microhorizons (Figure 10). The textural sequence is the reverse of the sedimentation sequence. At the top is a microhorizon of loose and very coarse sand, then a microhorizon of fine slightly cemented sand with high vesicular porosity; at the bottom is a plasmic seal with the same type of porosity.

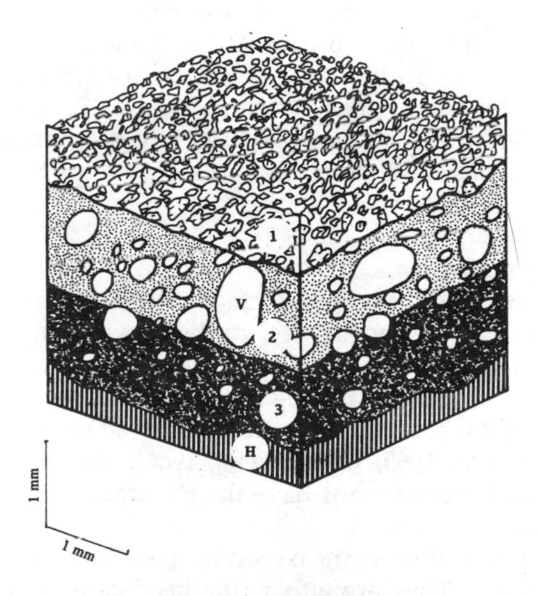

**Figure 10. Structural crust with three microhorizons: *1* - with coarse, loose sand; *2* - with fine, cemented sand,; and *3* - with plasma.
The cavities tend to group between *2* and *3*. The *H* horizon has not been affected by the direct action of wind or rain.**

- *Runoff crust.* This crust is characterized by sandy microhorizons alternating with plasmic seals. It can be several centimeters thick, especially between ridges. It is formed under flowing water.
- *Erosion crust* (Figure 11). This type of crust results from changes in the structural crusts that are subjected to water and wind erosion. Often, only a fine, very hard and impermeable plasmic seal remains on the surface. Porosity is frequently vesicular, whereas remainders of the sandy microhorizons are found in cracks and crevices.
- *Depositional crust* (Figure 12). This type of crust results from the deposition of sediments in a stagnant liquid. The difference in water tension between the fine microhorizons and the underlying sandy microhorizons leads to cracking, cleaving, and curling.

Formation processes

It is useful to characterize the initial state of a seedbed prior to the rains by determining the distribution of clod diameter. The diameter of the smallest clod that is still not embedded in the structural crust is monitored throughout the rainy season; it is thus possible to assess the percentage of the surface that is already occupied by the crust, without taking pictures in sequences, and using a surface integrator to determine the percentage of crusted area (Boiffin, 1986). The characteristic slope coefficient can be estimated from the linear regression of the diameter D_{min} in relation to cumulative rainfall or the kinetic energy of rainfall; it then serves as an indicator of the crusting rate. The crusting rate should, however, be distinguished from crusting intensity. The second paramenter is represented by mechanical and hydrodynamic properties and is related to crust thickness and therefore to the crust type (Valentin and Janeau, in press).

Resistance to penetration

One of the first constraints of a surface crust to a crop is its mechanical obstruction at plant emergence. Several studies have been carried out on these mechanical properties, including tables of pressure exerted by major crops at emergence (Goyal *et al.*, 1980). Crusting begins to have an effect on emergence at pressures exceeding 3-6 bars. Emergence is completely inhibited beyond 12-16 bars (Parker and Taylor, 1965; Taylor, *et al.*, 1966). As seed strength is related to seed weight, small-seeded crops have the maximum difficulty at emergence (Jensen, *et al.*, 1972).

Resistance to penetration is measured by penetrometers, both in the field and in the laboratory. They are either fine needles that are pierced into the soil at constant speed, and the force required is measured, or spring dynamometers. Due to the small cross section of the needles and penetrometer heads, the measurements are repeated a great number of times (up to 50 to obtain

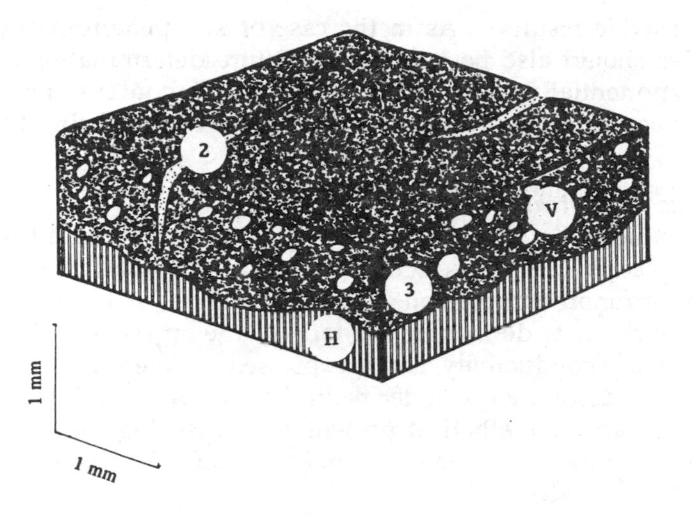

Figure 11. Erosion crust.
2 is the sand which subsists in the gaps and crevices; and *3* is plasma with cavities. The *H* horizon has not been affected by the direct action of wind or rain.

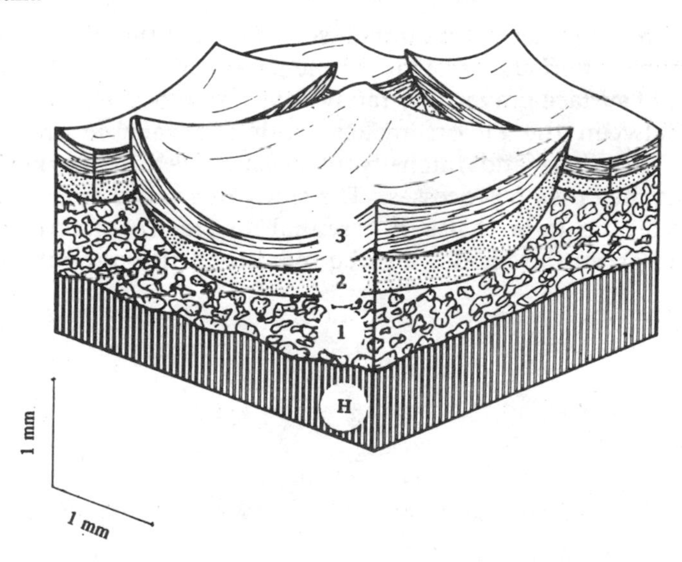

Figure 12. The depositional crust.
Note that the vertical distribution of the microhorizons is the reverse of the distribution for a structural crust with three microhorizons. *3* shows curling plasmic sand; *2* shows fine sand attached to *3*; *1* shows coarse sand which is independent of *2*. The *H* horizon has not been affected by the direct action of wind or rain.

statistically reliable results). As in the case of any penetrometric measurement, samples should also be taken for moisture determination. Resistance decreases exponentially in relation to moisture content, and increases exponentially according to the kinetic energy received (Valentin, 1986).

Resistance to infiltration

Several studies following that of Duley (1939) have revealed the key role of crusts and seals in the reduction of infiltration, and therefore in the development of runoff and erosion. This hydraulic resistance is exerted as an electric resistance. It is defined as the ratio of the crust/seal thickness to its saturated hydraulic conductivity, and is expressed in time units.

Apart from measurements under natural or simulated rainfall, there is no completely satisfactory method at present for measuring the infiltrability of soils prone to crusting. A method that could be useful for fine-textured soils is proposed in the appendix.

Conclusion

Improved management of mountain soils requires the identification of the factors governing runoff and erosion. Simple paramenters such as slope gradient or percentage of surface gravel only rarely have a univocal effect. Considerable interaction between the factors makes models unreliable and difficult to extrapolate. The type and intensity of crusting is often a key factor for understanding the erosion processes. The study of surface crust types, along a slope and according to land use, is essential for site characterization. Crust formation and evolution should also be monitored throughout the cropping season.

Appendix

Measuring infiltrability by the dropper method

Objective

To measure the infiltrability of a crusted soil.

Principle

A dropper with a very small diameter and which does not disturb the soil surface is used. A small shining spot is formed under the dropper and its diameter stabilizes if the initial moisture is close to saturation. If evaporation is not taken into account and no ponding appears, infiltration at saturation can be determined from the capillary flow and the area of the saturated spot (Figure 1).

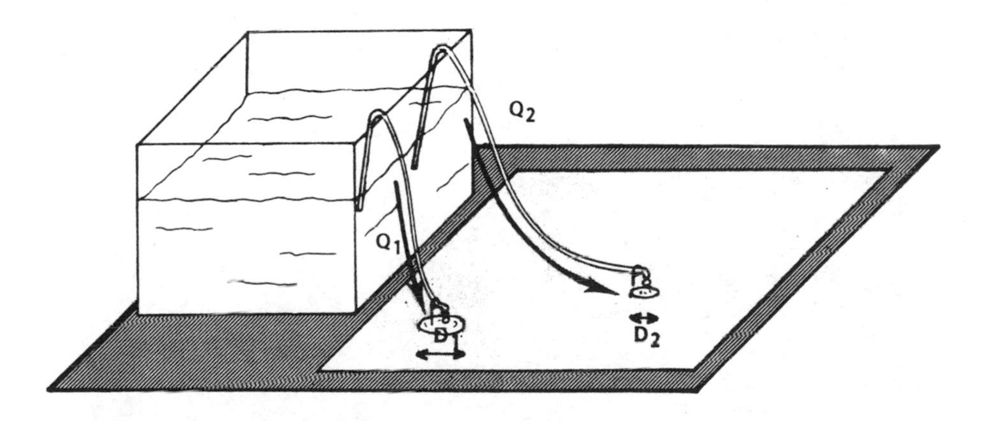

Figure 1. The principle of the device for measuring saturated spots.

Material

The instrument is made up of:

- about 15 spaghetti-type capillary tubes, of very small diameter (0.5-1.5 mm) and different lengths - varying, for example, from 1 to 3 m;
- one or several thermometers;
- three reservoirs of water;
- coloured pins;
- a chronometer or, if not available, a watch;
- sheets of transparent paper (optional).

Tubes with a larger diameter can be used if spaghetti tubes, commonly used in hospitals, are not available. But in this case, the tubes should be calibrated with Mohr clamps.

Procedure

Fifteen capillary tubes with flow rates ranging from 30 to 400 cm^3/h are set up on a plot. To obtain different flow rates, capillary tubes of varying lengths are used (Plate 1), and are fed from three reservoirs filled with water up to different levels. For each tube, corresponding to the given water level and tube length, the flow rate should be previously calibrated in the laboratory according to the water temperature. The diameter, and consequently the rate, is sensitive to variations in temperature. In the field, the temperature should be measured at the beginning and end of the operation for each reservoir, in order to calculate the flow from each capillary on the basis of preestablished regressions.

Plate 1. Infiltration measuring device, using "spaghetti" tubes. The surface crust developed on a seedbed. The surface area is 1 m^2.

In a simpler method, wider tubes of the same length are used. The flow rate is regulated by Mohr clamps. The tubes are calibrated in the field by counting the number of drops produced over a given period of time.

The following points should be considered while arranging the tubes:

- the surface under the dropper should be uniform and level;

- the droppers should be placed at an adequate distance from one another to avoid merging saturated drops;
- the dropper is kept in place with pins, and should be at an optimum height so that the drop is formed without acquiring kinetic energy.

The time at which the dropper is set up is noted. After a relatively long period of time, generally between 15 and 30 minutes, (depending on the moisture conditions and infiltrability), each saturated spot is marked with coloured pins. The same operation is repeated after 15 minutes. No change in the outline of the spot signifies that a stable state has been reached, and the diameter of the spot can be measured if it is circular. For irregular shapes, the outline is traced on transparent paper, and the area is measured with a surface integrator in the laboratory.

Expression of results

Infiltrability is calculated from the following equation:

$$F = \frac{10\,Q}{S}$$

where:

F : infiltration intensity (mm/h)
Q : flow rate of the capillary tube (cm^3/h)
S : area of the saturated spot (cm^3)

In practice, results that do not fulfil the application conditions of the equation are eliminated; these include:

- absence of the saturated spot: infiltrability is higher than the flow rate, or sandy microhorizons at the surface make it difficult to carry out the measurements;
- the appearance of puddles or runoff.

The value of F is the average or median of the remaining results. It can also be calculated with the regression $S = a.Q/F + b$, with the constraint of $b = 0$.

Accuracy

As the number of tubes used is high, it is possible to assess infiltrability at saturation, as well as its variability. The results are more accurate when the test area is not very permeable - for example on clayey or silty soils covered by an erosion crust.

Cost

This method costs very little.

Advantages

This is a low-cost apparatus that is light and easily transported in the field; it requires only one operator. The measurements can be normally completed within 1 h.

With this method, the surface state of the soil can be characterized without disturbing it. It can be carried out without any kinetic energy, which would modify the crust, or a deep water sheet, which would make the surface muddy and increase the hydrostatic charge. In this respect, it is better than the rainfall simulation and double-ring methods.

Disadvantages

The presence of lateral flow can prevent the effective use of the method for predictions. The method can only effectively predict the minimum rainfall intensity required for runoff occurence if there is no lateral water flow. It cannot therefore be applied to:

- soils with a surface structure, such as a seedbed;
- nonsaturated soils, or those very close to saturation;
- sandy soils, or those with sandy surface microhorizons;
- soils on steep slopes.

Remarks

At present, there is no completely satisfactory method for measuring crust infiltrability. The dropper method can be used to overcome this difficulty for fine-textured, crusted, and flat soils.

References

ALPEROVITCH, N. and DAN, J. 1973. Chemical and geomorphological comparison of two types of loessial crusts in the Central Neguev (Israel) *Israel Journal of Agricultural Research*. 23(1): 13-16.

BEN-HUR, M., SHAINBERG, I. BAKKER, D. and KEREN, R. 1985. Effect of soil texture and $CaCo_3$ content on water infiltration in crusted soil as related to water salinity. *Irrigation Science* 6: 281-294.

BOIFFIN, J. 1976. Histoire hydrique et stabilité structurale de la terre. *Annales Agronomiques* 27(4): 447-463.

BOIFFIN, J. 1984. La dégradation structurale des couches superficielles du sol sous l' action des pluies. Thèse, Docteur Ingénieur. Paris: INAPG. 320p.

BOFFIN, J. 1986. Stages and time-dependency of soil crusting. In: *Assessment of soil surface sealing and crusting*, ed. C. Callebaut, D. Gabriels and M. De Boodt, 91-98. Belgium: University of Ghent.

BOIFFIN, J. and MONNIER, G. 1986. Infiltration rate as affected by soil surface crusting caused by rainfall. In: *Assessment of surface sealing and crusting*, ed. C. Callebaut, D. Gabriels and M. De Boodt, 210-217. Belgium: University of Ghent.

BOX, J.E. jr. and MEYER, L.D. 1984. Adjustment of the Universal Soil Loss Equation for cropland soils containing coarse fragments. In: *Erosion and productivity of soils containing rock fragments*, 83-90. Madison, USA: Soil Science Society of America.

BRUAND, A. 1985. Contribution à l'étude de la dynamique de l'organisation de matériaux gonflants. Application à un matériau provenant d'un sol argilo-limoneux de l'Auxerois. Thèse, Université de Paris VII. 225p.

CASENAVE, A. et VALENTIN, C. 1988. *Les états de surface de la zone sahéliene.* Paris: ORSTOM. 202p.

COLLINET, J. 1988. Comportements hydrodynamiques et érosifs de sols de l'Afrique de l'Ouest. Evolution des matériaux et des organisations sous simulation de pluies. Thèse, Université Louis Pasteur, Strasbourg. 521p.

COLLINET, J. et VALENTIN, C. 1979. Analyse des différents facteurs intervenant sur l'hydrodynamique superficielle. Nouvelles perspectives. Applications agronomiques. *Cashiers ORSTOM, Série Pédologique* 17(4): 283-328.

COLLINET, J. and VALENTIN, C. 1985. Evaluation of factors influencing water erosion in West Africa using rainfall simulation. In: *Challenges in African hydrology and water resources,* 451-461. International Association of Hydrological Services. IAHS Publication no. 144.

COMBEAU, A. et QUANTIN, P. 1963. Observations sur les variations dans le temps de la stabilité structurale des sols en région tropicale. *Cahiers ORSTOM, Série Pédologie* 3(1): 17-26.

DE NONI, G., TRUJILLO, G. et VIENNOT, M. 1986. L'érosion et la conservation des sols en equateur. *Cahiers ORSTOM, Série Pédologie* 22(2): 235-245.

DE PLEUVRET, C. 1988. *Infiltration et réorganisations superficielles sous pluies simulées. Cas des sols ferrallitiques du Sud Togo.* D.E.A. de Pédologie. Paris: ORSTOM. 79p.

DE PLOEY, J. and MUCHER, H.J. 1981. A consistency index and rainwash mechanisms on Belgian loamy soils. *Earth Surface Processes and Landforms* 6: 319-330.

DE PLOEY, J. and BRYAN, R.M. 1986. Time-to-runoff as a function of slope angle. In: Assessment of surface sealing and crusting. ed. F. Callebaut, D Gabriels and De Boodt, 106-119. Belgium: Flanders Research Centre for Soil Erosion and Soil Conservation.

DE VLEESCHAUWER, D., LAL, R. and DE BOODT, M. 1978. Comparison of detachability indices in relation to soil erodibility for some important Nigerian soils. *Pédologie* 28(1): 5-20.

DESPHANDE, T.L., GREENLAND, D.J. and QUIRK, J.P. 1964. Influence of ion and aluminium oxide on the surface change of soil and clay materials. In: *Eighth International Soil Science Transactions* (Buchrarest, Roumania) 3: 1213-1225.

DULEY, F.L. 1939. Surface factors affecting the rate of intake of water by soils. *Soil Science Society of America Proceedings* 4: 60-64.

DULEY, F.L. and HAYS, O.E. 1932. The effect of the degree of slope on runoff and soil erosion. *Agricultural Research.* 45(6): 349-360.

DUMAS, J. 1965. Relation entre l'érodibilité des sols et leurs caractéristiques analytiques. *Cahiers ORSTOM, Série Pédologique* 3(4): 307-333.

FARRES, P.J. 1987. The dynamic of rainsplash erosion and the role of soil aggregate stability. *Catena* 14: 119-130.

FOURNIER, F. 1967. Research on soil erosion and soil conservation in Africa. *Sols Africains* 12: 53-96.

GABRIELS, D., PAUWELS, J.M. and DE BOODT, M. 1975. The slope gradient as it affects the amount and size distribution of soil loss material from runoff on silt loam aggregates. *Mededelingen Landbouw Gent* 40: 1333-1138.

GHADIRI, H. and PAYNE, D. 1977. Raindrop impact stress and the breakdown of soil crumbs. *Journal of Soil Science* 28(2): 247-258.

GOYAL, M.R., DREW, L.O., NELSON, G.L. and LOGANT, J. 1980. Critical time for soybean seedling emergence force. *Transactions of the American Society of Agricultural Engineers* 23(4): 831-835, +839.

GUMBS, F.A., LINDSAY, J.I., NASIR, M. and ANGELLA, M. 1986. Soil erosion studies in the northern mountain range, Trinidad, under different crop and soil management. In: *Soil erosion and conservation,* 90-98. Madison, USA: Soil Conservation Society of America.

HART, G.E. 1984. Erosion from simulated rainfall on mountain rangeland in Utah. *Journal of Soil and Water Conservation* 330-334.

HENIN, S. et MONNIER, G. 1956. Evaluation de la stabilité de la structure du sol. *Sixième Congrès International de Science du Sol* (Paris), vol B: 49-52.

HENIN, S. 1938. Etude physico-chimique de la stabilité structurale des terres. Thèse. Paris: Imprimerie Nationale. 70p.

HILLEL, D. and GARDNER, W.R. 1969. Steady infiltration into crust-topped profiles. *Soil Science* 109(2): 69-76.

HORVATH, V. and ERODI, B. 1962. Determination of natural slope category limits by functional identity of erosion intensity. International Association of Hydrological Services. IAHS Publication no. 59, 131-143.

HUDSON N.W. 1973 *Soil conservation,* ed. Batsford. London, UK. 320p.

JENSEN, E.H., FRELICH, J.R. and GIFFORD, R.O. 1972. Emergence force of forage seedling. *Agronomy Journal* 64: 635-639.

KHEYRABI, D. et MONNIER, G. 1968. Etude expérimentale de l'influence de la composition granulométrique des terres sur leur stabilité structurale. *Annales Agronomiques* 19(2): 129-152.

KUNKLE, S.H. and HARCHARIK, D.A. 1977. Conservation of upland wildlands for downstream agriculture. In: *Soil conservation and management in developing countries,* 133-149. FAO Soils Bulletin no. 33. Rome: FAO.

LAL, R. 1985. *Soil erosion problems and the role of mulching techniques in tropical soil and water management.* International Institute for Tropical Agriculture. Technical Bulletin no. 1. Ibadan: IITA.

LATHAM, M. 1988. Development on marginal lands. *IBSRAM Newsletter* 7: 1-2.

LE BISSONNAIS, Y. 1988. Analyse des mécanismes de désagrégation et de la mobilisation des particules de terre sous l'action des pluies. Thèse, Université d'Orléans. 195p.

LEMOS, P. and LUTZ, J.F. 1957. Soil crusting and some factors affecting it. *Soil Science Society of America Proceedings* 21: 485-491.

MONNIER, G. 1985. Recherches sur la stabilité structurale. Exemples d'applications. In: *Livre jubilaire du cinquantenaire,* 293-297. Paris: A.F.E.S.

MONNIER, G. et STENGEL, P. 1982. La composition granulométrique des sols: un moyen de prévoir leur fertilité physique. *Bulletin Technique d'Information* 370-372 : 503-512.

MOUGENOT, B. 1983. Caractérisation et évolution des états de surface en relation avec la dynamique des sols dans le delta du Sénégal. Méthodologie et résultats partiels de la campagne 1982-1983. Dakar: ORSTOM. 50p.

PALMER, R.S. 1965. Waterdrop impact forces. *Transactions of the American Society of Agricultural Engineers* 8(1): 69-72.

PARKER, J.J. and TAYLOR, H.H. 1965. Soil strength and seedling emergence. I. Soil type, moisture tension, temperature and planting depth effects. *Agronomy Journal* 57(3): 289-291.

POESEN, J. 1986a. Surface sealing as influenced by slope angle and position of simulated stones in the top layer of loose sediments. *Earth Surface Processes and Landforms* 11: 1-10.

POESEN, J. 1986b. Surface sealing on loose sediments: the role of texture, slope and position of stones in the top layer. In: *Assessment of soil surface sealing and crusting*, ed. F. Callebaut, D. Gabriels, and M. De Boodt, 354-362. Belgium: Flanders Research Centre for Soil Erosion and Soil Conservation.

POESEN, J. and SAVAT, J. 1981. Detachement and transportation of loose sediments by raindrop splash. II. Detachability and transportability measurements. *Catena* 8(1): 19-41.

RADWANSKY, S.A. 1968. Field observations of some physical properties in colluvial soils of arid and semi-arid regions. *Soil Science* (106) 4: 314-316.

RODIER et RIEBSTEIN. In press. Estimation des caractéristiques de la crue décennale pour les petits bassins versants du Sahel.

ROOSE, E. 1973. Erosion et ruissellement en Afrique de l'ouest. Vingt années de mesures en petites parcelles expérimentales. Thèse, Docteur Ingénieur, Université d'Abidjan. *Travaux et Documents*. Paris: ORSTOM.

SEBILLOTTE, M. 1968. Stabilité structurale et bilan hydrique du sol. Influence du climat et de la culture. *Annales Agronomiques* 19(4): 403-414.

SEGINER, I., MORIN, J. and SACHORI, A. 1962. Runoff and erosion studies in a mountainous terra-rossa region in Israël. *IAHS Bulletin* 7: 79-92.

SHAINBERG, I. 1985. The effect of exchangeable sodium and electrolyte concentration on crust formation. *Advances in Soil Science* 1: 101-122.

SINGER M,.J. and BLACKARD, J. 1982. Slope angle-interrill soil relationships for slopes up to 50%. *Soil Science Society of American Journal* 46: 1270-1273.

TAYLOR, H.M., PARKER, J.J. and ROBERSON, G.M. 1966. Soil strength and seedling emergence relation. A generalised relation for Graminea. *Agronomy Journal* 58: 393-395.

TESSIER, D. 1984. Etude expérimentale de l'organisation des matériaux argileux. Hydratation, gonflement et structuration au cours de la dessiccation et de la réhumectation. Thèse, Doctorat d'Etat. Université de Paris VII. 361p.

VALENTIN, C. 1986. Surface crusting of arid sandy soils. In: *Assessment of soil surface sealing and crusting*, ed. F. Callebaut, D. Gabriels, and M. De Boodt, 9-17. Belgium: Flanders Research Centre for Soil Erosion and Soil Conservation.

VALENTIN, C. 1988. Degradation of the cultivation profile: surface crusts, erosion and plough pans. In: *First Training Workshop on Site Selection and Characterization*, 233-264. IBSRAM Technical Notes no. 1. Bangkok: IBSRAM.

VALENTIN, C. et CAMARA, M. 1988. Evolution des états de surface, de l'infiltrabilité et de la détachabilité sous culture d'arachide, de manioc et de riz pluvial en zone de savane humide. Adiopodoumé: ORSTOM. 13p.

VALENTIN, C. et JANEAU, J.L. In press. Les risques de dégradation structurale superficielle en savane humide. *Cahiers ORSTOM, Série Pédologique*. Paris: ORSTOM.

WEBB, A.A. and COUGHLAN, K.J. 1989. Physical characterization of soils. This volume.

WISCHMEIER, W.H. and SMITH, D.D. 1960. An universal soil loss estimating equation to guide conservation farm planning. In: *Proceedings of the Seventh International Congress on Soil Science*, 418-425. Madison, USA: Soil Science Society of America.

YAIR, A. and KLEIN, M. 1973. The influence of surface properties on flow and erosion processes on debris-covered slopes in an arid area. *Catena* 1: 1-18.

YODER, R.E. 1936. A direct method of agregate analysis of soils and a study of the physical nature of erosion losses. *Journal of the American Society of Agronomy* 28: 337-351.

ZACHAR, D. 1982. *Soil erosion*. Developments in Soil Science no. 10. Amsterdam: Elsevier.

ZINGG, A.W. 1940. Degree and length of land slopes as it affects soil loss and runoff. *Agricultural Engineering* 21: 3-11.

Physical characterization of soils

A.A. WEBB and K.J. COUGHLAN*

Abstract

In soil management studies, physical characterization of soils is a major requirement for evaluation of data and the transfer of technology from site to site. Important soil characteristics can be measured or inferred from knowledge of soil morphology, bulk density, air porosity, soil strength, particle-size distribution, soil hydraulic properties, aggregate stability and plant available water capacity.

Comments are made on the use of the physical characteristics and the methods of measurement. An important requirement is that the methods used to measure these characteristics have relevance to the field situation.

Introduction

Considerable effort has been invested in various workshops organized by IBSRAM over the past two years in identifying the site and soil characteristics which should be taken into account and described at research and development sites, especially those where land management practices are to be evaluated or monitored. The rationale for this is that a knowledge of basic (or initial) site and soil characteristics is required for evaluation of subsequent data and for technology transfer.

The level of characterization can range from relatively simple to complex, but should be based on the degree of importance placed on the site. The number

* Queensland Department of Primary Industries, Indooroopilly, QLD 4068, Australia.

of sites at which detailed or complex characterization can be carried out will be determined by the resources and skills available. Pragmatism and budgetary limits often prevail, and the number of sites at which detailed data can be obtained is usually less than desirable. However, a minimum set of data must be collected for all sites to allow extrapolation of information from key sites at which the detailed studies are carried out.

In this workshop we are concerned particularly with the management of steeplands, and emphasis will be placed on those physical characteristics which are seen to be more important with respect to stability of the resource. The techniques of physical characterization listed here are applicable to rigid soils, which occur commonly in steeplands. If the soil exhibits significant volume change with wetting and drying, measurement of physical properties is much more difficult. In this case, the reader is referred to the paper by Coughlan *et al.* (1986).

Site and soil profile description

A site and soil morphological description is basic to any research and transfer of technology from site to site. Factors such as soil depth, thickness, texture, structure, consistence (strength) and colour of soil horizons and parent material can allow a considerable number of conclusions to be drawn on aspects such as water storage capacity, infiltration and drainage characteristics, and root growth. The methodology and criteria used for description are similar across most systems, but in the interests of uniformity within the Pacific region it seems logical to agree on one particular methodology. Those that seem suitable include the FAO methodology (FAO, 1968), the methodology used in Australia - the Australian soil and land survey (McDonald *et al.*, 1984), or the soil survey manual and taxonomy from the USA (Soil Survey Staff, 1951, 1975). Irrespective of the methodology used, of paramount importance is the need to describe the variability in soil morphology and physical properties across the site.

Bulk density and air porosity

Bulk density, the mass of soil per unit volume, is a measure of the compactness of soil and is a major determinant of soil water content, hydraulic conductivity, and root growth. Soil porosity is an alternative way of expressing a soil's structural condition. Air-filled porosity is the volume of air-filled pore space in a soil at a given moisture content and, as such, provides a guide to a soil's aeration status, which may be related to air diffusion and the potential for plant root respiration.

Values of bulk density and air-filled porosity can be derived from the one sample. Samples collected should be undisturbed and of known volume. A common and useful technique for collecting samples involves the use of thin-walled tubes hydraulically or manually pushed into the soil (Loveday, 1974, ch. 3). The trimmed wet soil core is weighed, dried at 105°C to constant weight, and reweighed after cooling. The calculation of dry bulk density and air-filled pore space, as well as the graviemtric and volumetric water contents of the sample, requires:

M = wet mass of the core (g)
M_s = mass of solids (that is, the oven-dry mass of the soil in the sleeve) (g)
V_b = bulk volume (that is, the volume of the metal sleeve) (cm^3)
AD = particle density, usually taken as 2.65 (g/cm^3)

to give:

- dry bulk density, $D_b = M_s/V_b$ (g/cm^3)
- gravimetric water content, $\theta_g = (M - M_s)/M_s$ (g/g)
- volumetric water content, $\theta_v = W.D_b$ (cm^3/cm^3)
- air-filled porosity, $f = 1 - (D_b/AD) - \theta_v$ (cm^3/cm^3)

Errors in estimation can occur if soil is compacted during sampling. This is a common problem in high clay content soils when they are near or above field capacity. Although tubės down to 50 mm diameter have been used for bulk density estimation in rigid soils, it is preferable if the diameter is 100 mm or greater. Care must be taken in controlling or measuring accurately the depth of the core sample. Details of soil samplers and some of the requirements are given by McIntyre (1974). In rigid soils, D_b remains constant with change in water content. In Vertisols (cracking clays), variation between maximum water content and wilting point may be greater than 20%.

Bulk density values of >1.7 for lighter-textured soils and 1.4 for clay soils are considered to be restrictive to root growth. Bulk density and air porosity are two characteristics which are often modified when land is developed for agriculture. However, bulk density is not a good measure of compaction since it is largely determined by the volume of pores formed by the packing of discrete particles. These pores are too small to be destroyed by the compressive forces of heavy equipment. Changes in air-filled porosity, or preferably macroporosity (calculated from the moisture characteristic - see later), are better indicators of compaction or soil structural degradation for two reasons: firstly, only large pores are affected by compaction; and secondly, air and water flow (which are of great importance to root growth) are proportional to (pore radius)^4, i.e. they are very dependent on large pores. The disadvantage of air-filled porosity measurement is the lack of information on pore continuity.

Soil strength

Penetration resistance

The weight of machinery used in land development and agriculture is seen to be a major cause of soil compaction. Soil resistance to the insertion of a penetrometer is a commonly accepted measure of soil strength, which many authors have attempted to relate to root exploration and expansion into the soil. For example, the use of penetrometers to assess compaction on sandy soils in Western Australia has been quite successful, showing strong correlations with wheat growth and yield (Henderson, 1987; Belford *et al.*, 1987). The ease and rapidity of data collection with penetrometers has ensured its popularity. However, uncertainty exists about the meaningfulness of the measure, both due to its interdependency with other soil factors, and to its interpretation as regards plant root exploration of the soil. Measurement is commonly made by means of a hand-powered penetrometer, though electric-powered recording penetrometers are becoming more widely used. The range of penetrometers available in terms of size, shape of tip, strength of internal spring, etc. is very large, but the uncertainties associated with interpretation of the values are common to all. If soil water content is not taken into account when penetration resistance is measured, misleading interpretations can be made, although this is less of a problem with rigid and lighter-textured soils than in swelling soils. Critical values of penetration resistance above which root growth stops may be found in the literature; a range of 2000 to 2500 kPa is common. However, such values cannot be directly related to root penetration as they cannot account for factors such as the lubrication of the soil by plant root exudates, or the ability of a plant to seek out the path of least resistance, especially in a structured soil. Hence there is considerable merit in making detailed visual assessments of root penetration from excavated pits.

Shear strength

Interest in cohesion and shear strength has increased recently because of the heightened concern by many over the degree of erosion in agricultural lands throughout the tropics and subtropics. There are several recent studies reporting increased erosion losses due to reduction in soil strength by tillage (Alberts *et al.*, 1980; Truong and Wegener, 1984), and there are encouraging studies showing the relationship between soil strength and entrainment parameters (Rauws and Govers, 1988; Bradford *et al.*, 1987).

Four methods have been promoted to measure shear strength of surface soil in the field: the vane shear tester, the torsional shear box, the drop-cone penetrometer, and the cone penetrometer. At this stage, there is no clear indication of the most suitable method(s) for field use in erodibility assessment, but the most favoured are the vane shear testers and the drop cone penetrometer. Considerable research is in progress attempting to determine the most appropriate method for estimating the strength-related parameter(s) related to soil detachment and entrainment.

Particle-size distribution

The relative proportions of particle sizes determine the observed texture of the soil, and these have controlling effects on other characteristics like structure, water-holding capacity, void volumes, soil strength, etc. Particle-size distribution is determined basically by parent material and the conditions under which the soils were developed. Highly siliceous lithologies like granite or sandstone commonly lead to the formation of coarse-textured soils. Lithologies like mudstone or basalt commonly lead to finer-textured soils. Particle-size distribution can be determined by a number of methods, but a requirement of each one is the separation and dispersion of the soil into discrete particles. Care has to be taken in this step not to break down the particles into smaller sizes. Thorburn and Shaw (1987) tested a range of procedures and recommended dispersion using a reciprocating shaker, and the measurement of fine particles by hydrometer. This technique is an acceptable compromise between accuracy of results and speed of determination.

Field experience has given rise to some useful generalizations about the effect of particle size on soil physical properties. For soils with relatively high clay content (>35-45%), a 'clay matrix' exists, and soil structure associated with clay dominates - e.g. strong aggregates associated with clay mineral bonds, 'self-mulching' in Vertisols, and clay dispersion. At lower clay percentages, the size of nonclay particles has a significant effect on soil properties - e.g. soils with a relatively high proportion of coarse sand tend to be friable; soils with a high proportion of fine sand tend to be dense, intractable, hard setting and have coarse aggregates; soils with a high proportion of silt tend to be crusting, cloddy, sealing soils.

In steeplands, one of the major factors of concern is soil erosion, and one very simple way of assessing or monitoring the degree of erosion in soils which show characteristic particle-size distribution down the profile is to monitor particle-size distribution. It is basic data for extrapolation of data across sites.

Soil hydraulic properties

Soil hydraulic properties determine both the capability and the stability of the soil. Factors of particular interest are the rate of water entry and the rate of movement through the soil. Water entry is determined by both surface processes, such as crusting and sealing, and the processes deeper in the profile that establish throttles to water movement. It is important that in the selection of a method for estimating hydraulic characteristics the purpose of the estimate is understood. For instance, is the measure for comparison of management effects to study infiltration of rainfall vs. (say) flood irrigation, or is it to provide parameters in an analytical solution?

One important use of hydraulic properties is to model runoff at a field scale. Information is best obtained at a scale sufficient to capture soil variability associated with soil macroporosity, and biotic factors such as root channels (typically, tens to hundreds of square metres in size). However, to allow extrapolation of data obtained on large plots, simpler techniques are required which can estimate soil hydraulic properties over time, and hence runoff. (Plot sizes for measurement are typically <1 m^2.)

In soils which surface seal, soil hydraulic properties change during the rainfall event. Therefore, we must use a rainfall simulator to obtain relevant data on the decay of saturated hydraulic conductivity (K_s) with time due to seal formation. K_s under nonsealed conditions is obtained on plots which are covered to protect the soil from kinetic energy.

Under conditions where surface seals do not form, profile K_s can be measured in ponded rings or large cores. Alternatively, K_s values for different layers of the soil profile can be measured using a suction infiltrometer (disc permeameter) for the surface layer, and a well permeameter for deeper layers. Information on antecedent water content before rainfall, saturated water content (θ_s) and K_s can then be used to model runoff under different rainfall events.

Information on K_s can also be used to:

- detect limiting layers where perched water tables may form, resulting in interflow or loss of cohesion and mass movement; and
- to detect soil degradation over time under different soil management systems.

In rigid soils, where volume change does not occur with water content, small cores (say 75 mm diameter x 50 mm deep) can provide useful information on soil characteristics, and notably:

- Saturated water content (θ_s) for use in modelling. The soil should be wet from the top down to retain entrapped air (as occurs in the field).
- Saturated hydraulic conductivity (K_s).
- The moisture characteristic, relating θ to soil water potential (ψ). Using a matching technique developed by Marshall (1958), the moisture characteristic can be used to give data on unsaturated conductivity for a range of water contents.

  Using the capillary rise equation (relating the size of the drained pore to the water potential for rigid soils), the volume of macropores can be measured. This is a sensitive measure of soil degradation. In soils which shrink during drying, the capillary rise equation cannot be used to calculate pore size because water removal involves both pore drainage and pore shrinkage.

We rely on two particular approaches for estimating water movement in the field in our soil conservation research and cropping systems studies. One involves using rainfall simulators capable of delivering high kinetic energy rain (1 x 1 m and 4 x 22.4 m), and the other is to use field plots ranging from 10 x 30 m to 10 ha. Both of these approaches have been used to compare the effects of tillage and cropping practices on infiltration. There are obvious disadvantages in using field plots under natural rainfall conditions, particularly in situations where rainfall is episodic, and when control over the surface conditions and soil water is desired. However, one major advantage is that variability in water movement at the small scale is removed, and the integrated effect is much more applicable to extrapolation in the field. One of the major difficulties in using large plot data to estimate infiltration has been the lack of a suitable procedure to account for the lag effects associated with the runoff. This problem has been addressed by Rose *et al.* (1983 and 1984), who have outlined the use of an approximate analytical procedure which allows the infiltration rate to be calculated where the runoff has been measured.

Aggregate stability

The stability of soil aggregates to wetting affects many aspects of soil behaviour, including infiltration, erosion, and crusting. Measurements of aggregate stability to wetting are commonly used in comparing soils, and in assessing the effects of various management practices on soil structure. Many methods have been developed to measure aggregate stability to wetting, and the two approaches most commonly used are to measure either wet aggregate size (by sieving), or dispersion of clay. However, there is widespread dissatisfaction with these methods, common criticisms being:

- the methods of wetting used are not relevant to the field situation;
- tests may not be applicable to a wide range of soils;
- the way in which aggregate breakdown is measured may not be relevant to field soil behaviour;
- tests have not been calibrated against field soil behaviour; and
- the soil samples tested are not representative of the range of aggregate sizes likely to occur in the soil surface under rain.

When we consider natural rainfall and a field soil, it is clear that a range of drop sizes (and impact forces) are applied to a range of aggregate sizes and strengths. The distribution of drop impact forces is therefore of major importance. It is impossible to equate the various energy inputs of many tests such as shaking, ultrasonic disruption and remoulding, with the sorts of forces found under natural rainfall, As a consequence, Loch and Foley (1987) have developed a simple test based on the degree of aggregate breakdown for a bed of soil subjected to simulated rain, and wet sieving to measure the degree of breakdown. The results compare well with the effects of rainfall under field behaviour. The strong relationship between the percentage of particles <0.125 mm in the surface crust and the final infiltration rates for a wide range of Queensland soils (Figure 1) has given us confidence in the use of this test. Soil clay varied from 8-70%, and Vertisols, Oxisols and Alfisols were included.

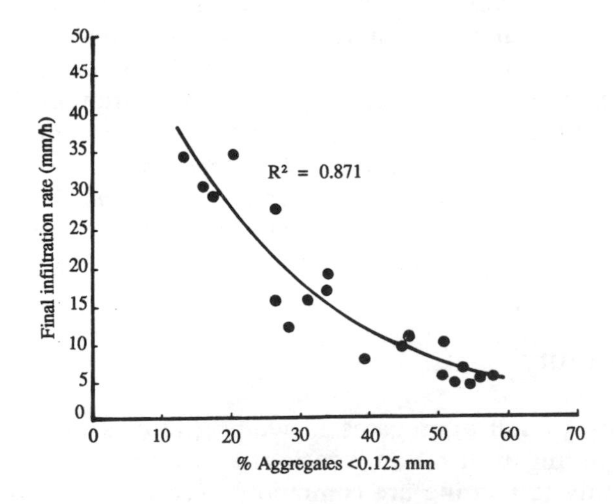

Figure 1. Relationship between final infiltration rates and percentage of aggregates <0.125 mm for a range of Queensland soils.

This test was compared with a number of others, including measures of clay dispersion, different methods of wetting and wet sieving, high-energy moisture curves and stability to successive water drops. For the soils used to derive Figure 1, results of other tests showed no correlations with measured infiltration rates, nor did they discern between soils that in the field have been observed to differ greatly in their aggregate stability.

Plant available water capacity

Characterization of plant available water capacity (PAWC) is important because PAWC is a measure of the soil's capability to store water that will be available for use by crops. Our concept of PAWC involves two factors: the soil water storage and plant water extraction. PAWC is defined (Gardner *et al.*, 1984):

$$\text{PAWC} = \sum_{z=0}^{z=RD} (\theta_{max} - \theta_{dry})\, BD\, \Delta z / DW$$

where: θ_{max} = the gravimetric water content at the upper storage limit in the field soil (g/g),

θ_{dry} = the final gravimetric water content after plant water extraction in the field (g/g),

RD = rooting depth,

BD = field bulk density at θ_{max} (measured or estimated),

Δz = depth increment,

DW = density of water.

PAWC may vary with crop, stage of growth, level of stress, root development restrictions, extent of soil water recharge, etc., but when measured using management appropriate for the intended application it has proved an extremely useful concept for comparison among soils (Shaw and Yule, 1978; Gardner and Coughlan, 1982).

W_{max} should be measured by sampling the fully wet profile (that is, two to three days after profile recharge) and W_{dry} by sampling when the crop shows specific stress symptoms. If suitable equipment is available, field bulk density can be measured in conjunction with W_{max} sampling. If no land under the crop of interest is available, it may be necessary to set up small protected plots to grow the crop under simulated conditions to provide the range of wetting and drying required. This approach has been used to assess the PAWC of potentially irrigable soils in Queensland (Shaw and Yule, 1978; Gardner and Coughlan, 1982). W_{dry} can be equated directly with the -15 bar water content for each depth interval, but it must be remembered that in the deeper subsoil crops might not remove water from the soil even though it is wetter than -15 bar (because of restrictions to root distribution and to the osmotic effects of salinity). A detailed discussion of soil water and its measurement and the estimation of PAWC has been given by Gardner (1985). This is an excellent coverage of the whole area of soil water measurement and some of the problems concerned with the estimation of soil water and bulk density.

References

ALBERTS, E.E., FOSTER, G.R. and MOLDENHAUER, W.C. 1980. Soil aggregates and primary particles transported in rill and interrill flow. *Soil Science Society of America Journal* 44:590-595.

BELFORD, R.K., HARVEY, J.R. and JARVIS, R. 1987. The use of a penetrometer to assess soil conditions and crop growth after direct drilling with an experimental seeder. In: *Effects of management on soil physical properties*, 182-185. QC 87006. Queensland, Australia: Department of Primary Industries.

BRADFORD, J.M, FERRIS, J.E. and REMLEY, P.A. 1987. Interrill soil erosion processes. I. Effect of surface sealing on infiltration, runoff, and soil splash detachment. *Soil Science Society of America Journal* 51:1566-1570.

COUGHLAN, K.J., McGARRY, D. and SMITH, G.D. 1986. The physical and mechanical characterization of Vertisols. In: *Management of Vertisols under semi-arid conditions*, 89-105. IBSRAM Proceedings no. 6. Bangkok: IBSRAM.

FAO (Food and Agriculture Organization of the United Nations). 1968. *Guidelines for soil profile description*. Soil Survey and Fertility Branch, Land and Water Division. Rome: FAO.

GARDNER, E.A. 1985. Soil water. In: *Identification of soils and interpretation of soil data*, 197-234. Queensland, Australia: Australian Society of Soil Science.

GARDNER, E.A. and COUGHLAN, K.J. 1982. *Physical factors determining soil suitability for irrigation crop production in the Burdekin-Elliot River area*. Agricultural Chemistry Branch. Technical Report no. 20. Queensland, Australia: Department of Primary Industries.

GARDNER, E.A., SHAW, R.J., SMITH, G.D. and COUGHLAN, K.J. 1984. Plant-available water capacity: concept, measurement and prediction. In: *Properties and utilization of cracking clay soils*, ed. J. McGarity, E.H. Hoult, and H.B. So, 164-175. Reviews in Rural Science no. 5. New South Wales, Australia: University of New England.

HENDERSON, G. 1987. Using penetrometer data to predict the effects of compaction of earthy sands on wheat yields. In: *Effects of management on soil physical properties*, 178-183. QC 87006. Queensland, Australia: Department of Primary Industries.

LOCH, R.J. AND FOLEY, J.L. 1987. Relationship between aggregate breakdown under rain and surface sealing. In: *Effects of management on soil physical properties*, 166-169. QC 87006. Queensland, Australia: Department of Primary Industries.

LOVEDAY, J. ED. 1974. *Methods for analysis of irrigation soils.* Technical Communication no. 54. Harpenden, UK: Commonwealth Bureau of Soils.

McDONALD, R.C., ISBELL, R.F., SPEIGHT, J.G., WALKER, J. and HOPKINS, M.F. 1984. *Australian soil and land survey: field handbook.* Melbourne, Australia: Inkata Press.

McINTYRE, D.S. 1974. Soil sampling techniques for physical measurements. In: *Methods for analysis of irrigated soils,* ed. J. Loveday. Commonwealth Agricultural Bureaux International. Technical Communication no. 54. Wallingford, UK: CABI.

MARSHALL, T.J. 1958. A relation between permeability and size distribution of pores. *Journal of Soil Science* 9:1-8.

RAUWS, G. and GOVERS, G. 1988. Hydraulic and soil mechanical aspects of rill generation on agricultural soils. *Journal of Soil Science* 39:111-124.

ROSE, C.W.R., FREEBAIRN, D.M. and SANDER, G.C. 1984. *Computing infiltration rate from field hydrologic data: GNFIL - A Griffith University programme.* School of Australian Environmental Studies. AES monograph 2/84. Brisbane, Australia: Griffith University.

ROSE, C.W., PARLANGE, J.Y., SANDER, G.C., CAMPBELL, S.Y. and BARRY, D.A. 1983. Kinematic flow approximation to runoff on a plane: an approximate analytical solution. *Journal of Hydrology* 62:363-369.

SHAW, R.J. and THORBURN, P.J. 1985. Prediction of leaching fraction from soil properties, irrigation water and rainfall. *Irrigation Science* 6:73-83.

SHAW, R.J. and YULE, D.F. 1978. *The assessment of soils for irrigation, Emerald, Queensland.* Agricultural Chemistry Branch. Technical Report no. 13. Queensland, Australia: Department of Primary Industries.

SOIL SURVEY STAFF. 1951. *Soil survey manual.* Soil Conservation Service, U.S. Department of Agriculture. Agriculture Handbook no. 18. Washington, DC: Government Printing Office.

SOIL SURVEY STAFF. 1975. *Soil taxonomy: A basic system of soil classification for making and interpreting soil surveys.* Soil Conservation Service, U.S. Department of Agriculture. Agriculture Handbook no. 436. Washington, DC: Government Printing Office.

THORBURN, P.J. and SHAW, R.J. 1987. Effects of different dispersion and fine-fraction determination methods on the results of routine particle-size analysis. *Australian Journal of Soil Research* 25:347-360.

TRUONG, P. and WEGENER, M. 1984. A review of results of trials with trash management for soil conservation. In: *Proceedings of the Australian Society of Sugar Cane Technologists* 6:101-106.

Section 7: Experimental design and monitoring

Experimental design for land clearing experiments on sloping lands

DENNIS F. SINCLAIR*

Abstract

The fundamental principles of experimental design are reviewed. The particular problems of the design and analysis of experiments on sloping land are discussed. Nearest-neighbour analysis and the analysis of covariance are suggested as appropriate ways to account for variability within experimental sites. Simple experimental designs are recommended, particularly for coordinated experiments in the ASIALAND network.

Introduction

As the title of this paper includes the words 'experimental design', a few introductory remarks on this important branch of statistical methods are warranted. Experimental design plays an important role in what is often called the 'scientific method' - the means by which we seek to obtain unequivocal answers to questions of interest with the minimum of effort. The scientific method can be simplified as follows:

Hypothesis->Experimental Design->Data->Conclusion->New Hypothesis

* Department of Statistics, The University of Newcastle N.S.W. 2308, Australia.

If the experimental design is good, often little in the way of statistical analysis is required. However, if an experiment is poorly designed, a statistical salvage operation usually needs to be launched, resulting in very complicated statistical analysis which may be difficult to understand and interpret. Thus to obtain the full value from the effort put into experimentation, considerable time should be spent in the planning stage of experimental design. It is at this stage, rather than at the data analysis stage, that a statistician can make the most valuable contribution.

We live in an uncertain world. No two things are exactly identical. If the same variety of maize is grown on two plots of ground, two different yields will be obtained. If the same amount of fertilizer is applied to two plots of potatoes, one will yield more than the other. Statistics has been described as the science of making decisions in the face of uncertainty. Statistical thinking has to do with understanding the concept of variability and making decisions based on data. In an experiment, one treatment will produce a higher average yield than the rest. Is this a real effect or simply a result of chance fluctuations? To answer this question statistical methods are needed. There is a need to reduce the influence of natural variation through sound experimental design, and to account for its presence in drawing conclusions from the experimental data.

There are three important principles of experimental design - replication, randomization, and blocking. Each of these principles will be discussed in turn. (For a more complete treatment of experimental design, see Snedecor and Cochran [1967], and Cochran and Cox [1957].)

Replication

Each treatment is applied to more than one plot, or experimental unit. This allows an estimate of experimental error. It can also increase the generality of results and avoid the problem of confounding where the apparent effect of one factor may in fact be due to another cause.

Randomization

Treatments should be allocated randomly to experimental units, preferably by means of random number tables or computer-generated random numbers. Randomization avoids the biases that can arise from subjective, or systematic, allocation of treatments to experimental units.

Blocking

Experimental units are often grouped into blocks (usually containing the same number of experimental units) with the aim of reducing the experimental error. This has the result of increasing the precision of comparisons of treatment effects. As the principle aim of experimental design is to compare treatments, successful blocking is clearly a desirable tool. Blocks are chosen to be homogeneous 'within', that is plots within blocks are as similar as possible. Differences in the natural variation in the experimental area should be reflected by differences 'between' the blocks. Usually blocks will consist of a complete replicate of experimental treatments (the completely randomized block design). In the running of an experiment, everything should be done 'by block' - fertilizing, mowing, harvesting, personnel, etc. Within a block everything should be uniform; it does not matter if conditions differ between blocks. In agricultural field trials in temperate climates, large gains in efficiency have resulted from blocking. However, in the tropical environment, particularly after land clearing, soil variability is high and the pattern of variability is very patchy (Sinclair, 1987). The opportunity to form blocks from uniform tracts of land is greatly diminished, and there may often be little advantage in choosing a randomized block design over a completely randomized design, in which treatments are randomly allocated to plots over the whole experimental area (Pearce, 1986).

Apart from the statistical design of an experiment, there are many other aspects which need to be addressed in the planning stage. These include the choice of treatments (including the possibility of one or more controls), the site(s) of the experiment, the size of the experiment (number of replications), the field layout, the budget, personnel, and equipment required. Thought needs to be given to the ongoing management of the experiment, the methods of data recording and transcription. Once the data has been collected, there is the question of data analysis, and the reporting of results and their interpretation. Of particular importance is the distinction which needs to be drawn between 'statistically significant' differences and practical differences. If, for example, an experimental treatment is found to be 1% better than the traditional method, but requires much more investment of time or money, this result has little practical significance. On the basis of this evidence there would be little sense in changing from the traditional method. Dyke (1988) and Preece (1982) are excellent sources of wisdom on the practical aspects of designing, running, and analyzing field experiments.

Sloping land

There are few references addressing the problem of experimental design on sloping land. Dyke (1988, p.33) argues in favour of experiments which reflect the conditions under which farmers grow their crops. For example, in areas where crops are grown on long narrow curved terraces. In areas where the traditional method is to farm small patches of land on irregular hillsides, experimental plots should be of similar size and laid out accordingly.

Experimental results from flat, relatively uniform, research stations are unlikely to transfer meaningfully to the 'real world' situation.

Pearce (1986) recommends nearest-neighbour analysis (Papadakis, 1937; Bartlett, 1978) to analyze an experiment in which plots are laid out along a terrace. Several terraces could be used, each regarded as a block, and the nearest-neighbour adjustment made within each block.

Dyke (1988) agrees that generally blocks would run along contours in experiments on steeply sloping land. However, he lists a number of factors such as varying soil depth, atmospheric conditions, degree of slope, and soil fertility which may influence the experimenter to adopt a different layout. This simply emphasizes the point that all sites are different, and there is no more a universal recipe for experimental design on sloping lands than there is for flat ground.

The degree of slope, combined with such factors as rainfall, soil type, and erosivity will determine whether terracing is necessary. A hydrological study of the experimental site, with a view to potential erosion, before deciding upon the final experimental design is highly recommended.

In keeping with the recommendation of the 1986 working group (see Appendix), the design should be kept as simple as possible. Where land-clearing treatments are involved, these would probably form main plots, with cropping system treatments being applied to the subplots in a split-plot design. Plot size will depend on the location, but should relate where possible to traditional farmers' plots.

Several general recommendations on the design and analysis of experiments in the ASIALAND network are given in Sinclair (1987), and it serves little purpose to repeat all of them here. However, in the context of experiments on sloping land, the much-underutilized statistical technique of analysis of covariance deserves special mention.

Site variability is expected to be high after land clearing. But when the cleared area is an uneven hillside, which even before clearing may vary greatly in soil characteristics, water movement, or vegetation, the resulting site variability can only be magnified. Too much site variability can mask the effect of a good treatment, or enhance the effect of a poor treatment. Thus there is a need to remove the influence of this underlying variability when we compare between treatments. The 'tried and true' way of achieving this in experimental design is through blocking (Sinclair, 1987). However, when the variability has little discernible pattern, and is generally patchy in nature, blocking may not remove much of the variability - other statistical tools are needed.

At the statistical analysis stage of the experiment, analysis of covariance (Snedecor and Cochran, 1967) can be used to remove variability due to underlying environmental factors, unrelated to the treatments being applied.

Covariates are subsidiary observations taken prior to the application of the treatments. Possible covariates in experiments involving clearing on sloping land are depth of soil, density of vegetation, available P, N, bulk density or moisture of soil, altitude, and aspect. These variables would be measured on each plot. Treatment means are 'adjusted' to allow for the effect of the covariates, which in turn reflect underlying site heterogeneity. The aim is to reduce error variance and increase the precision for comparing treatments.

Agrotechnology transfer

The aim of the ASIALAND network experiments is agrotechnology transfer rather than basic research. As such, they must relate to the end-user - the small farmer - and his likelihood of adopting improved methods. For this reason it would seem that recommended procedures should not be too far removed from familiar practices for the farmer. Otherwise, without a massive and successful education programme, the new technology is unlikely to be adopted. Besides, there is often considerable wisdom in the traditional farming methods.

With this in mind, intercropping and rotation crop treatments will be of particular interest. Statistically, intercropping and rotation experiments present difficulties in both analysis and interpretation. In intercropping the primary aim is usually to determine the most productive mixture of crops. As a consequence, effort has been concentrated on developing statistical analyses based on indices formed from various combinations of the individual crop yields. These techniques thus reduce to univariate analyses, which may be particularly useful in identifying optimal mixtures, but do not provide insight into the competition between components. The statistical aspects of intercropping experiments have been reviewed by Mead and Riley (1981). More recent contributions to the design and analysis of intercropping experiments are Wijesinha *et al.* (1982), Gilliver and Pearce(1983), and Mead *et al.* (1986).

Preece (1986) discusses some of the general statistical principles of crop rotation experiments. He makes the point that rotation experiments by their nature tend to be long term, and a major source of variation in a long-term experiment is management. Thus training and continuity of front-line workers should be given high priority. Thorough statistical analysis of rotation experiments can be quite complex (Yates, 1954; Patterson, 1964). Preece recommends the involvement of a statistician from the planning stage through to the analysis of results.

Concluding remarks

The basic principles of experimental design apply in land clearing experiments on sloping lands - to remove extraneous sources of variability so that treatments can be compared as precisely and efficiently as possible. This is more likely to be achieved through nearest-neighbour analysis and the analysis of covariance than the more traditional agricultural method of blocking (due to the expected patchy nature of the underlying variability over the experimental site).

The actual experimental design and layout should be kept as similar as practicable across sites in the network. The designs should be simple, probably no more complex than a split plot. The front-line workers should enter their own data on to a computer on-site and perform basic statistical operations, such as summary tables, graphs, and analysis of variance (or analysis of covariance). The results from all sites in the network should be centrally collated, stored, and analyzed. The more detailed statistical analysis can be carried out at this stage. Regular update reports should be circulated by the network coordinator to inform participants of the findings from their own experiments, and those of others in the network.

Appendix

Recommendations from the first ASIALAND seminar

At the first ASIALAND seminar at Khon Kaen and Phitsanulok in October 1986, I was the discussion leader of the working group considering 'Sampling and the design of experiments'. As they are relevant to this workshop, I summarize below the broad recommendations made by this working group for the ASIALAND network.

Socioeconomic survey. The ASIALAND network coordinator should be given the task of preparing a standard questionnaire. The questionnaire would be used to gather baseline information for the regions in which the experimental sites are to be located. Of particular interest will be information on `traditional' methods of farming.

Soil maps. These should be on several scales for the experimental sites, both before and after treatments.

Fixed point sampling. Where it is of interest to track changes, e.g. in vegetation, soil properties or erosion, through time, samples should be taken from approximately the same locations on each sampling occasion. This strategy is much more efficient than simple random sampling, especially on highly variable sites (see Gillman *et al*, 1985; Sinclair, 1987).

Measurements. We did not address the question of which data should be collected. There was strong opinion expressed, however, that forest assessment and biomass were important variables which should be closely monitored.

Observational studies. Supplementing designed network experiments, observational studies may be carried out to learn more about the environment and local conditions. It was recommended that proven sampling techniques such as stratification and pairing of locations (Gillman *et al*, 1985; Sinclair, 1987; Cochran, 1963) be utilized.

Relate design to objectives. The objectives of the ASIALAND network should be clearly stated and understood. Participating countries must agree with these common objectives, and see value in them. In addition, individual countries, and regions within countries, will have particular problems which are not common to all sites. The experimental design at a particular site should reflect both the common ASIALAND objectives and the special objectives related to that site. Thus the concept of one single experimental design for all sites is inappropriate.

Keep the design simple. The working group agreed in principle that in a network such as ASIALAND the experiment design should be kept simple - nothing more complicated than a split-plot design. But the actual design and layout, especially for experiments involving land-clearing treatments, needs to be site-specific.

Site selection. The working group recommended that specialist consultants be involved in the selection of suitable sites. Advice would be required on site classification, environmental variability, experimental design and layout, hydrological aspects, experimental management, and data collection.

Choice of treatments. A core group of treatments from the common ASIALAND network objectives needs to be identified. These core treatments would be supplemented by specific treatments of relevance at a particular site. The working group strongly recommended the inclusion of a 'traditional farmer' treatment at each site as a control. In many cases this would involve intercropping, as few traditional farmers crop monocultures. Recommendations on plot size were difficult to make, as these would be expected to vary from site to site. However, as a broad recommendation for land-clearing experiments, a minimum plot size of 0.025 ha, say 25 m x 10 m, was suggested. Experiments on steep slopes, as proposed, for example, for Malaysia, usually require terracing, and result in quite a different experimental layout to those on flat land. The choice of treatments should truly reflect agrotechnology transfer, and not simply duplicate applied agricultural research being done elsewhere in the world.

Coordination of experiments. The various experiments in the network need to be centrally coordinated and their progress closely monitored. The network coordinator was seen to have an important role in setting deadlines for participants, and ensuring these deadlines are met.

Data collection and analysis

Minimum data set. The working group saw merit in a basic set of variables being measured across the network. In addition, other measurements may be made which are more site- or experiment-specific. The group recommended that consultants be involved, at least in the early stages, to advise on the collection of the more technical measurements, such as runoff.

Standardized data forms. There are clear advantages to the coordination of the network if the data are recorded on standard forms, particularly for the minimum data set in the core experiments.

Standardized data storage. If the data are to be entered into a computer, on-site, then it was recommended that the same data base package be used at each site.

Standardized computing facilities. The working group felt it was desirable for computing facilities to be available on site (see Gomez, 1987). A minimum requirement would be an IBM-compatible personal computer with 64 Kbyte RAM and a 20 Mbyte hard disk. In addition there are advantages in performing some basic statistical analysis and data summary on site. There should be standardization of statistical computing packages to be used for these purposes.

Centralized compilation and analysis of results. While data can be collected, coded, summarized and even statistically analyzed by site personnel, there is still a need for centralized processing of all network results. The working group saw this as the network coordinator's responsibility. Updates and progress should be provided on a regular basis to network participants.

Other considerations

Available land. Some members of the working group felt that lack of available land could be a problem, particularly for land-clearing experiments which required relatively large plot sizes to be realistic.

Limited resources. Limited funds, lack of suitably trained personnel, and inadequate analytical facilities were seen as potential stumbling blocks to the successful implementation of the ASIALAND network.

Training needs. Training courses were seen as essential, especially in areas such as soil moisture measurement and statistical methods. The overwhelming view of the working group, however, was that these courses be held on-site, for front-line workers.

Regular meeting. The working group recommended that regular network meetings be held where participants discuss results and learn from others.

Field manual. Possibly the strongest recommendation from the working group was for an ASIALAND field manual, detailing field, laboratory and statistical techniques. While the production of such a manual would be a major job for the network coordinator, it could be of immeasurable value.

References

BARTLETT, M.S. (1978). Nearest neighbour models in the analysis of field experiments (with discussion). *Journal of the Royal Statistical Society* B40: 147-174.

COCHRAN, W.G. 1963. *Sampling techniques*. 2d ed. New York: Wiley & Sons.

COCHRAN, W.G. and COX, G.M. 1957. *Experimental designs*. 2d ed. New York: Wiley & Sons.

DYKE, G.V. 1988. *Comparative experiments with field crops.* 2d ed. London: Griffin.

GILLMAN, G.P., SINCLAIR, D.F., KNOWLTON, R. and KEYS, M.G. 1985. The effect on some soil chemical properties of the selective logging of a North Queensland rainforest. *Forest Ecology and Management* 12: 195-214.

GILIVER, B. and PEARCE, S.C. 1983. A graphical assessment of data from factorial intercropping experiments. *Experimental Agriculture* 19: 23-31.

GOMEZ, K.A. 1987. Data base for soil management network. *Soil management under humid conditions in Asia and the Pacific,* 433-445. IBSRAM Proceedings no.5. Bangkok: IBSRAM.

MEAD, R. and RILEY, J. 1981. A review of statistical ideas relevant to intercropping (with descussion). *Journal of the Royal Statistical Society* A144: 462-509.

MEAD, R., RILEY, J., DEAR, K. and SINGH, S.P. 1986. Stability comparison of intercropping and monocropping systems. *Biometrics* 42: 253-266.

PAPADAKIS, J. 1937. Méthode statistique pour les expériences sur champ. *Bulletin de L'Institut pour L'Amé lioration des Plantes, Salonique* (Grèce) 23.

PATTERSON, H.D. 1964. Theory of cyclic rotation experiments (with discussion). *Journal of the Royal Statistical Society* B26: 1-45.

PEARCE, S.C. 1986. Experimental design: the first sixty years. *Tropical Agriculture (Trinidad)* 63: 95-100.

PREECE, D.A. 1982. he design and analysis of experiments: what has gone wrong? *Utilitas Mathematica* 21A: 201-244.

PREECE, D.A. 1986. Some general principles of crop rotation experiments. *Experimental Agriculture* 22: 187-198.

SINCLAIR, D.F. 1987. Accounting for soil microvariability after land clearing. *Soil management under humid conditions in Asia and the Pacific,* 419-432. IBSRAM Proceedings no. 5. Bangkok: IBSRAM.

SNEDECOR, G.W. and COCHRAN, W.G. 1967. *Statistical methods.* 6th ed. Ames: Iowa State University Press.

WIJESINHA, A., FEDERER, W.T., CARVALHO, P. and PORTES, T.A. 1982. Some statistical analyses for a maize and beans intercropping experiment. *Crop Science* 22: 660-666.

YATES, F. 1954. The analysis of experiments containing different crop rotations. *Biometrics.* 10: 324-346.

Procedures for monitoring rainfall, runoff and erosion in the measurement of land-use effects

CYRIL CIESIOLKA, ADRIAN WEBB, TED GARDNER and CALVIN ROSE*

Abstract

Accelerated soil erosion is the result of interaction between a large number of physical, cultural, and economic variables. This paper summarizes techniques and problems associated with the measurement of the physical attributes of the erosional system.

Rainfall measurement is considered in relation to the spacing and location of instruments, instrument maintenance, recording times, and safety.

Instantaneous measurements that synchronize rainfall and runoff invlove the use of flumes/weirs/channel cross sections, drum and chart recorders, transducers and pressure-bubble recorders. Electronic advances have allowed the measurement of runoff height using capacitance-height recorders, while telemetry systems have linked field measurements to office computers.

Measurement of erosion is discussed under the headings of denudation, sedimentation, and instantaneous sampling techniques. Rainfall simulation is useful in studying component processes of land degradation under controlled conditions. Comments are made on the site requirements and the types of studies carried out using rainfall simulators.

The paper concludes with a discussion of analytical technology being developed at Griffith University. A computer programme, GUESS, is compared

* Queensland Department of Primary Industries, Indooroopilly, QLD 4068, Australia.

with USLE. Discussion of an emerging technology to be evaluated in ACIAR Project 8551 is discussed, along with its data requirements.

Introduction

"There is a lack of standard methodology for assessing the extent of erosion in tropical lands" (Lal, 1988). A perusal of the literature indicates that a wide range of soil-loss values have been reported from a variety of experiments within the tropics. Consideration of results from the studies indicates that it is seldom possible to make comparisons between experiments. Such a situation is a major deficiency in the use of the limited resources available for soil erosion research.

Soil erosion research involves the measurement of precipitation, infiltration, runoff, and soil loss in combination with farming techniques, especially tillage and agronomy. This paper considers instrumentation and methodology of these measurements as they apply to agriculture in tropical lands, based on experience in Queensland, and concludes with a section on erosion models and their data requirements.

Precipitation

The type of instruments used to measure precipitation are the 203 mm rain gauge (sometimes the 127 mm gauge) and siphoning and tipping-bucket pluviometers.

Voluminous literature exists on the problems of measuring rainfall. In the tropics, large volumes of intense precipitation occur over long durations. Tropical disturbances (typhoons) and localized thunderstorms are accompanied by strong winds. These conditions favour the use of large cylinder-siphon types of pluviometers or large sizes of tipping buckets. While the siphon type overcomes the errors involved in high rates of tipping, chart size must be large and therefore costly when short time intervals are needed in plot studies - for example, one minute time intervals.

In plots and small catchment studies (<1 ha), definition of the time interval may need to be as short as 0.5 minutes because the lag time of flows is only very short, in the order of 1 to 5 minutes. Interpretation of rainfall effects, in relation to hydrograph characteristics makes it desirable to have at least four time intervals in the rising limb of the hydrograph. Small plots respond to short bursts of rainfall, thus the requirement to measure at small time intervals. It should be noted that even with 0.2 mm tipping-bucket volumes, there is a discrimination of 12 mm per hour in the level of measurement.

Wind often accompanies the major rainfall/runoff/erosion events. Thus the number and location of gauges becomes important. For plot studies of less than 100 m^2, one pluviometer and two standard 203 mm rain gauges are recommended. Where groups of plots are used and the total length of the experimental area exceeds 100 m, it is recommended that one pluviometer in the centre of the area be supplemented by standard gauges based on the polygon technique, because of the extremely localized nature of tropical thunderstorms. The location of the pluviometer should take into consideration the height of the crop.

A range of pluviometers are available, some of which form an integral part of stage-height runoff recorders (e.g. AUSI and Leupold Stevens). Since Rimco pluviometers are no longer available, we have used a number of similar types manufactured locally in Queensland. In one type, 203 mm rain gauges with milled brass rings are used as a base for pluviometers. An inexpensive plastic tipping bucket developed by CSIRO has been fitted into the collector of the rain gauge, and the instrument has proven robust, accurate and cost-effective. The gauges are positioned 30 cm above ground level and surrounded by splash pads (green artificial turf). However the presence of livestock and wild animals sometimes makes it necessary to raise the pluviometers to a height of 1 m or more. If V-shape tipping-bucket pluviometers are used, it is necessary to clean and polish the stainless steel buckets regularly, because dust that deposits in them will allow water to siphon out by capillary action and the buckets will not tip in low-intensity rainfall.

Runoff

In many soil erosion studies, runoff has often been measured as a total volume, especially from plots. However, with increased emphasis on the study of physical processes, more effort has been concentrated on measuring rates of runoff, necessitating runoff measurement related to a time basis.

Soil erosion studies, because of their nature, immediately restrict the number of instruments that can be used. For example, in low sloping areas, pondage upstream of the measuring device radically alters the transporting capacity of flows, and deposited material must be estimated along with its bulk density. Thus it is necessary to assess the site characteristics when weirs and flumes are being installed. This requires a very careful topographic survey.

The types of instruments for measuring runoff can be classified in terms of small and large experimental catchments.

Small experimental catchments

For small plots, sunken tanks can be used to collect all of the runoff and sediment, which are then subsampled for sediment and nutrient analysis. This approach provides data on an event basis and is simple to construct - but has the disadvantage that the water must be disposed of before the next runoff event. Access to the field sites must be good at all times, especially during prolonged rainfall.

Flow splitters have been used widely, and a number of designs exist. Again, sunken structures are necessary along with skilled installation because the splitting device must be maintained in a perfectly level position. Trash collectors are necessary and these instruments need attention on a daily basis to check for insects building nests inside them.

In Queensland, runoff from small plots is being measured by tipping buckets. A slotted manifold at one end of the trough is used to distribute the water to the bucket under controlled flow conditions, thus enhancing the accuracy of the buckets. Buckets constructed of PVC and having volumes of up to 10 litres per side will service plot areas of up to 200 m^2 (assuming a 0.9 coefficient of runoff, and rainfall at 300 mm/h).

Tests have indicated that while the tip rate is below 10 tips/min, tip counters will provide an event or daily runoff volume. The design reduces the need for major excavation and considerable daily servicing. Above 10 tips/min, a calibration is required because of variations from the static tipping volume due to rates of filling and tipping.

The measurement of small flows by weirs and flumes necessitates constricting the flow so that stage height provides sufficient variation for accurate differentiation. Two main problems are that accuracy of rating curves diminishes considerably below depths of 3 cm, and that the hydrograph base is artificially lengthened. Consequently, in areas of high infiltration that give rise to protracted periods of low runoff, considerable error can occur.

Other problems found from field experience in Queensland indicate that the approach conditions of weirs and flumes must be carefully observed. In one of our studies, because of the considerable upstream ponding, 90° V-notch weirs altered peak discharges by up to 60% when installed in contour bay channels. By carefully selecting runoff measuring points, and shaping the upstream area, it is hoped that Cipoletti weirs will largely overcome this problem.

Weir failure can occur on dispersible soils. Where these soils occur, the ponding of water in upstream areas is definitely not advised, and structures such as Parshall and other cut-throat flumes are preferable.

Large experimental catchments

In Queensland during the 1960s, some large 160° V-notch weirs were constructed in 250 ha farm catchments. Sediment deposition rendered the weirs useless. However, construction of a rectangular sunken box upstream of the V-notch has achieved the necessary scouring of sediment from the channel upstream of the V-notch.

Because of the very high sediment loads, Parshall and San Dimas flumes have been used in contour bays and valley floors of small catchments. Large 1.2 m, 45° HL flumes are being used to measure runoff on clay soils where sediment load is not excessive. The HL flumes have been used in steeper catchments, where their ponding areas are small and their hydraulic approach conditions are stabilized by the construction of a stilling area immediately upstream of the structure.

To reduce the adverse effects of ponding, our approach has been to construct flumes with a throat width somewhat greater than the cross-sectional area of the approach channel, and design the flume surroundings so that when the structure is surcharged the adjacent wing walls act as a broad-crested weir. Failure of conservation works in the subhumid-semiarid Vertisol grainlands of the Darling Downs in Queensland has prompted measurement of runoff and soil movement at scales of 1000 to 5000 ha so that integrated 'nested' catchment studies can ascertain sources of runoff and sediment as well as tracing the movement of eroded material throughout the catchments. Flow measurements at channel cross sections are made by wading and gauging from a bridge because control structures are too expensive. Changes in vegetative cover and channel bed elevation must be controlled regularly so that reliable results can be achieved.

Three types of mechanical stage-height recorders have been used to measure runoff. They are:

Drum recorders

Charts are wrapped around a clock-driven drum and a pen mechanism is operated by a direct or pulley-supported float. Felt-tipped pens have overcome the problem of evaporation of ink in hot dry areas. Two important points about these recorders are:

- charts should be changed after each event, and times recorded on the chart as a measure for checking clock reliability;
- drum size should be sufficient to allow differentiation of time increments that correspond with lag times for the catchment areas under study.

In Queensland, a simplified version of this instrument for use in tillage and cropping system studies has been built. It is limited to measuring flows with less than 50 cm stage height.

Strip chart recorders

The most popular of these horizontal paper feed systems is the Leupold-Stevens recorder. An Australian adaptation of this instrument, the AUSl, which comes with variable speed drives, an electronic clock, and an attachment for registering pumped water samples is used. It is reliable and reasonably priced.

Pressure bubble recorders and pressure transducers

Sediment can be a problem with the transducers, and their use has been restricted to monitoring groundwater. In Queensland, there has been no experience with pressure bubble recorders.

Measurement of stage height using electronic equipment has advanced rapidly. Through the use of potentiometers driven by a float, voltages can be recorded and related to stage height. Capacitance-height recorders, using a teflon-coated wire in a stainless steel tube and connected to some simple circuitary, have revolutionized flow measuring. There have been problems in the drift of base levels, but regular cleaning and checking have reduced these.

Telemetry using radio and telephone connections with field instruments has improved communications and allowed almost instantaneous data processing.

Erosion

In establishing a soil erosion study, cognisance must be taken of the characteristics of the naturally occurring geomorphic units. Size and silting of the plots must be related to these units. For example, care should be taken to have similar types of slopes and depths of soil. Plots should be located at similar slope positions. Within every landscape there are specific sediment sources and sinks, and a particular land use at one location may serve to aggravate erosion and deposition at another. For example, the consequences of activities on a hillslope need to be assessed in terms of their effects on valley floors. This aspect has considerable importance when land management systems are being considered on a catchment basis. Therefore, land management research output must be integrated to the level where this can happen.

In process studies, it is usually forgotten that landforms are often the product of previous climates, and as such are not in equilibrium with contemporary climates. Therefore the design of an erosion study should endeavour to differentiate soil erosion in relation to relict geomorphic features and contemporary climate, as well as modern day land use. In densely populated countries, it is naive to assume that past land uses have not left an effect (for example, timber felling).

Measurement of soil erosion can be considered under three conditions:

Denudation
- erosion pins and small buried cylinders joined by a fine wire
- pedestalling associated with tree roots and grass tussocks
- photogrammetry
- tracers, caesium 137, beryllium 7

Sedimentation (deposition)
- deposition plates, especially for alluvial fans
- sediment collection tanks and troughs
- dams of various sizes

Instantaneous sediment sampling from runoff
- hand sampling
- stage samplers
- aliquot samplers and splitters
- pumping samplers
- turbity meters and infrared nephelometers

Denudation measurements

The technique of erosion pins and buried cylinders makes the assumption that erosion and deposition are point processes, when in fact they are spatially variant. Whether it be in row crops, broadacre grains or pasture lands, there are source and sink areas that are related to catchment size, hill and channel slopes, infiltration, surface roughness, antecedent soil moisture, vegetative cover, underlying geology, and runoff events of a particular recurrence interval. In seasonally humid lands that are characterized by one or two large events every year (typhoons, rain depressions and localized thunderstorms), there is evidence to suggest that erosion is episodic, and sediment moves in pulses from hillslopes to adjacent valley floors. Subsequently, it moves to the larger streams. As the level of vegetative cover increases, the microerosional systems become obliterated, especially with stoloniferous grasses.

Based on this observation, we have used sites with fixed pins, onto which a frame that contains drop pins is placed. In row crops, a series of traverses can be made at 2 to 5 cm spacings to characterize a cross section. This technique is not rigorous enough in situations where soils shrink and swell or where faunal activity is obvious (e.g. termites and ants). It measures only gross erosion and deposition, that is >50 t/ha.

The use of markers that provide a datum and a time have provided a relatively quick method of deriving mean annual denudation rates. Tree stumps and exposed roots have been used in Africa in conjunction with time-sequenced photography. Data from this technique is often more qualitative than quantitative. The major difficulty with this technique has been accurately dating a known surface. Similarly, ash and layers of flood debris in alluvial deposits can act as markers when dates can be ascribed to such events.

A photogrammetric method using an expensive quality camera on a tripod has been used to obtain a high resolution of surface denudation. Measurements can be made to within 0.5 mm, and interpretation of the photographs can be carried out in the office (Sneddon *et al.*, 1984). With the advent of radioactive markets, a whole new field of soil erosion measurement has been opened up. In Australia, Campbell *et al.* (1982) have shown that caesium 137 can be used to provide a good estimate of erosion over the past 30 years, as well as identifying source areas.

Sedimentation (deposition)

Following detachment, some of the material being transported is deposited within a short distance. Removal in this manner is a slow process of hillslope and channel side erosion especially in the case of contour channels. The channels act as major sinks for eroded hillslope materials, and sedimentation plates are used to estimate the volume of materials deposited. The plates are also useful in row crops such as pineapples, where they can be used to measure not only deposition in furrows between rows, but also the consolidation of the beds in which the plants grow. The plates are cheap, readily moved, and can provide a gross figure of soil movement quickly.

Runoff collection tanks are also used for collecting soil loss on a seasonal or on an event or daily basis. By thoroughly mixing the bed and suspended load, total solid loads can be calculated. Where events or daily data are required, tanks need to be washed out after each measurement, or a rigorous subsampling and depth measurement procedure adopted.

Many different types of troughs have been used to collect soil from plots and small hillslope catchments. These have ranged from tarpaulins that were used to line channels at the base of hillslopes to modified gerlach troughs.

Modified gerlach troughs have been used in grazing land and row crops in Queensland. The troughs collect bed load, while the suspended load is taken via a splitter inserted into the side of the manifold. The suspended load sample represents a mean concentration because the splitter takes a sample that is proportional to the flow.

Troughs of 1-, 2-, 5- and 10-m lengths have been used, but as yet no evaluation of the various widths have been carried out. Troughs are well suited to studying the effects of slope curvature on runoff and soil loss, both in plan and profile form, when they are combined with runoff tipping buckets.

While plot boundaries are well defined in row crops, in broadacre crops and grazing lands there arises the problem of the contributing area. Generally, researchers have inserted boundaries, often rectangular in shape, above collection gutters. This makes the assumption that water moves directly downslope, but field observation shows that water often flows downslope in a more tortuous path, especially in grazing lands. A careful inspection of a forested hillside will show that there are many small lobes of soil caused by fallen trees and tree stumps that will act as sources for ready removal when the cover has been cleared. Construction of plot sides on a convex hillside can increase the true contributing area. The University of New South Wales has used diamond-shaped plots, while there is the suggestion that triangular-shaped areas would more closely approximate the true contributing area. The construction of rectangular plots, especially on steep slopes, could well be increasing the sediment yield from the areas under study. Plot sides in particular develop rills along their edges, and bias results.

Small dams and reservoirs act as efficient sediment traps, especially in the monsoon lands where they dry up during the dry season. Where their contributing areas have a relatively homogeneous soil type and land use, sediment cores can be extracted and an average erosion rate calculated for the catchment. A series of larger dams situated progressively down a catchment can provide a useful picture of erosion and sediment delivery where landscape and land use are relatively homogeneous.

Instantaneous sediment sampling

The most reliable results are obtained by hand-sampling during runoff events. Hand-sampling allows variation in both the sampling period and the volume of the sample. The use of waterproof field note books and an accurate watch provide information necessary for linking sediment data to runoff hydrographs. This technique is recommended where field plots are situated on a research station.

Stage samplers have been used to collect water samples on the rising limb of hydrographs. The method provides very basic information because variation in sediment concentration may be quite small, depending on the rise time of hydrographs. Sediment bottles need to be changed after each event. Stage samplers are more suited to sites where flows exhibit a marked fluctuation of stage height. Consequently they are well suited to V-notch weirs and cut-throat flumes. Flow splitters that are slotted in the vertical direction can be used to sample the vertical profile of runoff and the resulting sample will give a mean concentration for the total flow. This system is being used in the ACIAR project. A slotted copper tube was attached to the side of the manifold that distributes water to the tipping bucket. The total sample was collected in a plastic drum, and three subsamples were taken for analysis.

Aliquot samplers, such as the Coshocton wheel, are a more sophisticated type of splitter that overcomes problems of blockage from debris in the flows. This method faces the same difficulties as other splitters in that hornets and spiders must be kept out of the instruments.

At ICRISAT, a siphoning system was attached to the side of a Parshall flume. The water flow was distributed to a set of bottles that allowed 6 minutes of sample to be put into each container.

Pumping samplers have been developed for the waste water industry and are being used for sediment sampling. Such pieces of equipment give options of sample time, number of samples per time interval, variations in sampling time with flow height, and length of purging time. Problems with the instruments concern the capacity of pumps to produce a reliable sample, particularly under varying sediment sizes and types. Generally automatic pumping samplers have not exhibited a high level of reliability, and vacuum systems offer a better chance of obtaining reliable samples.

In Queensland, we have developed our own sampler that used bilge pumps for taking samples. Its reliability has not been high, mainly because of the failure of the impellor pumps, which require a good power supply to start them. Sensing devices which control the pump are another source of unreliability.

Turbidity meters and infrared nephelameters have been used with varying degrees of success in relating turbidity to sediment concentration. The techniques appear reliable where suspended clay sediment loads are involved, but we have had no experience with them.

Denudation by solutes

In erosion studies of the tropical lands, the dissolved load is often overlooked. Under seasonal and annual high temperatures and rainfall, many ions are highly mobile, e.g. Cl, S, Ca, Na, Mg, K, F, Sio_2. In catchments developed on calcareous materials especially, up to 90% of denudation loss can be in the form of solutes. It is also important to understand that nutrient cycling in such cases involves the biota of the ecosystem.

Measurement techniques involve sampling surface runoff, hillslope through flow, and drainage to groundwater.

Rainfall simulators

Various sizes and types of rainfall simulators have been used to study the component process of land degradation under controlled conditions. Meyer (1988) takes the view that simulators are well suited to field studies that make comparisons and provide relative values rather than absolute rates. Long -term field studies should be operated in conjunction with simulations. Types of studies well suited to simulators are:

- evaluation of different crop covers,
- crop residue management,
- types and methods of tillage,
- slope length and angle,
- relative erodibility of different soils,
- cropping sequence and critical periods of erosion, and
- fundamental processes of erosion and runoff.

In deciding the suitability of sites for carrying out the above research the following points should be considered:

- previous land-use history,
- previous erosion on the plots,
- soil type,
- plot uniformity, size, shape, and slope, and access to the plots,
- availability and quality of water used,
- equipment for applying treatments to the plots, and
- disposal of water.

In Queensland, one-metre-square rotating disc simulators (Morin *et al.*, 1967) have been used to derive curve numbers for application to small catchments (contour bay and 10 ha grazed catchments), measure infiltration rates (including ponding and time to runoff) under various tillage treatments, and investigate surface roughness effects on infiltration and runoff.

A rainulator (4 m x 22 m) (McKay and Loch 1978) has been used to the study effects of:

- drop size and rainfall intensity on surface seal formation and infiltration,
- plot length on erosion processes and rates,

- relationships between soil and sediment properties and soil erodibility,
- soil tilth on surface sealing and erosion,
- stubble mulches on interflow rates, and
- soil type on organic carbon and clay enrichment in sediment.

Infiltration data from a one-square-metre rainfall simulator and an 88 m rainulator have been used to provide inputs for a distributed parameter model, ANSWERS (Beasley *et al.*, 1980), to predict runoff and soil loss from small catchments in grazing land and from contour bay catchments in cropped land.

Further information on techniques and instrumentation can be found in Abraham (1986), El-swarfy (1988), Goudie *et al.* (1981), Gregory and Walling (1973), Kirkby and Morgan (1980, Rapp *et al.* (1974) and Turner *et al.* (1984).

Erosion models and their data requirements

Using established methodologies

The universal soil loss equation (USLE)

The methodology which has been most extensively used is the USLE (Wischemeier and Smith, 1978). It is governed by data related to rainfall rate, and is based on the physical concept that "rainfall detaches, and runoff simply transports the soil detached by rainfall".

This methodology aims to provide an estimate of average annual soil loss, so data collection over a significant number of events (at least 5) is desirable in its use. The nomograms developed to infer K values from other basic soil data are now widely recognized as being limited by the data base from which the nomograms were developed (dominantly silty soils from the midwest of the USA). An advantage of this methodology is that it has been used extensively, is well documented, and has built up a useful empirical body of knowledge on the effectiveness of soil conservation methods. This is summarized in the non-point factors, LSCP, of the soil loss equation.

The GUESS programme

Another established methodology is the GUESS model, which is based on an erosion theory outlined by Rose (1988). GUESS is based on the physical assumption that erosion is governed by sediment entrainment in overland flow, and rain splash detachment has little effect. GUESS summarizes soil loss data in terms of two parameters η and ε. The η parameter is identified as the efficiency of net sediment entrainment, whilst ε is the threshold stream power for the entrainment process.

GUESS is available in microcomputer format. The theory manual (Carroll *et al.*, 1986) gives a full description of the assumptions underlying the model, whilst a user's manual (Crawford, 1986) describes the programme and its use.

Comparison of methodologies

Both USLE and GUESS are compatible in terms of the data collected. Thus, GUESS may be used to interpret the same erosion data which has been collected for use in the USLE. Lo *et al.* (1988) provide an illustration of this point where data on erosion from a range of soil types in Hawaii were interpreted using both the K value in the USLE and the GUESS programme. Soils were placed in the same sequence of erodibility using the USLE and GUESS.

GUESS is designed to interpret data from a single erosion event, which is an advantage over the USLE (which is limited to average annual values). The USLE uses data on rainfall rate and soil loss. GUESS uses data on runoff and soil loss (that is, sediment concentration).

Using an emerging methodology

A new emerging methodology which is more closely related to the processes involved in erosion is based on the assumption that entrainment as well as rainfall detachment is an important contributor to sediment concentration. The driving force causing sediment entrainment is the shear stress exerted on the soil surface by the overland flow runoff. This shear stress is parameterized by the stream power, ε, of overland flow (the rate of working of the shear stresses induced by it) and is calculated as the product of the surface slope and the specific discharge (m^3 of water per m width of slope per unit time) of overland flow. As specific discharge is the product of slope length (m) and has excess rainfall rate (m/unit time), it follows that the greater the slope and slope length, the more likely it is that entrainment will dominate over rainfall detachment.

A stumbling block in the application of this physically rigorous approach to erosion is the difficulty in representing the entrainment in any fundamental way. An example is describing the effect of soil strength on the amount of sediment entrained per unit of stream power. Until recently this phenomenon could only be allowed for by adjusting the bulk erosion parameter η (efficiency of entrainment) using a curve-fitting technique so that measured and predicted erosion data coincided.

Recent developments (Rose, 1988) have allowed a more fundamental description of processes by recognizing that the process of deposition continuously returns sediment to the soil surface to form a mechanically weak deposited layer of size-sorted sediment which covers some fraction of the surface. Thus a substantial fraction of the stream power is consumed in entraining original soil material against resistance provided by the strength of the soil with the remaining fraction used in reentraining sediment from the deposited layer. The work done in reentrainment of deposited sediment is done chiefly in overcoming the immersed weight of the sediment.

Data collection and measurement

The interpretation of erosion processes provided in this emerging methodology show collection of the following data to be desirable:

- *Sediment depositability*

    Whether the erosive agent is raindrops or runoff, the resulting sediment concentration depends on the settling velocity characteristics of the soil. This characteristic is the fraction of sediment with a settling velocity in water less than any specified value. If this characteristic is divided into an arbitrary number (I) of classes each of equal mass and defined by a settling velocity V_i for the class, then the 'depositability' of the sediment is defined as $\Sigma V_i / I$. The summation ΣV_i is strongly influenced by the higher values of V_i for larger aggregates.

    To determine the larger values of V_i in the presence of other sediment, the best method available is the modified bottom withdrawal tube, whose use is described by Lovell and Rose (1988a). For reentrainment, the soil characteristic of significance is its settling-velocity characteristic (Lovell and Rose, 1988a, b); for entrainment the soil strength is also important. The most suitable methodology for determining this strength-related parameter is currently under development.

- *Sediment concentration*

    Often it may only be possible to obtain average sediment concentration from total event soil loss and runoff. However, if it is possible to obtain sediment samples during a runoff event this is most desirable, as time-varying sediment concentrations can be used to obtain 'transport limit' and 'source limit' concentrations at any given runoff rate.

    Even if such time-varying data is limited, it can still be very valuable, especially for bare-soil plots, as it allows evaluation of a soil strength-related parameter.

Flow geometry

Rill frequency and rill geometry (if rills occur) can have a significant effect in increasing sediment concentration through their ability to concentrate flow, thus increasing stream power. These parameters can be measured using a combination of photographs and standard surveying techniques after runoff events.

Runoff rate as a function of time

Methods of measuring runoff are described in a previous section. One complication is that unless runoff is measured at the bottom of small rectilinear plots, there is a time lag between specific discharge of overland flow at the end of a plot/contour bay and discharge through the measuring device (for example, weir of a flume). This arises because of water storage effects, and transmission lags in the contour channel, as well as water banking up behind the weir. It follows that the sediment concentration measured at the weir at any time will not be time-synchronized with the bay runoff rate transporting this sediment.

The problem is conceptually similar to the one of deriving excess rainfall rate from measured runoff rates, and a number of computer-based numerical deconvolution techniques are available (Rose *et al.*, 1984).

References

ABRAHAMS, A.D., ed. 1986. Hillslope processes, *Binghampton symposia in Geomorphology*, 416ff. International Series no. 16. London: Allen and Unwin.

BEASLEY, D.B., HUGGINS, L.F. and MONKE, E.J. 1980. ANSWERS: a model for watershed planning. *Transactions of the American Society of Agricultural Engineers* 23:938-944.

CAMPBELL, B.L., LOUGHRAN, R.J. and ELLIOTT, G.L. 1982. Caesium as an indicator of geomorphic processes in a drainage basin system. *Australian Geographic Studies* 20: 49-64.

CARROLL, C., ROSE, C.W. and CRAWFORD, S.J. 1986. *GUESS - Theory manual*. Queensland, Australia: Department of Primary Industries.

CRAWFORD, S.J. 1986. GUESS - user's manual. Queensland, Australia: Department of Primary Industries.

EL-SWAIFY, S.A. 1988. Monitoring of weather, runoff and soil loss. In: *Soil management and smallholder development in the Pacific Islands*, 163-178. IBSRAM Proceedings no. 8. Bangkok: IBSRAM.

GOUDIE, A., ed. 1981. *Geomorphological techniques*.. London: Allen and Unwin.

GREGORY, K. J. and WALLING, D.E. 1973. Drainage basin form and process, London: Edward Arnold.

KIRKBY, M. J. and MORGAN, RP.C. 1980. *Soil erosion*. Chichester: John Wiley and Sons.

LO, A., EL-SWAIFY, S.A. and ROSE, C.W. 1988. Analysis of erodibility of two tropical soils using a process model. *Soil Science Society of America Journal* 52:781-784.

LOVELL, C.J. and ROSE, C.W. 1988. Measurement of soil aggregate settling velocities. I. A modified bottom withdrawal tube method. *Australian Journal of Soil Research* 26: 55-71.

LOVELL, C.J. and ROSE, C.W. 1988. Measurement of soil aggregate settling velocities. II. Sensitivity to sample moisture content and implications for studies of structural stability. *Australian Journal of Soil Research* 26:73-85.

LOUGHRAN, R.J., CAMPBELL, B.L. and ELLIOTT, G.L. 1982. The identification and quantification of sediment sources using C. *Proceedings of the Exeter Symposium*, 361-369. International Associotion of Hydrological Services. IAHS Publication no. 137.

LAL, R. ed. 1988. *Soil erosion research methods.* Iowa: Soil and Water Conservation Society of America.

MEYER, L.D. 1988. Rainfall simulators for soil conservation research. In: *Soil erosion research methods.* Iowa: Soil and Water Conservation Society of America.

McKAY, M.E. and LOCH, R.J. 1978. A modified Meyer rainfall simulator, In: *Conference on Agricultural Engineering* (Toowoomba), 78-81. Australia: Institute of Engineers of Australia.

MORIN, J., GOLDBERG, D. and SEGINER, I. 1967. A rainfall simulator with a rotating disc. *Transactions of the American Society of Agricultural Engineers* 10:74-79.

RAPP, A., BERRY, L. and TEMPLE, T. 1974. *Studies of soil erosion and sedimentation in Tanzania*, 379ff. Australia: University of Daves, Saloam and Uppasala University.

ROSE, C.W. 1988. Research progress on soil erosion processes. In: *Soil erosion research methods.* Iowa: Soil Conservation Society of America.

ROSE, C.W., FREEBAIRN, D.M. AND SANDER, G.C. 1984. GNFIL: *A Griffith University programme for computing infiltration from field hydrologic data.* School of Australian Environment Studies Monograph. Queensland, Australia: Griffith University.

SNEDDON, J., OLIVE, L.J. REIGER, W.A. and LUTZE, T.A. 1984. Erosion measurement using close-range photogrammetry. In: *Drainage basin-erosion and sedimentation*, ed. R. J. Loughran, 153-160. N.S.W., Australia: University of Newcastle.

TURNER, A.K., ed. 1984. *Soil water management*,. Canberra: I.D.P.

WISCHMEIER, W.H. and SMITH, D.D. 1978. *Predicting rainfall erosion losses.* Soil Conservation Service, U.S. Department of Agriculture. Agricultural Handbook no. 537. Washington DC: Government Printing Office.

Field measurements for soil management studies

TED GARDNER, CYRIL CIESIOLKA and ADRIAN WEBB*

Abstract

This paper discusses measurements carried out regularly during field experimentation to explain hydrological and soil erosion processes.

The paper commences with a brief discussion of the effect of plot size on the measurement of these processes. It is concluded that large plots (>100 m^2) are needed for tillage, erosion, and runoff studies. Small plots (<20 m^2) are appropriate for detailed studies such as are required for infiltration, surface sealing, and splash erosion.

The importance of drainage in the water balance equation is then discussed, and a number of direct and indirect methods for measuring it are considered. These range from lysimetry to soil hydraulic properties to the use of salt tracers.

The next section considers methods of measuring surface roughness using photographs, soil casts, and laser surveying. This is complemented with a brief discussion of the importance of cover measurements in erosion models. Methods of measuring cover range from photographs to transects with various degrees of sophistication. The importance of sample numbers in achieving statistical precision is illustrated by a worked example.

The paper concludes with a brief description of methods to measure sediment-size distribution, which is an important input to physically based erosion models.

* Agricultural Research Laboratories, Queensland Department of Primary Industries, Meiers Road, Indooroopilly, QLD 4068, Australia.

Introduction

This paper concentrates on measurements carried out regularly during field experimentation to explain hydrologic and soil erosion processes. The measurement of runoff and soil erosion are not covered in this paper.

Aspects covered include:

- a consideration of the effect of experimental plot size on soil and water processes.
- techniques for measuring the deep drainage component of the soil-water balance.
- measurement of surface roughness and cover for interpretation of both runoff and soil erosion data.
- measurement of settling-velocity distribution of soil particles to allow modelling of soil erosion (transport and deposition).

Size of experimental plots

The question most research scientists ask themselves when a research study is being planned is, "What size plots should/can we use?" Of course there is no perfect answer and the choice is usually a compromise based on available resources, efficiency/cost issues, the scale at which processes operate, the use of the data, and previous experience.

Small plots (less than 20 m^2) have been used widely to study processes such as surface sealing, aggregate stability, raindrop detachment, splash transport, infiltration, and soil strength - that is, most of the processes which are associated with the interrill area at the larger scale.

Advantages lie in the ease of establishment, the cost, and control of variables such as cover, moisture, slope, and soil condition. Problems associated with small plots relate to the relevance of the data to large areas - are the soils behaving in the same way? Variability in larger areas is not accounted for in small plots and edge effects are magnified (for example, splash losses, or water seepage around the edges). However, despite these problems small plots can provide much useful information.

Large plots (>100 m^2) have distinct advantages in field studies in that infiltration and the complete erosion process operating at the larger scale can be measured with greater confidence than in the small scale. Large plots are more appropriate for the study of tillage, soil water, slope length, slope gradient, and crop effects on runoff, erosion, and crop growth. The main disadvantages with large plots are the cost of installation and management.

There can be problems also in locating a sufficient area to have the number of plots required.

Drainage

Mass balance

The 'conservation of mass' statement for the water balance of a unit area of soil can be expressed as

$$P = R/O + \Delta S + ETR + D \qquad (1)$$

where:

P = precipitation (plus irrigation)
R/O = surface runoff
ETR = evapotranspiration
ΔS = change in soil water storage
D = drainage

Equation (1) may be rearranged to solve for D:

$$D = P - R/O - \Delta S - ETR$$

P and R/O are relatively straightforward to measure (rain gauges and weirs), whilst ΔS can be measured either gravimetrically or with a neutron moisture meter. The difficulty is to obtain an independent measure of ETR. This can be done using micrometeorological techniques ranging from a suitably calibrated potential ETR equation (for example, Penman-Monteith) to direct vapour flux measurement, such as the Bowen ratio, Eddy correlation, or lysimetry.

Whichever ETR measurement technique is used, it is important to realize that the D (a relatively small number) is being calculated as the difference between two relatively large numbers (P and ETR). As P and ETR can be measured with a precision of about ±5% and ±10% respectively, the precision of their difference (that is, drainage) is about ±11%. Hence, unless D is greater than about 20-30% of ETR, the uncertainty in calculated D can equal or exceed its mean value.

Lysimetry

Lysimeters are isolated monoliths of soil of known cross-sectional area (usually 2 m - 10 m) designed to keep all soil water flow within the confines of their container walls. They may be either of the weighing, draining, or constant water level type. They are usually used to measure potential ETR or actual ETR, in which case their area relative to plant row geometry, edge effects, and plant vigour relative to the surrounding area are the most important considerations.

However, the volumetric measurement of drainage from the bottom of such lysimeters will provide a direct estimate of drainage from the surrounding soil provided that:

- the soil monolith is undisturbed and retains its field structure;
- the profile soil water status within the lysimeter equals that of the surrounding area; and
- the soil suction at the bottom of the lysimeter approximates that of the surrounding soil.

The latter requirement ensures that an artificial perched water table does not build up at the lysimeter bottom/air interface. Suction at the interface can be manipulated via buried ceramic candles and a vacuum system.

Correlation with soil moisture

In uncropped soils of uniform texture protected from evaporation, the drainage rate from the lower part of the profile is often an exponential function of profile water storage above that depth (Black *et al.*, 1970). When profile water storage is divided by soil depth to calculate an average volumetric water content θ, there is often good agreement between measured drainage rates and the unsaturated hydraulic conductivity, $K(\theta)$, at corresponding water contents. In effect, this means the profile is draining under a unit potential gradient at a rate equal to $K(\theta)$.

The technique has been tested over a wide range of soils (sands to clays) and a useful equation is (Ritchie, 1981):

$$D = 0.1 \exp \beta(\theta - \theta d) \qquad (2)$$

where θd is the average profile water content when $D = 0.1$ cm/day and β is related to the slope of the log linear relationship between θ and D. As 0.1 cm/day is a convenient working definition of "field capacity" conditions, it follows that θd is an estimate of the field capacity moisture content.

Estimates of β and θd are obtained using measurements taken during drainage of a profile protected from evaporation. Subsequently, D from a cropped soil is obtained by substituting measured θ values into equation (2) and solving for D.

If extended moisture data from a draining profile is not available, β can be estimated as:

$$\beta = \frac{\ln K_s - \ln 0.1}{\theta_s - \theta d}$$

where subscript *s* refers to field saturated conditions and 1n means natural log.

Darcy's Law

The verticle form of Darcy's equation is

$$q = -K(\theta) \left[\frac{-\Delta\psi}{\Delta Z} - 1 \right] \tag{3}$$

where q is the soil water flux (mm/day); $K(\theta)$ is the hydraulic conductivity mm/day (that is, K is a unique function of soil wetness), and ψ is soil matric suction (cm). For uniform soil moisture conditions, $\Delta\psi/\Delta Z$ is approximately zero , hence equation (3) can be written as

$$q = -K(\theta) * (-1) = K(\theta) \tag{4}$$

where -1 is the gradient due to gravity alone.

In layered soil, matric potential gradients are not zero, and hence ψ must be either measured directly using tensiometers or calculated indirectly from measured θ and the soil moisture characteristic ($\psi - \theta$).

Provided q is calculated at the bottom of the root zone (beyond the influence of root water uptake), it can be equated with the profile drainage flux D (mm/ day).

The largest practical problem in applying equation (3) in the field is obtaining the appropriate $K - \theta$ function. This is best achieved using the instantaneous profile method (Hillel *et al.*, 1972), where q and ψ are measured on a draining soil profile protected from evalporation, and $K(\theta)$ is then calculated by inverting equation (3). Alternatively, $K - \theta$ can be calculated from the measured soil moisture characteristic, and a relationship based on the capillary rise equation (that is, ψ and pore radius are uniquely related [Hillel, 1980]) with measured K_s used as a matching factor.

Use of Darcy's equation is considerably easier in soils with Campbell-type hydraulic properties. By this we mean:

$$\frac{K}{K_s} = \left[\frac{\theta}{\theta_s} \right]^{2b+3} \tag{5}$$

where b is the slope of the linear log-log soil moisture characteristic.

Whilst not universal, equation (5) has been shown to apply to a wide range of soil textures (Talsma, 1985).

Salt balance

The 'conservation of mass' statement for the slat balance of a soil profile under steady state conditions can be expressed as

$$Q_i * C_i = Q_o * C_o \qquad (6)$$

where Q_i is the net flux of water into the soil profile (rainfall and irrigation mm/yr) of solute concentration C_i whilst Q_o is the flux of soil water below the bottom of the root zone of solute concentration C_o.

As Q_o is equal to D, equation (6) can be restated as

$$Q_o = D = \frac{Q_i * C_i}{C_o} \qquad (7)$$

Equations (6) and (7) are strictly only applicable to steady state conditions, but may be extended using simple theory to soil profiles undergoing gradual change from one steady state to another (Rose *et al.*, 1979).

Q_i and C_i are straightforward to measure. However C_o is the solute concentration of the drainage water below the root zone, and special care is required to convert 1:1 or 1:5 extracts into soil solution concentrations. Many of the dilution-precipitation problems can be overcome if chloride concentration is used in place of total solute concentration.

The technique is least useful in very permeable soil profiles with relatively small Q_i/Q_o ratios because of the error in measuring small C_o values (say, electrical conductivity of 0.05 dS m^{-1} or 10-20 ppm Cl).

Salt tracers

Tracer techniques are based on the assumption that a conservative non-reacting solute will move nearly as piston flow in soil profiles, that is the solute is carried along with the vertical soil water flow (Biggar and Nielsen, 1976).

After introducing a solute (for example potassium bromide) to the soil surface, the solute peak is tracked as a function of time by regular destructive soil sampling. If the average water content during solute movement is θ, then the Darcy flow, V_D, can be calculated as the product of water content and tracer velocity, V_s, that is,

$$q = V_D = V_s * \theta \qquad (8)$$

Points to consider when applying this technique are:

- The introduced solute must be easily distinguishable from naturally occurring solutes.

- Sufficient samples are taken on each occasion to account for variability in the solute peak.
- Sampling methodology is sufficiently precise to recover most of the solute (on a unit area basis) at each sampling.
- The technique is only applicable in uncropped soils or at soil depths below the root zone, because of the requirement to specify a θ associated with drainage.
- If the tracer moves preferentially through macrovoids, there can be a large deviation from the piston flow assumption. Hence the effective θ may be much smaller than the θ calculated from gravimetric sampling. This bypass flow phenomenon can cause up to a fivefold difference in calculated V_D. Unfortunately, there appears to be no easy way to correct for it (Rice *et al.*, 1986).

Surface roughness

The hydrologic processes of infiltration and surface runoff are closely related to, and affected by, microrelief surface storage (Moore and Larson, 1979). Burwell *et al.* (1966) identified oriented roughness produced by tillage equipment, and random roughness which is unrelated to the direction of tillage. Roughness is quantified by measuring surface elevations at a specified density and computing detention volumes or roughness indices such as those related to the standard deviation of the soil surface heights (Moore *et al.*, 1980).

Many microrelief measuring techniques are described in the literature, or developed for individuals needs. They fall into two types; contact and non-contact methods.

Contact methods

The simplest method involves a sheet of white-painted masonite fitted with a grid. The board is pressed into the soil, and either photographed or the outline of the surface is marked onto the board. The roughness coefficient is calculated from this line.

Profile meters are boards of suspended pins, typically 5 cm apart, which are gently lowered to the soil surface. The shape of the profile contact is measured manually (very tediously), photographically, or automatically at the top of the pins. Automatic versions (Podmore and Huggins, 1981) use either a single pin contact or a multiple pin contact (Grevis-James, 1984), and dump the spot heights directly to a microprocessor.

Noncontact methods

One of the latest methods involves a low-power laser which is used to focus an intense spot of light on the soil surface. The reflected light is detected through a modified 35-mm camera in which is built a 256-pixel array. The angle of incidence and distance from the laser to the ground spot are related geometrically, and calculated as spot heights in a microprocessor. Resolutions of 0.1 mm can be achieved (Huang *et al.*, 1988).

Welch and Jordan (1983) described a photogrammetric method to obtain spot heights. A cartographic digitizer is used to analyze photographs from a standard 35-mm camera. Resolution of 6.5 mm were obtained vertically, but the camera was positioned 9.5 m above the ground. Generally, higher resolutions require very high-precision cameras.

Others

Aggregate structure has been measured by pouring white-dyed araldite (for example) onto soils to embed and record the surface geometry. Samples are then retrieved from the ground, cut with a diamond saw, polished, and scanned using an image analyzer. These methods are generally unsuitable for depression storage measurements (Dexter, 1976).

Depressions can be stabilized by bitumen, grout, resin, or plastic sheet and their volumes calculated by pouring on water. Such methods are fast but imprecise. Plastic sheet has the advantage of being nondestructive to the plot.

Comments on techniques

Resolution is limited by the density of measurement. Profile meters are typically used in 5 x 5-cm grids (Moore *et al.*, 1980). Laser scanners can produce almost infinite resolution to less than 1 mm in the vertical axis. Random roughness, which is related to the standard deviation of height measurements is a useful comparator of roughness treatments. Linden and van Doren (1986) devised a statistical technique which used two surface configuration parameters (liming slope and limiting elevation difference) to model erosion processes. Moore *et al.* (1980), developed a model for estimating flow into and out of miniature subbasins during ponding and runoff. Volumes of depression storage can be calculated from spot height (X, Y, Z) coordinates using computer programmes such as SUPER by Golden Software.

Measurement of cover

The amount and rate of runoff from a catchment as well as the sediment concentration of that runoff has been shown to be strongly influenced by the amount and type of cover (Freebairn and Boughton 1985; Freebairn and Wockner, 1986; Rose and Freebairn, 1983).

Recent experimental evidence and theoretical considerations suggest there is a need to distinguish between two types of cover.

C_e The fraction of soil surface exposed to the splashing action of rain drops. (This influences the amount of soil detached by rainsplash as well as the formation of surface hydraulic throttles.)

C_r The fraction of the soil exposed to the shear stress from overland flows. (This cover is strongly related to the efficiency of sediment entrainment.)

Although both cover parameters have a direct effect on sediment concentration, the soil surface exposure factors are different. For a crop such as sunflower, with a high leaf area but relatively few stems per unit area $(1 - C_e)$ the cover will be much larger than $(1 - C_r)$. When the surface is covered by plant residue or stone, then $C_e = C_r$.

Measurement techniques

There are five common methods for measureing the amount of vegetative and residue cover.

Quadrat

Aboveground vegetation and residue are clipped from within a randomly placed quadrat, usually 1 m^2. The collected material is washed free of soil, dried and then weighed. Alternatively, the C_e and C_r values are estimated visually by the operator, who refers regularly to a set of photographic standards. The technique is rapid and the results are immediately available. It is best to have two operators who estimate the same quadrats independently of each other, but who then check themselves and the photographic standards regularly.

Overhead photographs

A camera is attached to a pole and suspended vertically over the area to be sampled. The camera shutter is activated by a self-timer or an extension cable. The colour slides are projected onto a grid pattern, and the percentage cover is equated to the proportion of hits the grid makes with the cover. Parallax error occurs at the edges of the slide film, and this may be corrected in part by having a reference scale in the photograph. However, as the cover is "seen" obliquely at the edge of the photo, rather than vertically as occurs at the centre of the field of view, the ground cover at the edges is overestimated. It is for these reasons that Elwell and Gardner (1975) recommended a quadrat sighting frame in place of photographs. An improved sighting frame is described by Stocking (1988).

Metre stick

A metre stick is placed on the surface, either perpendicular to the plant rows or at three different angles (at each measuring site), and the number of intercepts at 1 cm intervals recorded. In leafy crops such as shrubs, cotton, and young trees, the intersect method requires sighting through foliage, and errors usually occur. Canopy cover can be measured as the fraction of shadow over the stick on sunny days. As the percentage of shadow changes with the sun's elevation angle, the same time of day should be used during a crop cycle.

Line transect

A tape or cable is stretched across the area to be sampled. Beads spaced along the tape are judged to be in contact with the residue. A modification of the technique can be used to assess gap frequency in tall crops such as kenaf, cassava, and trees. Here a right-angle prism with cross hairs replaces the "beads" of the tape.

Wheel point

The wheel point consists of a number of wheel spokes pushed along line transects by the operator. Cover is recorded if plans or residue are struck by the point. The technique is used widely in pasture surveys for both canopy cover and basal area measurement. As the method can be destructive (because of operator foot traffic), it is not suitable for frequent use on small plots.

Bias and precision

Measurement bias occurs when the statistical true value differs from the scientific true value. Precision refers to the dispersion of observations around the mean, regardless of whether or not the mean value approximates the scientific true value. Precision can be increased by replication, whilst bias can only be removed by a change in experimental technique. The edge effect of overhead photos is an example of bias. Bias may be checked by using different measurement methods on the same (residue) sample. Laflen *et al.* (1982) compared the photographic, metre stick and line transect methods on surface residue. There was no difference in the mean value, but precision did vary between methods, with the line transect being the most statistically efficient.

Statistical efficiency for this situation can be judged by the sample number required to achieve a target percentage error (PE) of the mean at a given confidence interval. This statement can be expressed mathematically as:

$$n = t^2_{(n-1)} \frac{CV^2}{PE^2}$$

where t is students t at n-1 degrees of freedom at the $(1-\alpha)100\%$ confidence interval, CV is the coefficient of variation of the population, and PE is the sampling error expressed as a percentage of the mean. PE is set by the user.

Table 1 shows the application of this equation to assess the precision of measuring sorghum crop cover using overhead photographs. It is important to appreciate the strongly nonlinear relationship between sample number and PE at a given confidence interval, and between sample number and confidence interval at a given PE.

Table 1. The number of photographs required to satisfy a range of error levels and confidence limits for two field methods.

Error as a percentage of the mean	Field method used	Confidence limit ± %			
		99	95	90	80
2	Random walk	294	168	168	71
	Monitoring site	306	175	123	75
5	Random walk	50	29	20	12
	Monitoring site	52	29	21	13
7	Random walk	27	15	11	7
	Monitoring site	27	16	11	7
10	Random walk	15	9	6	4
	Monitoring site	15	9	6	4

A clear appreciation of the minimum acceptable precision of cover estimates for subsequent inputs into hydrology-erosion models will avoid unnecessary field effort.

Particle settling velocity

Particle settling velocity is a function of the size, density and shape of a particle, the viscosity and density of the fluid, and the concentration and distribution of other particles within the fluid.

It is of fundamental importance to the process of sediment transport and sediment deposition, and as such is an important parameter in the modelling of soil erosion by water (Rose *et al.*, 1979). Eroded sediment consists of primary soil particles and soil aggregates, and can be measured directly using settling columns.

Two simple methods commonly used in Queensland are described by Hairsine and McTainsh (1986) and Lovell and Rose (1988). The latter uses a modified bottom withdrawal tube.

References

BIGGAR, J.W. and NIELSEN, D.R. 1976. Spatial variability of the leaching characteristics of a field soil. *Water Resources Research* 12:78-84.

BLACK, T.A., GARDNER, W.R. and TANNER, C.B. 1970. Water storage and drainage under a row crop on a sandy soil. *Agronomy Journal* 62:48-57.

BURWELL, R.E., ALLMARAS, R.R. and SLONEKER, L.L. 1966. Structural alteration of soil surfaces by tillage and rainfall. *Journal of Soil and Water Conservation* 21:61-63.

DEXTER, A.R. 1976. Internal structure of tilled soil. *Journal of Soil Science* 27:267-268.

ELWELL, H.A. and GARDNER, S. 1975. *Comparison of two techniques for measuring percent canopy cover of rice crops in erosion research programmes.* Research Bulletin no. 19. Harare, Zimbabwe: Department of Conservation and Extension.

FREEBAIRN, D.M. and BOUGHTON, W.C. 1985. Hydrologic effects of crop residue management practices. *Australian Journal of Soil Research* 23:23-35.

FREEBAIRN, D.M. and WOCKNER, G.H. 1986. A study of soil erosion on vertisols of the eastern Darling Downs 1. The effects of surface conditions on soil movement within contour bay catchments. *Australian Journal of Soil Research* 24:159-172.

GREVIS-JAMES, I.W. 1984. Applications of single-board microcomputers. *Proceedings of the Conference on Agricultural Engineering* (Bundaberg), 185-189.

HAIRSINE, P.B. and McTAINSH, G. 1986. *The Griffith tube: a simple settling tube for the Griffith University.* Queensland, Australia: Griffith University.

HILLEL, D. 1980. *Fundamentals of soil physics.* New York: Academic Press.

HILLEL, D., KRENTOS, U.D. and STYLIANOU, Y. 1972. Procedure and test of an internal drainage method for measuring soil hydraulic characteristics *in situ. Soil Science* 114: 395-400.

HUANG, C., WHITE, I., THWAITE, E.G. and BENDALI, A. 1988. A non-contact laser system for measuring soil surface topography. *Soil Science Society of America Journal* 52:350-355.

LAFLEN, J.M., AMEMIYA, M. and HINTZ, E.A. 1981. Measuring crop residue cover. *Journal of Soil and Water Conservation* 36:341-343.

LINDEN, D.R. and VAN DOREN, D.M. 1986. Parameters for characterizing tillage-induced soil surface roughness. *Soil Science Society of America Journal* 50:1560-1565.

LOVELL, C.J. and ROSE, C.W. 1988. Measurement of soil aggregate settling volocities. 1. A modified bottom withdrawal tube method. *Australian Journal of Soil Research* 26:55-71.

MOORE, I.D. and LARSON, C.L. 1979. Estimating micro-relief surface storage from point data. *Transactions of the American Society of Agricultural Engineers* 22:1073-1077.

MOORE, I.D. LARSON, C.L. and SLACK, D.C. 1980. *Predicting infiltration and micro-relief surface storage for cultivated soils.* Water Resources Research Center. Bulletin no. 102. Minnesota, USA: Graduate School, University of Minnesota.

PODMORE, T.H. and HUGGINS, L.F. 1981. An automated profile meter for surface roughness measurement, *Transactions of the American Society of Agricultural Engineers* 24:663-665, 669.

RICE, R.C. BOWMAN, R.S. and JAYNES, D.B. 1986. Percolation of water below an irrigated field. *Soil Science Society of America Journal* 50:855-859.

RITCHIE, J.T. 1981. Soil water availability. *Plant and Soil* 58:327-338.

ROSE, C.W., DAYANANDA, P.W.A., NIELSEN, D.R. and BIGGAR, J.W. 1979. Long-term solute dynamics and hydrology in irrigated slowly permeable soils. *Irrigation Science* 1:77-87.

ROSE, C.W. and FREEBAIRN, D.M. 1983. A new mathematical model of soil erosion and deposition processes with applications to field data. In: *Soil erosion and conservation*, ed. S. A. El-Swaify, W.C. Moldenhauer and A. Lo, 549-557. Iowa, USA: Soil Conservation Society of America.

ROSE, C.W. WILLIAMS, J.R., SANDER, G.C. and BARRY, D.A. 1983. A mathematical model of soil erosion and deposition processes. I. Theory for a plane land element. *Soil Science Society of America Journal* 47:991-995.

STOCKING, M.A. 1988. Prediction of hydraulic conductivity from soil water retention data. *Soil Science* 140:184-188.

TALSMA, T. 1985. Prediction of hydraulic conductivity from soil water retention data. *Soil Science* 140: 184-188.

WELCH, R. and JORDAN, T.R. 1983. Analytical non-metric close range photogrammetry for monitoring stream erosion. *Photogrammetric Engineering and Remote Sensing* 49:367-374.

Soil biomass evaluation: its quantity and composition in relation to cropping systems in Thailand

PRASOP VIRAKORNPHANICH*

Abstract

Soil organic matter composed of nonliving components and soil biomass comprising the living cells of microorganisms have an important role in soil fertility. The most important components of soil biomass are the carbon and nitrogen contents. Their relationship is largely dependent on the soil environment and C/N ratios, and also on the amount and nutrient composition of the organic materials available in the soil.

CO_2 formation on mulching with different materials was determined by the chloroform technique. This showed that flush carbon, and hence the activity of aerobic bacteria, in a reddish brown lateritic soil, incubated with additions of cronstalk and ammonium sulphate, reached a peak after two weeks of incubation. The maximum value was about 13 mg CO_2-C/g soil. On the other hand, when a mulch of thornless minosa was incorporated, a much higher CO_2 formation, equivalent to the release of 108 kg N/ha of soil, was observed.

The flush in decomposition rate was more rapid from freshly added organic residues than from native soil humus. The amount of native soil humus (except that which was newly accumulated) was related to the clay content. The use of residues of thornless mimosa was an effective way of increasing the carbon content in the soil.

* Soil Science Division, Department of Agriculture, Bangkhen, Bangkok 10903, Thailand.

The paper also presents data to show that mulching with Stylosanthes humata *L. (as living plants and thus with only their litter playing a mulching role) or prunings from* Leucaena leucocephala *improved the yield of cassava and rice on poor soils in the ustic moisture regime in Thailand.*

Introduction

During the decade from 1965 to 1975, food production increased at a slightly faster rate than population in food-deficient countries (IFPRI quoted by Sanchez and Salinas, 1982). This achievement is due to a number of factors, which include the development and adoption of high-yielding varieties of several crops with improved agronomic practices. However, in several tropical developing countries, the main intrinsic factor for increasing food production is the opening and expanding of new lands and irrigation systems. The expansion through increased areas is very limited nowadays, and the use of irrigation is limited to relatively small areas, as it is a costly agricultural investment. In Thailand, statistics (Office of Agricultural Statistics, 1982) show that the yield of upland crops such as corn, cassava and sugarcane, which are mostly grown on soils classified as Ultisols and Alfisols, has not increased - but has actually shown a gradual decrease, despite the introduction of improved agrotechnology. Thus there is little doubt that increasing productivity on land already under cultivation is the principle way to increase food production.

Studies on soil fertility in Thailand, which have been carried out by Igarashi *et al.* (1980), Kubota *et al.* (1979), Suzuki *et al.* (1980), Inoue *et al.* (1984) and Nakaya *et al.* (1986), have shown that soil productivity tends to decrease through the depletion of soil organic matter. Organic matter in soils mainly serves as a supplier of three important nutrient elements, namely nitrogen, phosphorus, and sulphur. Through its effect on the physical properties of soils, organic matter also increases the amount of water a soil can hold. Thus to obtain the most economic return for any input, soil organic matter is important, as it is needed to keep a balance between several functions of the soil which are related to connected crop production (Flaig, 1977).

In general, plants supply organic materials to soil in the form of leaves, stems, twigs, seeds, trunks, roots, etc. Although a large part of plant biomass is decomposed by soil organisms on and in the soil, a part of its decomposition products goes into the microbial cells, which are called soil biomass. Some parts of such newly formed biomass are humified and remain there. The soil microbial biomass is considered to be a small but labile pool of plant nutrients (Jenkinson and Ladd, 1981). A decrease of soil organic matter has unfavourable effects on soil microbial biomass, and thus on soil productivity, if some provisions are not made by man.

The soil biomass and its evaluation

Generally, biomass can be defined as the amount of living matter per unit area or per volume of habitat (Mulongoy, 1988). The soil biomass is estimated by making a microbial count, which requires a knowledge of cell weights. The components of microbial biomass to be measured are present in all living cells but absent from all nonliving material. It exists in fairly uniform concentration in all cells, regardless of environmental stress.

The number of bacteria is extremely high, often reaching one billion per gram (Stevenson, 1982). Actinomycetes are the next largest group, and usually occur in the order of several hundred million per gram. The number of fungi varies widely, typically ranging from as few as 20 000 to as many as 1 000 000 fungal propagules per gram (Alexander, 1977). The propagule is considered as any spore, hypha or hyphal fragment capable of giving rise to a colony. Algae are abundant in habitats in which moisture is adequate and light can penetrate, but their numbers are somewhat lower (between 10 000 and 3 000 000 per gram), and they can be observed with the naked eye. Protozoa are found in populations as large as 100 000 to 300 000 cells per gram, though values between 10 000 and 100 000 are more typical.

It is well known that the microorganisms of each soil depend on soil properties, while the environment can bring about large differences in population and types within the same soil. Araragi and Tangcham (1974) collected paddy soil samples to examine and report the numbers and cause of microbial fluctuation between fresh and brackish water alluvial soils in the central plain of Thailand. The numbers of aerobic bacteria, including both ammonifiers and nitrifiers, actinomycetes and azotobacters, were higher in the inner areas of fresh water alluvial soil than in those near the roadside, while the opposite was the case with denitrifying bacteria. Such differences in microbial numbers between two sites was not found in brackish water alluvial soils. For low humic gley soils in the northeast of Thailand, only actinomycetes and nitrobacters showed higher numbers in the inner areas than near the roadside, whereas the number of denitrifiers was similar to the number in the two soils in the central plain. The spots of higher and lower conglomerations of microorganisms were attributed to differences in elevation, and to the supply of organic matter by buffaloes and ducks. The differences in the order of numbers and types of microorganisms was: fresh water alluvial soils > low humic gley soils > brackish water alluvial soils.

The numbers were usually higher in the rainy season than in the dry season, except in marine alluvial soils. The numbers of each microorganism were generally higher in the topsoil than in the subsurface soil in the rainy season. This was not the case with the non-spore-forming groups of anaerobic bacteria (ammonifiers and nitrifiers).

Araragi *et al.* (1979) also examined the numbers of microorganisms in upland fields (reddish brown lateritic soil), and showed the following sequence: aerobic bacteria (9×10^6/g) > actinomycestes (7×10^6/g) > denitrifiers (5×10^5/g) > fungi (7×10^4/g) > ammonifiers (2×10^4/g) > azotobacters (6×10^3/g). Chairoj *et al.* (1988) also incubated reddish brown lateritic soil at 30°C for 20 weeks and found that the numbers of microorganisms were highest after two weeks of incubation under soil with cornstalk, with and without ammonium sulphate applied. An exception was anaerobic bacteria. The decreasing sequence of microbial numbers per gram of soil was aerobes > actinomycetes > fungi > anaerobes. The numbers of aerobic bacteria, actinomycetes and fungi in soil treated with cornstalk and ammonium sulphate were 30×10^6, 9×10^6 and 1.4×10^6/g. This was more than that in soil treated with only cornstalk (25×10^6, 7×10^6 and 6×10^5/g) at two weeks of incubation, and more than that in untreated soil (13×10^6, 2×10^6 and 3×10^5/g), even after six weeks of incubation. The results of the above incubation experiment can be interpreted in terms of slow nitrogen mineralization when cornstalk was applied as mulching material in the field. Thus plants may show nitrogen deficiency symptoms in the early stages of growth.

Soil biomass nitrogen and its components

The nitrogen suppliers in the soil may be separated into four components:

- inorganic nitrogen, including nitrogen fixed or trapped in the clay micelle;
- biomass nitrogen, being nitrogen synthesized within the living cells of microbes;
- microbial cell mass and other easily decomposable organic matter (non-biomass nitrogen); and
- slowly released soil organic matter such as humus or organomineral complexes showing resistance to microbial degradation.

The amount of nitrogen supplied from slowly released soil organic matter may be negligible. Therefore only three types of nitrogen (inorganic nitrogen, biomass nitrogen, and nonbiomass nitrogen) were considered by Watanabe *et al.* (1989) as being pertinent in relation to the available nitrogen under corn cultivation in the reddish brown lateritic soils of Thailand.

Fumigated soils from the Ap horizon (0-15 cm) were collected from Phraputthabat Field Crop Experiment Station, Lopburi Province. Cornstalk, with or without alcohol-free chloroform, had been applied to the samples. The Jenkinson and Powlson (1976) method was used to measure the amount of cell carbon mineralized to carbon dioxide in the succeeding ten days from the fumigated soils. Watanabe *et al.* (1989) obtained a k value of 0.45 for a range of organisms, a value the same as that obtained by Jenkinson *et al.* (1976). However, the Anderson and Domsch (1978) method uses a k value of 0.41, and this now seems to be the most appropriate value to use.

The biomass B (mg C 100/g oven-dry soil) in duplicated samples can be calculated from:

$$B = \frac{X - x}{k}$$

where X is the carbon dioxide produced by fumigated soil in ten days (mg CO_2/ C), and x is that produced by unfumigated soil in twenty days. The biomass nitrogen (mg N 100/g) can be calculated from: $B/^{1/}6.7$, where 6.7 is the C/N ratio of microorganisms (Anderson and Domsch, 1978). Ammonium and nitrate nitrogen in soil extracted with 2N KCl were determined by steam distillation with MgO and Devarda's alloy in accordance with the Bremner (1965) method.

The effect of mulching on soil biomass nitrogen under corn cultivation, with and without nitrogen fertilizer application, is shown in Figure 1. It was observed that an increase in the rate of nitrogen fertilizer gave a negative effect on the amount of soil biomass nitrogen, which after gradually increasing reached a maximum at two weeks after planting (26 June 1986) when 100 kg/ha of nitrogen was applied to the corn.

The nitrogen present in soil as a source of nutrient for microorganisms, together with the application of organic material as an energy supplier for microorganisms, are conventionally considered to be the most important sources for increasing soil biomass. Thus increasing the amount of mulching materials with a high C/N ratio can delay the development of the maximum amount of soil biomass nitrogen. The effect of applying organic material to the soil on soil biomass nitrogen (Figure 1) may be attributed to the suppression of microbial activities by changes in the soil C/N ratio during organic material decomposition.

A seasonal variation of the available nitrogen obtained after 3 weeks of incubation of the same soil samples (Watanabe *et al.* 1989) is shown in Figure 2. The amount of available nitrogen can be accelerated by the application of nitrogen fertilizers. The more organic materials the soil contains, the less nitrogen will be available from the nitrogen fertilizer which is applied, because the absorption of nitrogen by microbes will be higher. However, to

obtain a vigorous growth and high yield of corn, it is necessary to apply nitrogen fertilizer before planting the corn, particularly when mulching is practiced (Inoue *et al.*, 1984; Watanabe *et al.*, 1989).

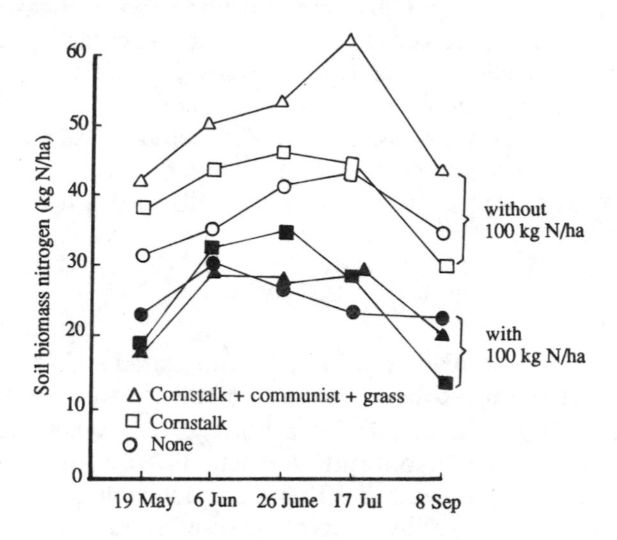

Figure 1. The effect of the application of organic materials and nitrogen fertilizer on soil biomass nitrogen in Oxic Paleustults in 1986 (adapted from Watanabe *et al.*,1989).

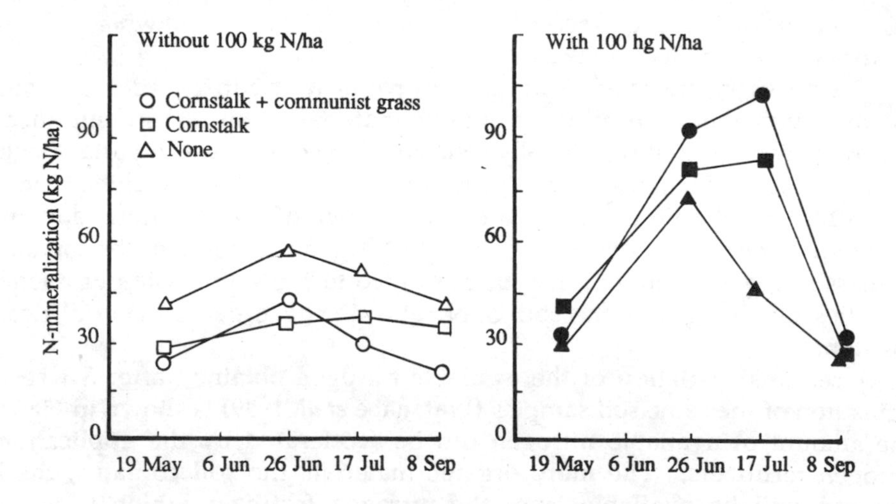

Figure 2. The effect of the application of organic materials and nitrogen fertilizer on available nitrogen in Oxic Paleustult in 1986 (adapted from Watanabe *et al.*, 1989).

CO_2 *formation*

CO_2 formation is more accelerated in fumigated soil (both paddy and upland soils) than in unfumigated soil (Watanabe *et al.*, 1989, Inubushi and Wada, 1988). A comparison of the results reported by Inubushi and Wada (1988) and Watanabe *et al.* (1989) shows that the incubation period when flush carbon reached a maximum value was longer under anaerobic conditions in paddy soils than under aerobic conditions in upland soils. Flush carbon (or the difference between the amount of CO_2 in the fumigated soil and that in the unfumigated soil) reached a maximum value (about 20 mg CO_2-C/g soil) after 14 -21 days of incubation of the paddy soil (Inubushi and Wada, 1988). With the upland soils, on the other hand, a maximum value of about 13.1 mg CO_2 - C/g was obtained after 14 days of incubation (Watanabe *et al.*, 1989). If the value of flush carbon refects the amount of microbial biomass carbon, then the decomposition of the microbial biomass may be slower under anaerobic conditions than under aerobic conditions.

The amount of soil biomass carbon is in parallel with that of soil biomass nitrogen. This was shown by Watanabe *et al.* (1989), who measured flush carbon from several treatments of corn intercropped with legumes and with mulch applied in the following year (Figure 3). The results were similar to those obtained for soil biomass nitrogen reported earlier. However, a high amount of soil biomass has been observed in mulching cornstalk with mimosa, even without fertilizer application. This exceeded 48 mg C/100 g soil, corresponding to 108 kg N/ha of soil biomass nitrogen (Figure 3).

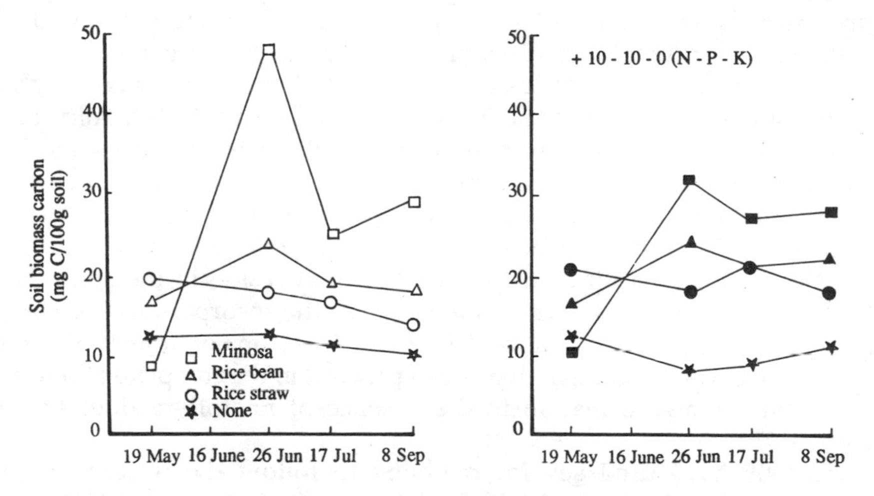

Figure 3. The effect of rice straw and leguminous crops used as mulching materials on the soil biomass carbon in Oxic Paleustults in 1986 (adapted from Watanabe *et al.*, 1989).

Soil organic-matter decomposition

Soils in tropical America, Africa and Southeast Asia largely belong to the order of Ultisols. The major soils occurring in Thailand also belong to this order, and many of the regions they occupy are being rapidly developed. The rapid depletion of soil organic matter (at yearly rates of 5 to 10%) can result in markedly detrimental effects to plant production and agroecosystems (Sanchez, 1976). The organic matter in tropical soils decreases at alarming rates due partly to inadequate soil management practices, and partly to the short supply of biomass. Some workers consider that the content of organic matter is lower in tropical soils than in temperate soils, because of the higher soil temperature and higher decomposition rate of organic matter by soil microorganisms in tropical soils. This view has, however, been opposed by Sanchez (1976). He compared the organic carbon contents of randomly chosen Oxisols, Ultisols, and Alfisols from the United States. The organic carbon contents of the top one metre of the Oxisols were not significantly different from those of the temperate Mollisols, and there was also no difference between the tropical and temperate Ultisols and Alfisols.

The terminology for soil organic matter and humus, as discussed at the 8th International Congress of Soil Science (ICSS) meeting in Bucharest in 1964 indicated clearly a diversity of opinions. No further discussions have been held on this matter, and no unified opinion yet seems to have been reached.

The statements of Hayes and Swift (1978) based on the grouping of soil organic matter proposed by Kononova (1966 and 1975) are that the complete soil organic fraction is made up of living organisms and their undecomposed, partly decomposed, and completely transformed remains. Soil organic matter is the term used to refer more specifically to the nonliving components, which are a heterogenous mixture composed largely of products resulting from microbial and chemical transformations of organic debris. These transformations, known collectively as the humification process, give rise to humus, which is a heterogenous mixture of substances having a degree of resistance to further microbial attack. Thus the decomposition of organic matter depends on the activities of numerous different populations that reside in the soil. Though earthworm populations are very important in the incorporation of the plant residues into the soil, bacteria and fungi play the major role in the mixing process. The decomposition of debris can proceed in the complete absence of all animals, but humus formation in the absence of microflora does not occur (Allison, 1973).

Scientists have used several methods to follow the decomposition of organic matter in the field and in the laboratory. Jenkinson (1971) reported that the proportion of the added residual carbon remaining in field soils after one and five years was about one-third and one-fifth respectively. He noted that this proportion was not affected by climatic conditions or plant materials, and that even fresh green manure behaved in this way - contrary to the wide-spread opinion that such residues decompose rapidly and completly in soil. Similar results have been obtained in the laboratory (Marumoto *et al.*, 1972).

Virakornphanich *et al.* (1987) incubated 5% alfalfa meal in Thai Ultisols, Oxisols, Alfisols and Inceptisols in a short-term experiment in the laboratory, with the water content of the soil adjusted to about 50% of its water-holding capacity. The incubation was carried out for up to 30 days at about 23-25°C. Their report showed a relationship between soil organic-matter contents and soil texture, except in two Oxisol samples and one Ultisol sample (Figure 4).

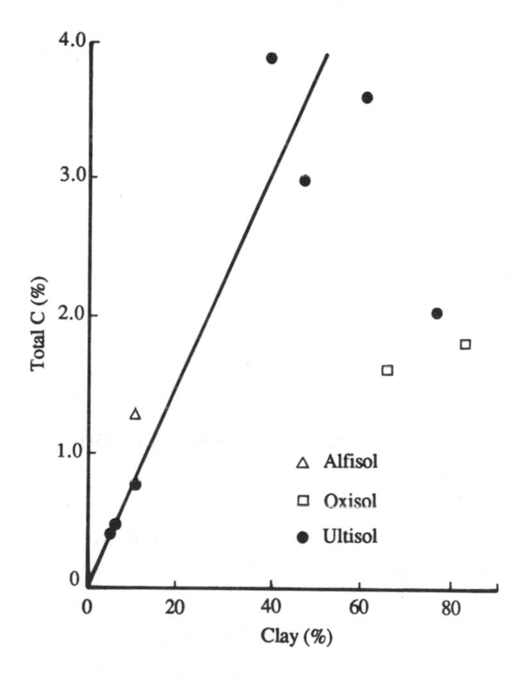

Figure 4. Relationship between total C and clay content for Ultisol Oxisol, and Alfisol samples (Virakornphanich *et al.*, 1987).

This was also found applicable for larger numbers of Thai and Korean Ultisols and Alfisols (Virakornphanich *et al.*, 1988). In one month, the rate of loss of the added carbon was in the range of 46 to 64%. The authors suggested that the higher the pH of the incubated sample, the lower was the retention of added organic matter, irrespective of differences in soil type and treatments. The two Oxisol samples and one Ultisol sample used earlier showed a high increase in soil pH due to the addition of alfalfa. The relatively weak pH buffering by clay in these soils, together with the observed effect of pH on the retention of newly formed humus, may explain why the two Oxisols are exceptions to the general correlation between the retention rate of added carbon and the clay content (Figure 5).

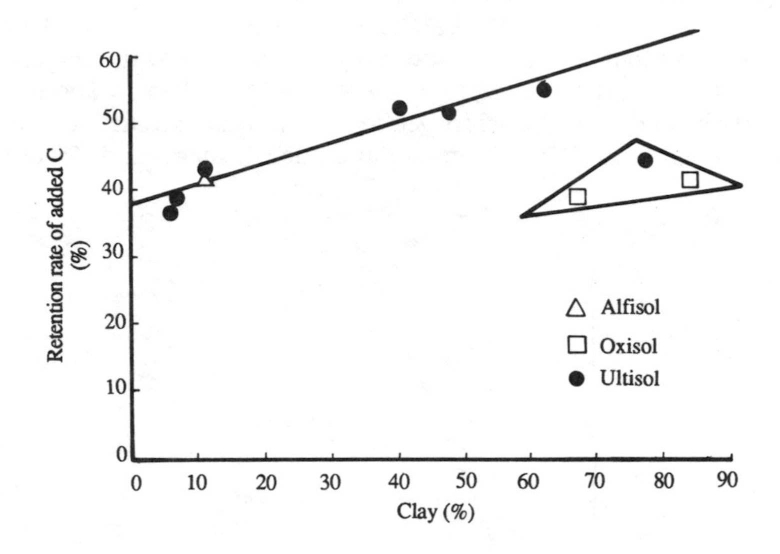

Figure 5. Relationship between the retention rate of added C and clay content for Ultisol, Oxisol, and Alfisol samples (Prasop Virakornphanich *et al.*, 1987).

In the field, there is a rapid flush of decomposition which accompanies the addition of residues. This flush rapidly decreases in a few months, but several years later the added organic carbon will still be decomposing at a more rapid rate than the native soil humus (Jenkinson, 1971). This supports the idea, as the line in Figure 4 indicates, that the humus present in the soil is zero at 0% clay, whereas newly formed humus is retained in large amounts, even in soils without clay (Figure 5).

Watanabe *et al.* (1989) used the kinetic equation to calculate the biodegradation speed of cornstalk added to Thai reddish brown lateritic soil, with and without nitrogen fertilizer, and incubated at 30°C in an incubator. The degradation coefficient k was calculated from:

$$\ln = \frac{A_t}{A_o} = k_t$$

where ln in the natural log, A_t is organic carbon at a given time, and A_o is organic carbon at the commencement of the incubation.

The degradation coefficient of cornstalk was, on an average, about 0.02565/ day during the first 20 days of incubation. The decrease of the k value was nearly half (0.01292/day) after two to four weeks in a preincubated soil. Soil added with cornstalk and nitrogen fertilizer gave a higher CO_2 evolution than that without nitrogen fertilizer in the first 20 days of incubation. The different CO_2 evolution was attributed to the increased microbial population, accelerated by the addition of nitrogen. However, at each period of termination of incubation (2, 4, 6, 8, 10 and 20 weeks), the CO_2 evolution from soils with and without cornstalk, but with nitrogen fertilizer, was lower than that from soil with cornstalk but without nitrogen fertilizer. Finally, the total C of the cornstalk remained at about 48-50% of that added for both the soils (with and without nitrogen application).

Nakaya *et al.* (1986) estimated the decomposition ratio of cornstalk which was mulched on the bare land of Thai reddish brown lateritic soil after harvesting in the early dry season. They showed that approximately 20% of the initial organic matter in the cornstalk was decomposed in two months, and it remained the same for up to six months. Suzuki *et al.* (1980) studied the decomposition rates of several plant residues, including corn leaf and stem, either surface applied or incorporated into soils similar to those used in the experiments of Nayaka *et al.*, (1986) in the wet season. Their results showed that plant residues applied to the surface of the soil showed lower loss in weight han those buried in the soil. The decomposition rates of buried samples were nearly double those of residues applied to the surface.

Under anaerobic conditions, the decomposition rate is dependent on anaerobic bacteria, which operate at a much lower energy level and are less efficient than aerobic bacteria. Thus both mineralization and immobilization rates are considerably ratarded. Snitwongse *et al.* (1988) have used tracers to follow the decomposition of rice straw in paddy soil (low Humic gley) in northeastern Thailand (Figure 6). Under submerged conditions, 3.3 t/ha of straw incorporated into the soil in planted plots decomposed slightly faster than that in the bare plots. The bare and planted submerged plots showed losses of ^{14}C-carbon of about 21.1 and 29.4% respectively, while the nonsubmerged plots gave a lower loss of ^{14}C-carbon (about 16.8%) within 4 weeks. Scharpenseel (1987), quoted by Snitwongse *et al.* (1988), reported 80-90% ^{14}C losses during the first year of rice straw incorporated into aerobic and submerged bare rice soil in the IRRI farm (Figure 6). The differences in decom-position between submerged planted soils and bare soils was small. They explained that this could be due to the fact that during the rice growth the degree of aeration was enhanced, as the rice root excrete oxygen and oxidize the reduced substances.

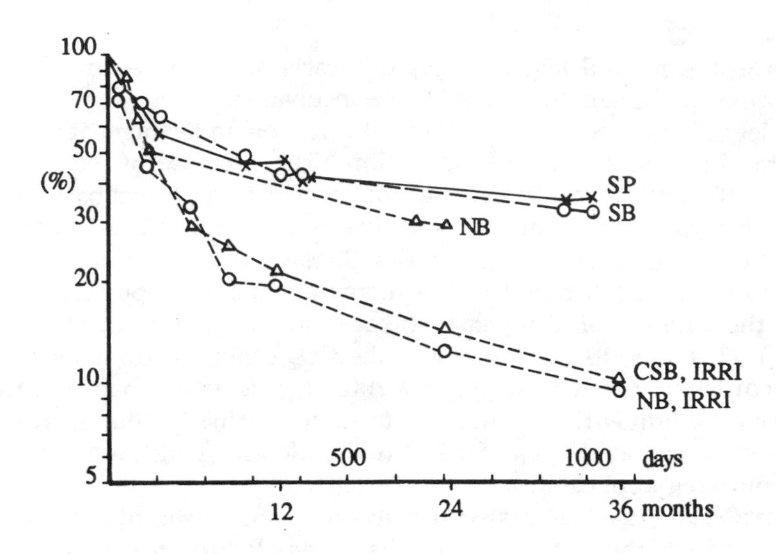

Figure 6. Decomposition of rice straw in the paddy soil of a low humic gley soil, northeastern Thailand (Snitwongse *et al.*, 1988).

Effect of crop management

In crop production systems, better soil cover protection can be obtained by the use of mulches and green manures, as well as by the incorporation of organic materials.

It has been demonstrated that mulching with plant residues is very advantageous in maintaining soil physical properties under low-input technology in the subhumid (ustic) forest region of West Africa (Lal, 1975). Impressive results have been obtained not only in maintaining soil physical properties, but also in improving chemical properties, especially with regard to organic-matter and nitrogen content in upland soils in Thailand. This demonstrates the advantages of mulching for sustained crop production.

Data on the effect of mulching on organic-matter contents under corn cultivation obtained from Thai and Japanese reports are shown in Table 1. Thornless mimosa is the most effective plant for increasing the amount of soil carbon content (with and without nitrogen fertilizer). The organic carbon content in mimosa plots (with and without nitrogen applied), after 8 years of intercropping with corn, was 134.33% and 98.57% respectively for the control plot. Soil organic matter under mimosa was more than that under rice straw, which in turn was greater than that under *Crotolaria*. The lower supply of biomass made it unsuitable for intercropping with corn, and showed the least increase in soil organic matter among the nonliving mulch.

Table 1. Effects of mulching on the organic carbon contents in reddish brown lateritic soil at Phraputthabat Field Crop Experiment Station, Lopburi, Thailand.

Treatment	Carbon content (%)								
	Year								
	1979	1980	1981	1982	1983	1984	1985	1986	1987
10-10-0	0.66	0.53	0.61	0.57	0.72	nd	0.77	nd	0.67
+ Rice straw[1]	0.76	0.63	0.72	0.67	0.80	nd	1.08	nd	1.06
+ *Crotaralia*[2]	0.75	0.64	0.73	0.68	0.83	nd	0.88	nd	0.87
+ Rice bean[3]	0.70	0.61	0.71	0.66	0.83	nd	1.04	nd	1.05
+ Mimosa[4]	0.66	0.59	0.64	0.75	0.93	nd	1.42	nd	1.57
0-0-0	0.65	0.48	0.59	0.57	0.63	nd	0.64	nd	0.70
+ Rice straw	0.77	0.67	0.71	0.66	0.76	nd	0.94	nd	0.88
+ *Crotaralia*	0.68	0.53	0.71	0.64	0.72	nd	0.88	nd	0.81
+ Rice bean	0.66	0.50	0.67	0.67	0.80	nd	0.95	nd	0.96
+ Mimosa	0.70	0.61	0.64	0.71	0.83	nd	1.49	nd	1.39

nd = not determined.

1 Rice straw was applied as mulch (4 tons/ha/yr) from 1976.

2 Cornstalk was incorporated (4 tons/ha/yr) from 1976-1979, and has been changed to *Crotaralia* by intercropping with corn and applying it as mulch since 1980.

3 Vinyl sheet was used as mulch from 1976-1978 and was substituted with water hyacinth (4 tons/ha) in 1979. Water hyacinth was in turn substituted with rice bean (*Vigna umbellata*) by intercropping with corn from 1980 onwards. Its residue was used as mulch in the following year before corn planting.

4 Cornstalk was used as mulch (4 tons/ha/yr) from 1976-1979, and was substituted with thornless mimosa by intercropping with corn from 1980. Its residue was used as mulch, and the remaining seed could germinate in the following year, before and after corn planting.

Mulching with living plants, or live mulch, may be another possibility with managing soil moisture for low-input technology in the ustic region in Thailand. Mulching with legumes may be the best method of improving the soil fertility. Growing legumes and nonlegumes together results in the transfer of nitrogen to the nonlegumes due to the sloughing off of nodules from legume roots. Mulching crops with *Stylosanthes hamata* L. cv. Verano has introduced low-input technology into the upland soils of Thailand. Verapattananirund *et al.* (1988) used Verano as a live mulch in cassava crops with tillage practices and herbicide control (Table 2). The cassava crop (Rayong-1) was planted at a density of 10,000 plants/ha on a fine loamy sand (Oxic Paleustults). The increase in root and shoot fresh weight in the system, with no tillage (but mulched), was 40 and 80% more respectively than in the traditional system. The increase in shoot-root ratio of the no-tillage system decreased the harvest index. Though no-tillage with herbicide use did not show the highest yield of root fresh weight, its harvest index was superior to that of other treatments (Table 2).

Table 2. Effects of four management systems on the fresh weight of roots and shoots and the harvest index of Rayong-1 cassava cultivar grown on Warin soil at the ADRC in northeast Thailand, 1987-1988.

Treatment	Fresh weight (ton/ha) Roots	Fresh weight (ton/ha) Shoots	Harvest index (%)
S1	16.6	14.2	54
S2	20.1	16.7	55
S3	14.1	21.0	40
S4	23.2	26.2	47
LSD at 0.05	5.2	7.2	4
C.V. (%)	17.7	23.1	4.7

Source: Verapattananirund *et al.* 1988.

S1: conventional system of plough-harrow and three hand-weedings.

S2: no-tillage, with one postemergence herbicide application before planting of cassava, plus two hand-weedings.

S3: no-tillage with *Stylosanthes hamata* L. cv. Verano as living mulch.

S4: no-tillage with living mulch, but with four mowings of Verano, maintaining the height at 7-10 cm from the ground surface during the growing season.

There is still a need for more intensive research, particularly on the soil biology and chemistry of living mulch in Thailand. The study should include the nature and components of crops, crop ratios, crop density, and cropping geometry (Suwanarit, 1987).

Green manure is usually beneficial as it involves nitrogen fixation, humus formation, and an improvement of the nutrient supply. Today farmers seldom grow green manure crops, even for maintaining soil organic matter, because the subsequent crop yields show very small increases, and there is no appreciable increase in income. Thus the introduction of nitrogen-fixing trees for use as green manure - but which would also give some products for the farmers - would be more acceptable.

Rathert *et al.* (1988) transplanted two months old *Leucaena leucocephala* at the beginning of the wet season on paddy field bunds in northeastern Thailand. The double hedgerow spacing was 50 by 50 cm. After three months of growth, the *L. leucocephala* was lopped at 1 m and the fresh biomass (branches and leaves) incorporated into the submerged paddy soils (low humic gley) as green manure at rates of 1875 and 3750 kg/ha. After seven weeks, the trees were lopped again and applied to the young rice at the same weight as before, giving a total of 3.75 and 7.5 t/ha of biomass. The results (Table 3) show that the yield of paddy grain was increased by both leucaena and mineral fertilizer. However, 50 kg of fertilizer N/ha could be replaced by 3.75 t/ha of *L. leucocephala* biomass.

Table 3. Effect of mineral fertilizer and fresh green biomass of *L. leucocephala* (LEU) on paddy grain and straw yield.

Treatment		Yield			
N-P_2O_5-K_2O (kg/ha)	LEU (t/ha)	Grain (t/ha)		Straw (t/ha)	
00-00-00	–	1.243	e	1.199	cd
00-37.5-37.5	–	1.286	de	1.130	d
50-37.5-37.5	–	1.990	c	1.842	b
00-37.5-37.5	3.75	2.385	bc	2.343	ab
50-37.5-37.5	3.75	2.826	ab	2.857	a
00-37.5-37.5	7.50	3.073	a	2.913	a
LSD 1%		623		578	
CV %		14.0		13.5	

The means followed by a common letter are not significantly different at the 1% level.

Source: Ràthert *et al.* (1988).

Other intercropping systems can be even more efficient, but little research has been carried out in Thailand. Leihner 1979; CIAT 1980, quoted by Sanchez and Salinas (1982) reported that when cassava was interplanted with cowpeas or peanuts at their normal planting densities, neither crop suffered significant yield declines. This was apparently due to less interspecific competition between the early-maturing grain legumes and the later-maturing cassava. These approaches provide an avenue for improving crop productivity with low inputs.

Conclusions

In order to maintain soil biomass activities or to improve soil organic matter, the application of several kinds of organic residue - on the surface or incorporated into the soil - is recommended. Various low-input technology components may be appropriate for Thai Ultisols. An understanding of the changes in soil biological, chemical, and physical properties over a period of time is helpful in designing and reducing the need for chemical fertilizers, especially nitrogen fertilizer, for improving soil fertility in cropping systems.

Acknowledgments

The author wishes to thank Dr. Prapai Chairoj for supplying several sources of soil biomass data, and Miss Srisuda Hirunyaphrerk for typing the manuscript.

References

ALEXANDER, M. 1977. *Introduction to soil microbiology*, 2d ed. New York: Wiley. 467p.

ALISON, F.E. 1973. *Soil organic matter and its role in crop production.* Developments in Soil Science no. 3. Amsterdam: Elsevier. 673p.

ANDERSON, J.P.E. and DOMSCH, K.H. 1978. Mineralization of bacteria fungi in chloroform-fumigate soils. *Soil Biology and Biochemistry* 10: 207-213.

ARARAGI, M. and TANGCHAM, B. 1974. Volatilized loss of soil nitrogen and microflora related to the nitrogen cycle. Thai-TARC Report. Bangkok: Department of Agriculture. 305p.

ARARAGI, M. , TANGCHAM, B., CHOLITKUL, W. and PHETCHAWEE, S. 1979. *Studies on microflora in tropical paddy and upland farm soils.* Tropical Agricultural Research Center. Technical Bulletin no. 13. Tokyo: Takayama Inc. 89p.

BREMNER, J.M. 1965. Total nitrogen, In: *Methods of soil analysis*, ed. C.A. Black, 1149-1255. Agronomy Publication no. 9. Wisconsin: USA.

CHAIROJ, P., CHOLITKUL, W., TANGCHAM, B. and ONJAN, A. 1988. *Effect of organic material and nitrogen fertilizer on soil biomass in Pakchong soils*, 18-31. Soil Chemistry and Fertility Report no 11. Bangkok: Department of Agriculture. (In Thai).

HAYES, M.H.B. and SWIFT, R.S. 1978. The chemistry of soil organic colloid. In: *The chemistry of soil constituents*, ed. D.J. Greenland and M.H.B. Hayes, 179-320. Chichester, UK: Wiley.

FLAIG, W. 1977. The function of soil organic matter in the environment. *Proceedings of the International Seminar on Soil Environment and Fertility Management in Intensive Agriculture*, 65-77. Tokyo: The Society of the Science of Soil and Manure.

IGARASHI, T., VIBULSUKH, N., CHAIROJ, P., PHETCHAWEE, S., CHOLITKUL, W. and ISHIDA, H. 1980. Behavior of nutrient in upland soils and effect of mulching on soil fertility and growth of upland crop in Thailand. Thai-TARC Report. Bangkok: Department of Agriculture. 56p.

INOUE, T., MORAKUL, PL, VIRAKORNPHANICH, P., CHONSPRADITNANT, P., PHETCHAWEE, S. and CHOLITKUL, W. 1984. Dynamic behavior of organic matter and available nutrients in upland soils of Thailand. Thai-TARC Report. Bangkok: Department of Agriculture. 377p.

INUBUSHI, K. and WADA, H. 1988. Mineralization of carbon and nitrogen in chloroform fumigated paddy soil under submerged conditions. *Soil Science and Plant Nutrition* 2: 287-291.

JENKINSON, D.S. 1971. Studies on the decomposition of ^{14}C-labelled organic matter in soil. *Soil Science* 3: 64-70.

JENKINSON, D.S. and POWLSON, D.S. 1976. The effects of biocidal treatments on metabolism in soil: a method for measuring soil biomass. *Soil Biology and Biochemistry* 8: 209-213.

JENKINSON, D.S., POWLSON, D.S. and WEDDERBURN, R.W.M. 1976. The effects of biocidal treatements on metabolism in soil. III. The relationship between soil biovolume, measured by optical microscopy, and the flush of decomposition caused by fumigation. *Soil Biology and Biochemistry* 8: 189-202.

JENKINSON, D.S. and LADD, J.N. 1981. Microbial biomass in soil measurement and turnover. In: *Soil biochemistry*, ed. E.A. Paul and J.N. Ladd, vol. 5, 415-471. New York: Dekker.

KONONOVA, M.M. 1966. *Soil organic matter.* 2d English ed., Oxford: Pergamon. 544p.

KONONOVA, M.M. 1975. Humus of virgin and cultivated soils. In: *Soil components*, ed. J.E. Gieseking, 497-526. Berlin: Springer Verlag.

KUBOTA, T.,VERAPATTANANIRUND, P., PIYASIRANOND, T. and PIYAPONGSE, P. 1979. The moisture regimes and physical properties of upland soils in Thailand and their improvement by soil managements. Thai-TARC Report. Bangkok: Department of Agriculture. 118p.

LAL, R. 1975. *Role of mulching techniques in tropical soil and water management.* International Institute of Tropical Agriculture. Technical Bulletin no. 1. Ibadan, Nigeria: IITA. 38p.

MARUMOTO, T., FURUKAWA, K., YOSHIDA, T., KAT, H. and HARADA, T. 1972. Effect of the application of rye-grass on the contents of individual amino acids and sugars contained in the organic nitrogen in soil. *Journal of the Faculty of Agriculture, Kyushu University* 17: 37-47.

MULONGOY, K. 1988. Evaluation of plant and animal biomass. In: *First Training Workshop on Site Selection and Characterization*, 195-205. IBSRAM Technical Notes no. 1. Bangkok: IBSRAM.

NAKAYA, N., CHUEYSAI, D., HANSAKDI, N., MORAKUL,P. and NANAGARA, T. 1986. Studies on the improvement of soil physical properties through the utilization of organic matter in upland soils of Thailand. Thai-TARC Report. Bangkok: Department of Agriculture. 278p.

OFFICE OF AGRICULTURAL STATISTICS. 1982. *Agricultural statistics of Thailand.* Bangkok: Ministry of Agriculture and Cooperatives.

RATHERT, G., NAMMUANG, C., SONGMUANG, P., KANAREUGSA, C., ROJANAKUSON, S. and PODISUK V. 1988. Nitrogen fixing trees as an alternative nutrient source for paddy rice cultivation in northeast Thailand. In: *Proceedings of the First International Symposium on Paddy Soil Fertility* (Chiang Mai, Thailand), 831-840. Bangkok, Thailand: Paddy Soil Fertility Working Group.

SANCHEZ, P.A. 1976. *Properties and management of soils in the tropics*, 162-183. New York: Wiley.

SANCHEZ, P.A. and SALINAS, J.G. 1982. Low-input technology for managing Oxisols and Ultisols in tropical America. *Advances in Agronomy* 34: 280-406.

SNITWONGSE, P., PHONGPAN, S. and NEUE, H.U. 1988. Decomposition of ^{14}C labelled rice straw in a submerged and aerated rice in northeastern Thailand. In: *First International Symposium on Paddy Soil Fertility* (Chiang Mai, Thailand), 831-840. Bangkok: Paddy Soil Fertility Working Group.

STEVENSON, F.J. 1982. *Humus chemistry*, 1-26. New York: Wiley.

SUZUKI, M., THEPPOOLPON, M., MORAKUL, P., PHETCHAWEE, S. and CHOLITKUL, W. 1980. Soil chemical studies on rotting process of plant remains in relation to fertility of upland soils in Thailand. Thai-TARC Report. Bangkok: Department of Agriculture. 225p.

SUWANARIT, A. 1987. Soil, water, and fertilizer management for corn, sorghum and corn or sorghum-based intercrops with extra consideration for the northeast soils. *Thai Journal of Soils and Fertilizers* 9: 95-121.

VERAPATTANANIRUND, P., NANAGARA, T., TONGYAI, C. and NUALLA-ONG, S. 1988. A promising low-input management to sustain high cassava yield in northeast Thailand. Paper presented at the 8th International Society for Tropical Root Crops, 30 October to 5 November, Bangkok, Thaialnd.

VIRAKORNPHANICH, P., WADA, K. and WADA, S.I. 1987. The amount and nature of humus formed from alfalfa meal in Ultisol, Oxisol, Alfisol and Inceptisol samples in short-term incubations. *Bulletin of the Institute of tropical Agriculture, Kyushu University* 10: 1-15.

VIRAKORNPHANICH, P., WADA, S.I. and WADA, K. 1988. Metal-humus complexes in A horizons of Thai and Korean red and yellow soils. *Journal of Soil Science* 39: 529-537.

WATANABE, M., CHAIROJ, P., MASANGSAN, W., PHETCHAWEE, S. and CHOLITKUL, W. 1989. Studies on the improvement of soil productivity through incorporation of organic matter into upland soils of Thailand. Thai-TARC Report. Bangkok: Department of Agriculture. 197p.

Methods for monitoring agricultural sustainability

CYRIL CIESIOLKA and CALVIN ROSE*

Abstract

In many parts of the world, advances in agricultural technology and genetic resources have made it very difficult to identify reductions in crop yields with land degradation. However, in highly weathered soils in the tropics, reductions in productivity striplands is more readily apparent than in temperate lands.

Quantification of declining productivity requires meticulous experimentation because the interrelationships are complex. The major approaches have been to consider: (i) measurement of agronomic changes - comparative studies are relatively slow and costly, but measure reality; (ii) computer modelling of agricultural systems - the approach is rapid when a sizeable data base exists, and has the advantage of holding variables constant; (iii) changes in soil physical and chemical properties, and the remittant effect on crop yield; (iv) weathering rates of parent material - an approach which has highlighted the slow rate of replacement of soil, and therefore the depletion of the soil storehouse.

* Queensland Department of Primary Industries, Indooroopilly, QLD 4068, Australia.

Introduction

It is very difficult to establish a simple correlation between soil degradation and reduced crop yields because of the gains derived from better hybrid seed, applications of weedicide insecticide and fertilizer, and more efficient harvesting systems. In Western agriculture, while one can find examples of poorer yields where rill erosion has stripped away the A horizon of a soil, or prolific growth where an old fence has been removed, farmers seldom leave their land because of the severity of erosion. Rose and Dalal (1988) present evidence for decreases in yield associated with erosion. As cost-push inflation rises, agricultural yields seldom rise as quickly, and farmers without sufficient capital reserves are never able to catch up. Often, there are other social factors that encourage farmers to leave their industry.

In old weathered soils of steep tropical lands, especially where a good vegetative cover existed, soil degradation results in rapidly declining production. Thus the lack of sustainability in agriculture is more readily evident in the tropics than in many more temperate lands.

Intuitively, one would expect that the potential for yield would decline as soil depth decreased because of the associated reduction in available soil moisture, rooting depth, organic-matter content, and other biological factors affecting plant nutrient storage. Surface crusting or exposure of laterite and plinthite could be other factors reducing yield. Reduction could be expected in rates of infiltration, causing greater runoff that transports soil and nutrients from small areas. The process is part of a physical positive feedback system characterized by nonlinear functions.

Lal (1988) categorized techniques of investigation of sustainability into four classes:

- measurement of agronomic changes,
- measurement of changes in soil properties,
- quantification of weathering rates, and
- modelling.

Measurement of agronomic changes

Where yields have not been influenced by increasing technology, and well-documented site information exists, regression analysis can be used to compare yields of different soils and ascertain significant changes. The assumptions of homogeneity can even be relaxed to some extent, and the same analyses as above used.

In all comparative studies, two basic assumptions are made. Firstly, it is assumed that spatially all soil properties were similar before cultivation and secondly, as a consequence, productivity was uniform before erosion began.

At sites where soil loss has been measured over a number of years, a single crop planted on all of the measurement sites can quantify the effects of the previous history. Such an experiment requires careful analysis of characteristics of the sites, and can prove to be a very difficult task.

Measures of plant vigour at important stages of growth and tissue testing are possible areas of fruitful investigation. As leaf area index data at certain crop stages correlates well with yield, the measure can act as a surrogate indicator for productivity.

Measurement of changes in soil properties

Much research and prediction of the effects of erosion has centred around such soil depth-related variables as:

- plant available water capacity,
- soil rooting depth and soil compaction,
- organic-matter content, and
- plant nutrient storage and chemical oxygen demand.

Yield-soil depth relationships have received much attention, and nonlinear and polynomial relationships have been found.

Each of the four characteristics above can be considered graphically and compared with virgin sites found in protected areas or even along roadsides.

An interesting development is that aerial concentration of caesium 137 can be correlated with the depth of soil that has been removed by erosion (Loughan, *et al.*, 1982; Longmore *et al.*, 1983). Consequently, data from the eroded areas can be used to estimate losses when comparison with a virgin site has been made. Where the virgin sites can be cropped in close proximity to the trial sites, high quality data can be obtained.

The most common type of experiment has been to scalp soils to various depths and grow crops on this unnatural surface. Yields from such experiments do not simulate real-world erosion because of selective transport and deposition both in sheet and rill flow. The enrichment ratio of deposited sediments can explain variable yields across some plots.

Where a long-term project is planned, it would be worthwhile to begin with new plots where all variables could be monitored carefully and a precise plot history developed.

In different soils, various factors may become important. One instance is the clay content in the A horizon. Under some tropical soils, clay can be eluviated or transported by runoff, and surface soil will become sandier. However, in other cases, clay from the B horizon may be brought to the surface and mixed into the A horizon by cultivation. The clay-enriched horizon enhances runoff and further erosion.

Researchers should note that soil physical properties can deteriorate simply by keeping all cover from growing on level plots where there is minimal runoff and erosion. Cultivation on such plots can have a very deleterious effect on soil physical characteristics such as aggregate size, pore space, surface crusts and bulk density, and lead to the development of surface crusts.

Quantification of weathering rates

Weathering studies have been made possible by the development of laboratory equipment whereby weathering cycles can be generated using specially designed ovens. Results from such studies indicate that soil losses usually regarded as acceptable are far too high. Kirkby and Morgan (1980) also point out that by increasing overland flow, there will be less water in the soil profile for weathering action.

There is further evidence of the slow rate of soil formation where weathering of rocks has been recorded in historical times. Depending on rock type, the weathered zone and B horizon can represent a store capable of being converted into reasonably productive soil - for example, basic igneous and calcareous rocks.

Modelling

"Assessment of the impact of erosion on long-term productivity requires an estimate of long-term average annual erosion" (Foster, 1988), and the advent of computers has made such analyses posible.

Soil erosion, soil compaction, fertility and yield

The causes of yield variation are many and various, as experimentation testifies.

An area of research of direct relevance to this question is the modelling of crop growth and yield. Use of some such models, which hope to reflect the consequences of above- and belowground environmental variables and their calibration with genotypic factors would seem a very desirable complement to this area of research. Their use would assist in discerning the effects of erosion.

There is also a case for taking such research in the opposite direction from yield variation to possible causes. This "opposite direction" is to measure the consequences of soil erosion, which will certainly include loss of nutrients and no doubt a range of other possible consequences referred to earlier (lowered infiltration, storage of plant available water, increased bulk density, and surface sealing).

The development of USLE and the concept of a tolerable soil loss grew out of the disastrous erosion of the 1930s in the USA. Previously, construction of water storages in semiarid/subhumid lands had highlighted the enormous quantities of sediment that was being moved, especially in newly settled lands. Modified versions of the USLE, such as MUSLE and SLEMA, have been developed, attempting to remove some of the limitations of the USLE.

The idea of simulating physical processes, as opposed to the use of empirical methods, opend up alternative avenues for predicting not only erosion but also its effects.

CREAMS, a field-scale model for Chemicals, Runoff and Erosion for Agricultural Management Systems has been a fairly widely used model (Knisel, 1980). Such models endeavour to integrate the major physical processes in the environment. EPIC - Erosion Productivity Impact Calculator - takes analyses a step further in that it includes management and economic analyses (Williams *et al.*, 1983). The P.I. model - Productivity Index - focuses on root growth and water depletion using available water capacity, resistance to root penetration, soil texture and structure, bulk density, aeration, pH, and electrical conductivity (Kiniry *et al.*, 1983).

A methodological problem common to all experimentation in a variable environment is how to extrapolate expected long-term behaviour from a limited set of experiments. This problem remains even if the experiments have yielded adequate definition of relevant characteristics. A method which has proved fruitful, for example, in defining the characteristics of rainfall rate or duration (Julien and Frenette, 1985) has been to describe such characteristics in probabilistic terms. A probability density function is a mathematical expression describing the relative likelihood of occurrence of the variable, such as a given runoff rate. From the measurement of runoff rate over an adequate number of runoff events, this probability characteristic can be determined. It is likely that, as for rainfall rate, an experimental type of probability density function may describe the distribution of runoff rate, though the shape factor in such a distribution would be expected to vary seasonally and with surface cover and management.

Since runoff rate plays an important role in soil loss (Rose and Hairsine, 1988), then this probabilistic description of runoff rate can be combined with a deterministic model of soil loss to yield the probability characteristics of soil loss in any particular context of climate, soil, and land management. Such an approach has been outlined by Ward and Rose (pers. comm.), and will be used in ACIAR Project 8551 (The Management of Soil Erosion for Sustained Crop Production).

One outcome of such an analysis is an expected value of soil loss over any given period (such as a year), together with an estimate of the uncertainty in such an expected soil loss. This expected soil loss could be compared with available data to ascertain a "tolerable" rate of soil loss (that is, a rate of soil loss for which there is no evidence of yield decline, or a loss rate where such potential decline can be overcome by other tolerable economic or management inputs).

The type of methodology briefly outlined above holds promise as an efficient way of generalizing gathered data and aiding exploration of the likely sustainability of alternative management systems.

Sampling for sustainability

The concept of 'sustainability' has economic and management dimensions to it, as well as biophysical aspects. Thus consideration of the density and time span of sampling, needs to take such factors into account, probably largely as a matter of judgment.

Thus what is thought to be long-term variable economic and management inputs to, and economic outcomes from, any particular agricultural system, is part of the input to a judgment on 'sustainability'.

In biophysical terms the duration of experimentation required to make a judgment on sustainability would appear to be somewhat inversely related to the potential rate of degradation - ignoring such potential problems as disease build up.

However, a proper objective of research should be to discover, in different contexts, guidelines to answer such sampling and location questions. At the same time, it is not strictly possible or useful to lay down such guidelines *a priori* before experimentation has given some guidance. Experience on the actual rate of degradation in particular contexts for particular systems would be very valuable in providing such guidelines. This could be a reason for including 'worst management' options in experiments designed to ascertain the sustainability of different management options.

References

FOSTER, G.R. 1988. Modelling soil erosion and sediment yield. In: *Soil erosion research methods*, 97-118. Ankeny, Iowa: Soil and Water Conservation Society.

JULIEN, P.V. and FRENETTE, M. 1985. Modelling of rainfall erosion. *Journal of Hydraulic Engineering* 3: 1344-1359.

LAL, R. 1988. *Soil erosion research methods.* Ankeny, Iowa: Soil and Water Conservation Society.

LONGMORE, M.E., O'LEARY, B.M. ROSE, C.W. and CHANDICA, A.L. 1983. Mapping soil erosion and accumulation with the fallout isotope Caesium-137. *Australian Journal of Soil Science* 21: 373-385.

LOUGHAN, R.J., CAMPBELL, B.L. and ELLIOTT, G.L. 1982. The identification and quantification of sediment sources using $137C_s$. *Proceedings of the Exeter Symposium* (July 1982), 361-369. IAHS Publication no. 137.

KINIRY, L.N., SIRIVNER, C.L. and KEENER, M.E. 1983. *A soil productivity index based upon predicted water depletion and root growth.* Research Bulletin no. 1051, Columbia: University of Missouri.

KIRKBY, M.J. and MORGAN, R.P.C. 1980. *Soil erosion*, 312ff. Chichester, England: John Wiley and Sons.

KNISEL, W.F., ed. 1980. A field scale model for chemicals, runoff, and erosion from agricultural management systems. Soil Conservation Service, U.S. Department of Agriculture. Conservation Research Report no. 26. Washington DC: Government Printing Office.

KNISEL, W.F. and FOSTER, G.R. 1981. CREAMS: A system for evaluating best management practices. In: *Economics, ethics, ecology: roots of productive conservation*, 177-194. Ankeny, Iowa: Soil Conservation Society of America.

ROSE, C.W. and HAIRSINE, P.B. 1988. Processes of water erosion. In: *Flow and transport in the natural environment*, ed. W.L. Steffen, O.T. Denmead and I. White. New York: Springer-Verlag.

ROSE, C.W. and DALAL, R.C. 1988. Erosion and runoff of nitrogen. In: *Advances in nitrogen cycling in agricultural ecosystems*, ed. J.R. Wilson, 212-233. Wallingford, UK: CAB International.

WILLIAMS, J.R., RENARD, K.G. and KYKE, P.T. 1983. EPIC - A new method for assessing erosion's effect on soil productivity. *Journal of Soil and Water Conservation* 38: 381-383.

Appendixes

Appendix I

Methodological guidelines for IBSRAM's soil management networks*

* This is an updated version of the guidelines published in IBSRAM's Technical Notes no. 1, 275-291.

INTRODUCTION

The methodological guidelines presented here are intended as a guide to the various methodological procedures which are regarded as desirable for each of a succession of steps normally involved in carrying out agronomic field experiments. They have been compiled primarily to guide cooperators working within the framework of IBSRAM soil management networks who wish to select and characterize sites for experimental trials on soil management.

The five steps normally envisaged are:

(a) Site selection at the regional level, aimed at locating the general area in which the experiments are to be located and examining it in enough detail to ensure that it is representative of the soils and climate desired.

(b) Site selection at a local level within the area already examined under (a). The area examined in more detail will normally cover between 50-200 ha, and this exercise is designed to facilitate the final choice of the project site itself, usually about 2 ha.

(c) The detailed characterization of the selected project site, which involves soil survey on a regular grid system and a consideration of the climate and socioeconomic characteristics of the site.

(d) The design of the agronomic experiment to be carried out.

(e) The monitoring, during the course of the experiment, of the growth and yields of the crop in relation to the treatments selected, of the weather, and of changes in soil morphology, soil physical properties, and soil fertility. A description of socioeconomic parameters is also required.

In order to assist in this sequence of operations, the guidelines incorporate a number of separate forms relating to the five steps outlined. Additional relevant information is given in appropriate appendices.

"Guidelines", as the name suggests, should serve as a general guide and not as a rigid set of requirements. They should be used and interpreted bearing in mind the frequent need for modification in order to adapt them to local conditions and the particular nature of the treatments tested. In some cases additional information will be needed. On the other hand, cooperators should not be discouraged if they are not able to respond fully to all the questions given.

These guidelines, which have arisen out of discussions at a series of meetings, are themselves subject to modification in accordance with future experience.

General Sequencing

Forms A:	*Site Selection at the Regional Level*
Forms B:	*Site Selection at the Local Level*
Forms C:	*Site Characterization*
Forms D:	*Experimental Design*
Forms E:	*Experiment Monitoring*

FORM A: SITE SELECTION AT THE REGIONAL LEVEL

Form A1: **Type of Land Selected**
Form A2: **Farming Systems**
Form A3: **Relevance of Site Selected**

Form A1: Type of Land Selected

- Location. country
  . province
  . latitude
  . longitude
- Altitude (range)
- Geological unit
- General landform unit*
- Soil unit** . FAO
  . *Soil Taxonomy*
  . Local
  . Agroecoclimatic zone
- General climatic*** and climatic zone****
- Land use . present
  . past

* Refer to FAO Guidelines for soil profile description
** As per existing maps in U.S. *Soil Taxonomy* or FAO legend and the complementary site selection survey
*** As per existing climatological and agrometeorological documents and/or existing local climatic classification
**** FAO - agroecological zone

Form A2: Farming Systems

1. General information
   - Density of rural population ________________ people/km²
   - Household size a. average ______________ people
     b. range (min. - max.) _______ people
   - Farm size a. average ______________ ha
     b. range (min. - max.) _______ ha

- Land tenure

Categories	% of household*	% of land*
Owned		
Partially-rented		
Rented		

* Please fill up both, if possible.

2. Farming systems
   - Cropping pattern
     a. How many major cropping systems exist ____________________
     b. Minor crop(s)
        - vegetables ______________________
        - fruit trees ______________________

     Cropping system 1: (type of crop)

Year	Variety/type	J	F	M	A	M	J	J	A	S	O	N	D
1	1st crop												
	2nd crop												
2	1st crop												
	2nd crop												
3	1st crop												
	2nd crop												

If more than one major cropping system please do the same as above.

- Soil management practices used for major cropping systems
  a. Land preparation by
     __ hoe __ animal __ traction __ mechanized
  b. Level of monetary inputs (fertilizers, insecticide, etc.)
     __ low (0-20% of total farm expense)
     __ moderate (21-50% of total farm expense)
     __ high (>50% of total farm expense)
  c. Irrigation __ yes __ no
  d. Planting
     __ flat __ ridge __ mounds __ if any others (specify)

3. Livestock

Total number of livestock (number)

a) Poultry ____________________

b) Cattle ____________________

c) Others ____________________

4. Marketing of crops

a) Proximity of local markets ________ km

b) Level of activeness and competition ______________________________
(volume of trade or business, and number of buyers)

c) Major crops

Crops	Volume (tons)	Value (US$)
1.		
2.		
3.		
4.		

d) Minor crops but having potential are ______________________________

Form A3: Relevance of the Site Selected

a. Extent of the type of land selected

- ha ______________________________
- % in country ____________________

b. Extent of farming system

	System 1	2	3
- area (ha) cultivated on this land type	___	___	___
- % on country	___	___	___

c. National priorities

	High	Medium	Low
- to develop this land type	___	___	___
- to improve this farming system			
System 1	___	___	___
2	___	___	___
3	___	___	___

FORM B: SITE SELECTION AT THE LOCAL LEVEL

B1: Selection of Experimental Site

B2: Accessibility and Security

B3: General

Form B1: Selection of Experimental Site

On 50-200 ha available
- Characterize catena (if any)

	Homogeneous	Heterogeneous
* upper part	(___)	(___)
* middle part	(___)	(___)
* lower part	(___)	(___)

- Select one representative area relevant to
  * farmers
  * the project
  * agronomic potential

Form B2: Accessibility and Security of Experimental Site

- Distance from a major city (km) ________________________
- Distance from a laboratory (km) ________________________
- Access roads All weather __ yes __ no
- Is water available? __ yes __ no
  If yes, indicate source ________________________
- Is electricity available? __ yes __ no
- How secure is the site? __ yes __ no

Form B3: General

- Total number of livestock (type/number)
  a. Poultry ________________________
  b. Cattle ________________________
  c. Others ________________________
- Marketing of crops:
  a. Proximity of local market _______ km
  b. Level of activeness and competition ________________________
     (volume of trade or business and number of buyers)
  c. Major crops:

	Crops	Volume (tons)	Value (US$)
1.			
2.			
3.			
4.			

  d. Minor crops but having potential are ________________________

FORM C: SITE CHARACTERIZATION

C1: Assessment of Site Homogeneity
C2: Soil Characterization
C3: Climatic Characteristics
C4: Socioeconomic Characterization
- **4.1 Socioeconomic questionnaire**
- **4.2 Farmers' practices questionnaire**

Form C1: Assessment of Site Homogeneity (2-5 ha)

1. Topographic survey (minimum contour interval of 1 m)
2. Very detailed survey of surface features and soils based on 10 m grid
   - 2.1 Surface features with location of major features
     1) slope (percentage, aspect, form)
     2) vegetation/land use (cover % for basal and aerial, main species, history, deficiency symptoms)
     3) rock outcrops/surface stoniness, indicate percent cover
     4) biological activity (termite mounds, worm casts, give nature and density)
     5) microrelief
     6) cracks
     7) erosion features (type eg. crust, rill, sheet, deposition, etc.), also indicate location, size etc.
     8) left-over tree stumps and standing trees - give location and extent of influence
     9) man made structures or disturbance - include, burnt patches, ash heaps etc.
   - 2.2 Soil characteristics (using spade for first 20-30 cm and auger up to at least 1 m)
     1) parent material
     2) succession of horizons (depth)
     3) on each horizon
        - colour
        - mottles
        - texture (field texture)*
        - structure (moist) and consistency (moist, wet) of upper horizons
        - coarse material (size, amount in volume %, nature)
        - concentration of carbonates, sulphates, etc.
        - pH value by pH test kit of portable pH meter
        - depth and type of root-restricting layer

* By training with samples of known texture, it is possible to develop manual texturing skill adequate to determine soil textures on the field.

3. Preparation of a site map using the previous information to draw fairly homogeneous land units based on natural groupings of soils and surface features.

4. Experiments, or at least experimental blocks, have to be located in relatively homogeneous land untis.
5. Detailed description of soils and surface features should be kept for future comparisons and stored in a common data base for the IBSRAM network.

Form C2: Soil Characterization

- Description of two to three representative soil profiles* in the homogeneous land units selected for the experiments using USDA Soil Survey Manual, 1981, chapter 4.
- Soil sampling on profile for standard chemical, physical and mineralogical analyses and for classification according to the following systems:
  Local soil classification
  U.S. *Soil Taxonomy* (family level)
  Fertility capability classification
  FAO/Unesco soil map legend (optional)
- Field measurements

* Pits to be located nearby and not in experimental blocks. The whole soil profile should be characterized.

Chemical parameters to be analyzed:

Essential

pH (water, KCl), % C, CEC (NH_4OAc, pH 7), base saturation

Optional

Total P, available. P (Bray II for acid soils and Olsen for others)
Exchangeable cations, extractable acidity ($BaCl_2$- triethyl at pH 8.2)
Extractable Al (KCl extraction) for soils with pH <5.0

Soil physical measurements to be performed:

Essential

1. Particle-size analysis
   - ISSS size limits should be used.
   - pipette or hydrometer method
2. pF curve
   - to be measured after drainage of excess water
   - using soil cores for low suctions

Desirable

1. Particle density
2. Saturated hydraulic conductivity (core method)
3. Aggregate density - by paraffin coating method
4. Aggregate stability - to be conducted on naturally moist samples (at present there is no method that can be recommended).
5. Atterberg limits

Field measurements

1. Infiltration rate - double-ring method; in soils prone to crusting, the capillary tube method should be used.
2. Bulk density - core method (excavation lining method for gravelly soils): Diameter of core should be at least 50 mm; larger diameter is preferable.
   Should be conducted throughout the whole profile, in relation to soil horizons.
   Particle density is assumed to be 2.65 g/cc. Total porosity to be calculated.
3. Compaction - using mass penetrometer at 5 cm intervals
   - to be conducted together with a moisture profile by the auger method.
4. Field capacity - using the field method
   - to be measured after drainage of excess water

Form C3: Climatic Characteristics

To be obtained from a nearby long-term weather station (record period preferrably 30 years or longer) located within the same agroclimate zone as the experimental site.

Agroclimatic zone*: ______________________________
Climatic station name: ______________________________
Distance from site: ______________________________
Address of responsible organization: ____________________
Latitude (deg. min.): ______________________________
Longitude (deg. min.): ______________________________
Elevation (m): ______________________________

* According to national agroclimate classification, otherwise FAO agroecological zones (FAO, 1981).

	J	F	M	A	M	J	J	A	S	N	D	YEAR
No. of years __________												
Temp. av.												
Temp. min.												
Temp. max.												
Precipitation av.												
Precipitation min.												
Precipitation max.												
Precipitation 25% prob.												
Precipitation 75% prob.												
Rainy days												
Solar radiation												
Relative humidity (%)												
Wind speed												
Max. rainfall intensity												
PE (Penman method)												

Diagram showing monthly T. av., P, PE and 0.5 PE values.

Form C4: Socioeconomic Characterization

Form C4.1: Socioeconomic questionnaire

To be undertaken by a team which preferably includes the project leader, who is usually expected to be a soil/agricultural scientist and a socioeconomist or extension officer. The questionnaire is to be used on a sample of 20 farmers from the selected area. Interviews should not be conducted at the village only, but should be accompanied by field visits, during which discussions could also take place. The interviewers should be familiar with approaches to such surveys and adjust their mode of obtaining the information based on the sensitivities of individuals. Those not familiar with such socioeconomic surveys should seek cooperation and assistance from socioeconomists in other organizations in their countries.

Farmer: ____________________ Address: ____________________
Date: ____________________ Survey no.: ____________________
Interviewers: ________________

I. Social background information

1. What ethnic group do you belong to? ________________
2. What is your religion? ________________
3. How large is your household?
   a. Total number ________________
   b. Male ________ Female ________
   c. Children under 14 years old ________________
4. What is the highest level of education of your household members? ________

II. Land and land utilization

1. Is farming your most important occupation?
   a. in time spent? _ yes _ no
   b. in income? _ yes _ no
2. What are your most important crops
   a. for food? ________________
   b. for cash? ________________
   c. for your livestock? ________________
3. Fragmentation: How many plots do you have? ________________
4. Land specification and utilization

Plot no.	Size (ha)	Location km from home	% slope	Rent/ owned	Soil condition*	Cropping pattern**

* Shallow soil, badly eroded, etc.

** Please provide details if there is intercropping or more crops grown in the same season on the same plot.

5. When land is short, how do you obtain more? ________________
6. What do you raise? (number)
   a. Cattle ________________
   b. Poultry ________________
   c. Other (specify) ________________

III. Labour

1. Have you hired labour in the past 12 months?
   _ yes _ no

2. Labour utilization

Activities	J	F	M	A	M	J	J	A	S	O	N	D
Level of farm activities*												
- on-farm**												
- off-farm												

* Busiest 3
Moderate 2
Not busy 1

** On-farm includes farm activities on other farms for exchange.
Please note also the festivals or ceremonial occasions (used only for local level).
There is no need to ask all the farmers this question. One or more from each ethnic group or religion is sufficient.

IV. Capital and inputs

1. Please list the crops with level of cash and credit spending (most to least).

 1. ____________________ 4. ____________________
 2. ____________________ 5. ____________________
 3. ____________________ 6. ____________________

2. How do you pay for the inputs to produce all your crops?

Input	Cash $	Credit $	In kind[1]	Source of input[2]
Seeds				
Seedlings				
Fertilizers				
Insecticides				
Fungicides				
--				
--				
--				
Farm tools				
Equipment				
--				
--				
--				
Labour				

1 Your own inputs, or what you have to pay in kind, exchanged labour, etc.
2 Relatives, friends/landlord/merchants/cooperative/commercial banks, etc.

Potential source, amount, and cost of credit.

Source	Maximum US$	Interest rate (%)

V. Yield and returns

Crop or livestock	Total products (kg/ha)	Quantity sold (kg)	Price received ($/kg)	Total expenses (own labour excluded)

VI. Problems and potential expansion

1. What are the major production problems of the following crops? (Please describe)

Crops	Production problems	Marketing and other problems
_____	_______________	_______________________
_____	_______________	_______________________
_____	_______________	_______________________

2. Do you feel you lack of technical knowledge?
__ yes __ no
If yes, how do you solve this problem? ____________________________

3. Where do you normally obtain marketing information? ______________
4. How often do you meet extension officers ______________
5. If the farmer does not mention extension officers; ascertain his attitude to extension officer i.e. positive or negative. Provide details if possible.

---

6. Which crop do you like to produce most? ______________
   Why?

---

7. Do you observe your yield reducing with time?
   __ yes __ no
8. Are you aware of soil conservation practices?
   __ yes __ no
   If yes, what kind of soil conservation are you currently practicing?

   If no, why? ______________
   (Note: interviewers should observe farmers' actual practices)
9. Do you earn enough just to fulfil the requirements of your household consumption?
   (include food, cloth, medical care)
   __ yes __ no
   If no, how do you solve this problem? ______________
10. What do you think is the most important change you could make to improve productivity? ______________
11. How much is the annual household expenditure? ______________ $
    How much is your household income, including off-farm income? ________ $

Form C4.2: Farmers' practices questionnaire

To be completed
- (i) at the same time as the socioeconomic questionnaire
- (ii) for individual crops

(Part of farmers' practices are included in C4.1)

1. What type of land preparation do you use? ______________
   (tillage implements, land shaping)
2. What spacing do you use? (by crops) ______________
3. What type of weeding do you perform and when? ______________
4. Do you apply fertilizer?
   __ yes __ no
   If yes, what kind/amount/time of application (please specify): __________
5. What do you do with the crop residues? ______________

6. Do you fallow your land?
   __ yes __ no
   If yes, how long? (years)
7. Do you use fallow for grazing?
   __ yes __ no
8. Do you irrigate? __ yes __ no
   If yes, what is the irrigation technique? ____________________
9. Do you control insects and diseases?
   __ yes __ no
   If yes, what do you use to control them? ____________________
   Is it successful?
   __ yes __ no
10. Farmers' general comments

FORM D: EXPERMENTAL DESIGN

1. The experiment will be laid out according to site characteristics and with regard to the treatments and crops to be employed (annual crops, perennials).
2. Randomized block designs with three to four blocks are favoured. Each block must, as far as practicable, be on a homogenous soil.
3. Due to the long-term maintenance of the experiment, the size of the plots should be at least 100 m^2 and should be suitable for the cropping systems and soil pattern.

FORM E: EXPERIMENT MONITORING

E1: Treatments Tested
E2: Weather Monitoring
E3: Soil Morphological Features
E4: Monitoring Soil Physical Properties
E5: Sampling and Analysis of Soil and Plant Samples for Monitoring Soil Fertility

Form E1: Treatments Tested

1. Treatment no.

----

2. Cropping system tested (crops, type of association, cropping period): __________
3. Field preparation
   a. Method of vegetation clearing:
      _____ slash-and-burn   _____ plough in
      _____ herbicides   _____ other (specify)
   b. Soil preparation:
      _____ conventional tillage (describe)   _____ minimum tillage
      _____ zero tillage   _____ other (specify)
   c. Surface shaping:
      _____ ridges   _____ beds
      _____ other (specify)   _____ nil
   d. Lime input: _____ kg/ha
   e. Fertilizer input before planting:
      _____ N kg/ha   _____ P kg/ha
      _____ K kg/ha   _____ other kg/ha

   Specify form, quality of the fertilizer and mode of placement.
4. Planting and establishment
   Planting date (in relation to rains)
   - Planting method:
     _____ manual   _____ mechanical
     _____ other (specify)
   - Spacing between plants within one crop, and spacing and arrangement between different crops in multicropping, and seeds/plants per point
   - Variety and seed rate - if possible both in weight/ha and number per 100 g seeds, and for tubers weight per 10 units of planting material
   - Seed germination or plant count can be based on count per metre in each plot . Percent germination normally observed when 50% of plants with some points visible aboveground, i.e. often emerge about 10 days after sowing
   - Cause of poor germination e.g. lack of moisture, not sown to proper depth, eaten by birds, due to surface crust, etc.
5. Cultural practices and crop development
   - Weeding
     * Date:
       1st weeding _____ days after planting
       2nd weeding _____ days after planting
       3rd weeding _____ days after planting

     Note: different types and percentage cover prior to each weeding round. If weeding is irregular, note weeds at third week and at harvesting.
     * Method: manual, chemical, etc. If chemical, specify chemical and rate.

- Crop development observations (phenology), deficiency symptoms, etc.
- Supplementary fertilizer input (nature, quantity, time, mode of placement)
- Pests and extent of damage
- Pesticide, insecticide, herbicide (date, nature, dose, effectiveness)
- Diseases, aboveground and roots
- Supplementary irrigation (if any) (date, dose)
- Special tillage practices

6. Harvesting
   - Date
   - Relation with rains
   - Yield (dry weight, gross, and marketable by grades), and (where applicable) components of yield
   - Crop damage
   - Crop residues (dry weight)
   - Total biomass (total of crop yields, crop residues, weeds, prunings, etc.)

Form E2: Weather Monitoring

Essential

1. Daily rainfall amount
2. Relative humidity (nearest weather station)
3. Max., min., average temperature
4. Solar radiation
   (Automatic weather stations to be installed at site)

Desirable

1. Open-pan evaporation
2. Soil temperature
3. Rainfall intensify
   a) Pluviometer and rain gauge should be of the same height.
   b) Weather station should be centrally located.
   c) A few rain gauges should be placed around the experimental area.

Form E3: Soil Morphological Features*

Essential observations to be performed during and at end of cropping and fallow period.

1. Erosion features - type, location, size, density, estimated through semiquantitative assessment (e.g. grades 0 to 5)
   - after every major erosion event

2. Crop cover - aerial and surface covers
- type and percentage
- visual assessment and quadrat method
- presence of gravel on the soil surface

3. Surface crusting - type, thickness, percentage of soil surface
- minimum clod diameter
4. Faunal activity - nature, density
5. Surface roughness - nature, depth, orientation
- semiquantitative assessment (e.g. grades 0 for smooth to 5 for extremely rough)

After harvest

1. Erosion and depositional features - also deposition from one treatment plot to another.
2. Plough pan
3. Macro- and microstructure
4. Waterlogging
5. Rooting

* A semiquantitative approach should be adopted.

Form E4: Monitoring Soil Physical Properties

Essential

1. Infiltration rate: - conducted within the plot
- double-ring or capillary tube method

   This is to be done after major rainfall events
2. Erosion and runoff measurements to be done in runoff plots** - after every major erosion event
   - plot length should be in relation to land use
   - plot width should be in relation to land use (preferably at least 10 m)

** If it is not possible to have runoff plots, permanent pegs and deposition plates should be used. Where one or more replicates have runoff plots, such pegs could be placed in the remaining replicates.

Desirable

1. Grain size of sediment
2. Use of water level recorder to measure runoff rate
3. Nutrient loss in runoff and sediment
   - 2 or 3 events per crop
   - org. C, N, P, K, exch. cations
4. Moisture profile

After harvest

Essential

1. Bulk density - 0-5 cm, plough layer, subsoil (30-35 cm)
2. Compaction - mass penetrometer method (together with a moisture profile)

Form E5: Sampling and Analysis of Soil and Plant Samples for Monitoring Soil Fertility

I. Soil Samples

1. Sampling method

For each plot, one composite sample (which is made up of a minimum number of 10 cores for each 100 m^2) should be taken. One composite sample should be taken from the surface to a depth of 15-20 cm. Additionally, a sample at a greater depth may be taken in order to account for deep-rooted crops (for example 50 cm or more). Preferably, each of these samples should be collected over the entire depth. Each core sample should be kept in a plastic bag and mixed thoroughly in the field if moisture content permits. Then a composite sample should be taken and stored in a properly labelled plastic bag.

For each plot, samples should be collected according to a systematic (regular) sampling scheme with random start. In the case of mixed-cropping systems, the number of samples taken from each crop area should be proportional to the total area occupied by each crop. The interval between samples within a transect (row) should be set so that the entire crop is represented by the sample.

Sampling should be performed before planting and at harvesting time. If two successive crops are grown, sampling should be performed at each harvesting time in addition to the preplanting time.

2. Chemical parameters and analytical methods

A distinction is made between the minimum number of parameters (obligatory) to be analyzed and optional ones.

Essential parameters	*Analytical methods*
- pH	- H_2O and N KCl (1:1)
- Total N	- Kjeldahl
- Available P	- Bray II (acid soils), Olsen (others)
- Exchangeable bases (Ca, Mg, K and Na?)	- NH_4OAc, pH 7
- Organic C	- Walkley and Black

Optional parameters	*Analytical methods*
- Micronutrients:	
B	- Azo-methine method
Mo	- Zinc dithiol method
Cu, Fe, Mn and Zn	- DTPA - extraction method
- Exchangeable acidity	- N KCl
- S	
Total	- Na_2CO_3 - $NaNO_3$ extraction
Available	- Extraction with ammonium acetate and measured by light transmittance or absorbance with colourimeter or spectro-photometer

II. Plant Samples

1. Sampling methods

In addition to fertility assessment by soil chemical analysis, input from crop and weed residues also has to be estimated. Only the residues of the crops and weeds and clippings should be analyzed per plot.

Quadrat sampling will be employed using 1 x 1 m quadrats. There will be a minimum of 5 quadrat samples per plot according to a systematic sampling scheme. A composite sample from these 5 samples should be used for the determination of water content and chemical parameters, and for estimating the dry weight of the combined quadrat above ground biomass. The biomass per hectare may then be obtained by multiplying by the appropriate conversion factor.

2. Chemical parameters and analytical methods

A similar distinction as with soil chemical analyses is made between obligatory and optional parameters to be analyzed. Apart from dry matter and ash contents, the following parameters are suggested:

Essential parameters	*Analytical methods*
- Organic C	- Walkley and Black
- Total N	- Kjeldahl

Optional parameters	*Analytical methods*
- K, Ca and Mg	- Atomic absorption spectrometer
- P	- Modified ammonium-vanodo-molybdate method (after dry ashing and measuring colourometrically)

III. Quality Control

It is recommended that the IBSRAM laboratories of this network should participate in the LABEX programme (interlaboratory trial) of 1988/89. Details on the provisional timetable and organization are described in *LABEX Newsletter* no. 4, issued in November 1988.

Appendix II

SOIL-MAPPING EXERCISE*

Introduction

In the tropics, large variability in soils occurs even within a small area. Such variations could be large enough to affect plant growth and hence exert an influence on the results of treatments in the experiment laid out on the site. The potential of the soil can be easily assessed by examination of visible features, such as surface features (including vegetation), slope, depth, texture, colour, concretions, stoniness, roots and faunal activity. These evaluations will enable the preparation of a relatively useful map to assist in laying out the trial. The field observations can then be supplemented by physical and chemical analyses.

In view of the above, an important activity of the training course was a field exercise. The objectives of the exercise were: to demonstrate the large soil heterogeneity even within a small area; to improve the participants' ability to recognize differences in soil properties in the field; to identify the soil heterogeneity; and to prepare a soil map of the area.

The site

A site of about three hectares adjoining the Land Development Department's training unit at the Regional Office in Chiang Mai was selected. The site was on a rolling topography, and was formerly under dense forest, but was cleared about 10 years ago. Currently it is partly under planted eucalyptus and partly under naturally regenerating secondary forest.

Design of the field exercise

An area of about one hectare was chosen for the exercise. The participants were divided into five groups. Initially they observed the geology, landform, drainage, and erosion features. Subsequently soil mapping was carried out. This was achieved by augering on a 10 x 10 m grid with a barrel auger. However the first 15-20 cm of soil was studied by digging a minipit with a spade. As the field exercise was done during the dry season, the soil was dry, and in a number of observation points it was not possible to auger to the stone line. A further factor was that due to time constraints each team observed only 10 cores, resulting in a coverage of only 500 m^2.

---

* The preliminary maps done at the workshop were finalized for publication by Ms. Pratumporn Funnpeng.

The following features were observed:

- slope
- erosion features
- vegetation
- soil colour
- soil texture and structure
- stoniness
- number of roots (visual estimate of intensity)
- location of termite mounds
- pH - using field pH kit.

The data from the different groups was pooled to produce the final report and maps.

Results

Geology

The underlying rock and surrounding hills were composed of granite. This was deduced from the rock fragments on the surface and from the semiweathered rock in the lower part of the C horizon.

Parent material

The parent material in the site was mainly alluvium deposited by streams, but was influenced (on the surface) by piedmont material from the adjoining hills. The lower part of the C horizon had some admixture of residuum from the granite. This was indicated by the presence of saprolite in the semiweathered materials.

Landform

The area consisted of a level plain and a dissected plain with slopes ranging from 3 to 20%.

There were five landforms:

- gently undulating erosion surface
- strongly undulating erosion surface
- moderately steep hillslope
- toe slope, and
- gulley floor.

Erosion was very evident on the sloping area. The surface features included rill and gulley erosion, crusting, deposition of gravels, and coarse sand.

Drainage

Currently the area is well drained. The presence of a gleyed horizon and distinct reddish brown mottles indicated saturated conditions in the profile, and fluctuation of the water table some time in the past.

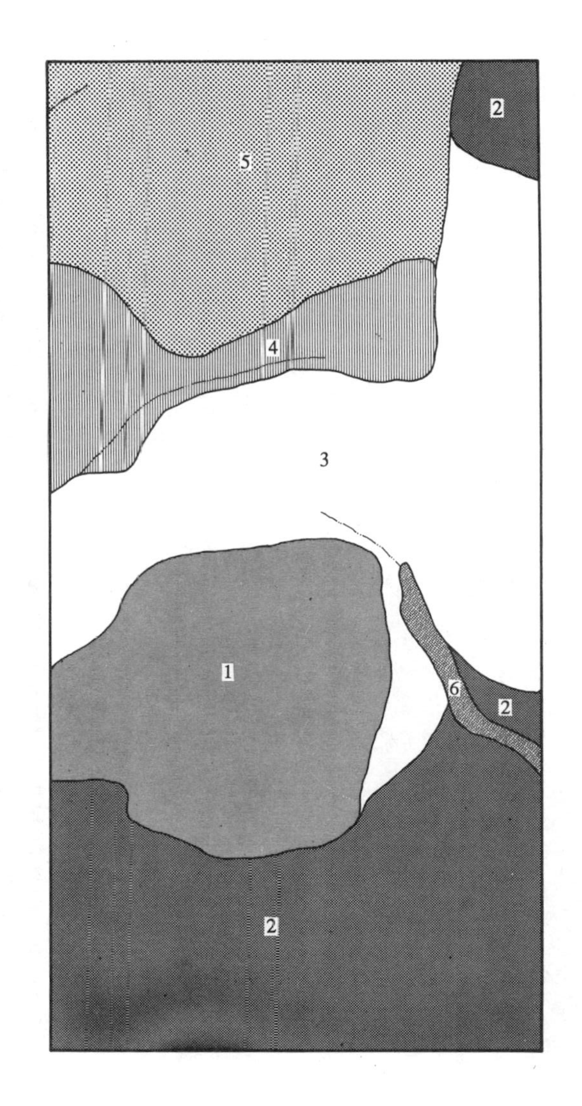

Legend

Eucalyptus plantation

1 — poor eucalyptus, few grass cover (almost bare soil) with phosphorus deficiency symptom on trees.

2 — poor eucalyptus in association with bush and/or grasses with phosphorus deficiency symptoms.

Regrowth vegetation of dry dipterocarp forest after slash and burn

3 — sparse trees and grasses

4 — sparse trees but slightly dense grass cover

5 — dense trees

Natural grass

6

incised gully

Figure 1. Map of vegetative cover.

Erosion

The following erosion features were observed:

- splash erosion (indicated by the presence of crust and vesicular pores on the surface)
- sheet erosion
- rill erosion
- gulley erosion

The degree of erosion could be gauged by the presence of rills and gullies, and was also supported by the truncation of the profiles in the dissected plain.

Vegetation and land use

The vegetation ranged from sparse grass, through regeneration of dry dipterocarp secondary forest to established eucalyptus plantation (Figure 1). Phosphorus deficiency was evident in the leaves of eucalyptus and grasses. This indicated that the area is low in phosphorus. In fact the growth of the planted eucalyptus was poor.

Termite mounds

A number of termite mounds were observed. The mounds varied in size and extent; the extent (particularly in the subsurface layers) was deduced by pH measurements. The location of the termite mounds is shown in Figure 2.

Soils present

On the basis of the surface texture and the slope of the landscape, the participants identified 12 mapping units. Subsequently it was decided to reduce these units to soil units, with regard to units exerting an influence on management practices. As the network was designed to evaluate management practices on sloping land, the degree of slope was considered to be a major factor. On this basis, the number of soil units was reduced to eight (Figure 2).

The general profile characteristics of these units as displayed by a transect (northwest to southeast) is shown in Figure 3. This shows that the soil colour in the surface soil ranged from 10YR to 2.5YR. In the latter (2.5YR) there was no topsoil; the surface soil being the C horizon. This indicates that this section had possibly been graded with a bulldozer, and could have been the remnants of a timber storage yard during the earlier logging. As this soil unit (A_3) forms a faily large portion of the area surveyed, it was concluded that this section of the site would not be suitable for the trial in mind.

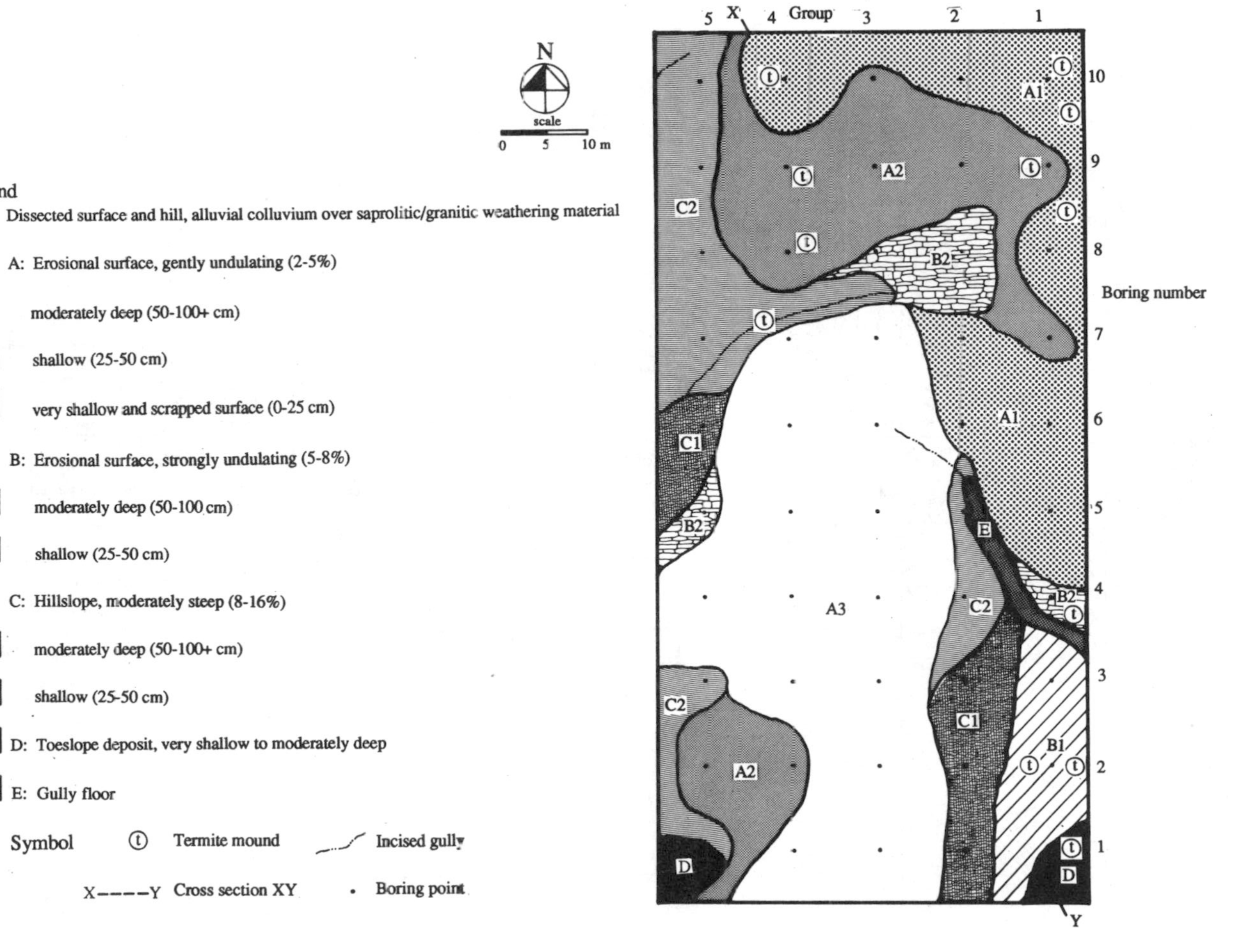

Figure 2. Soil map.

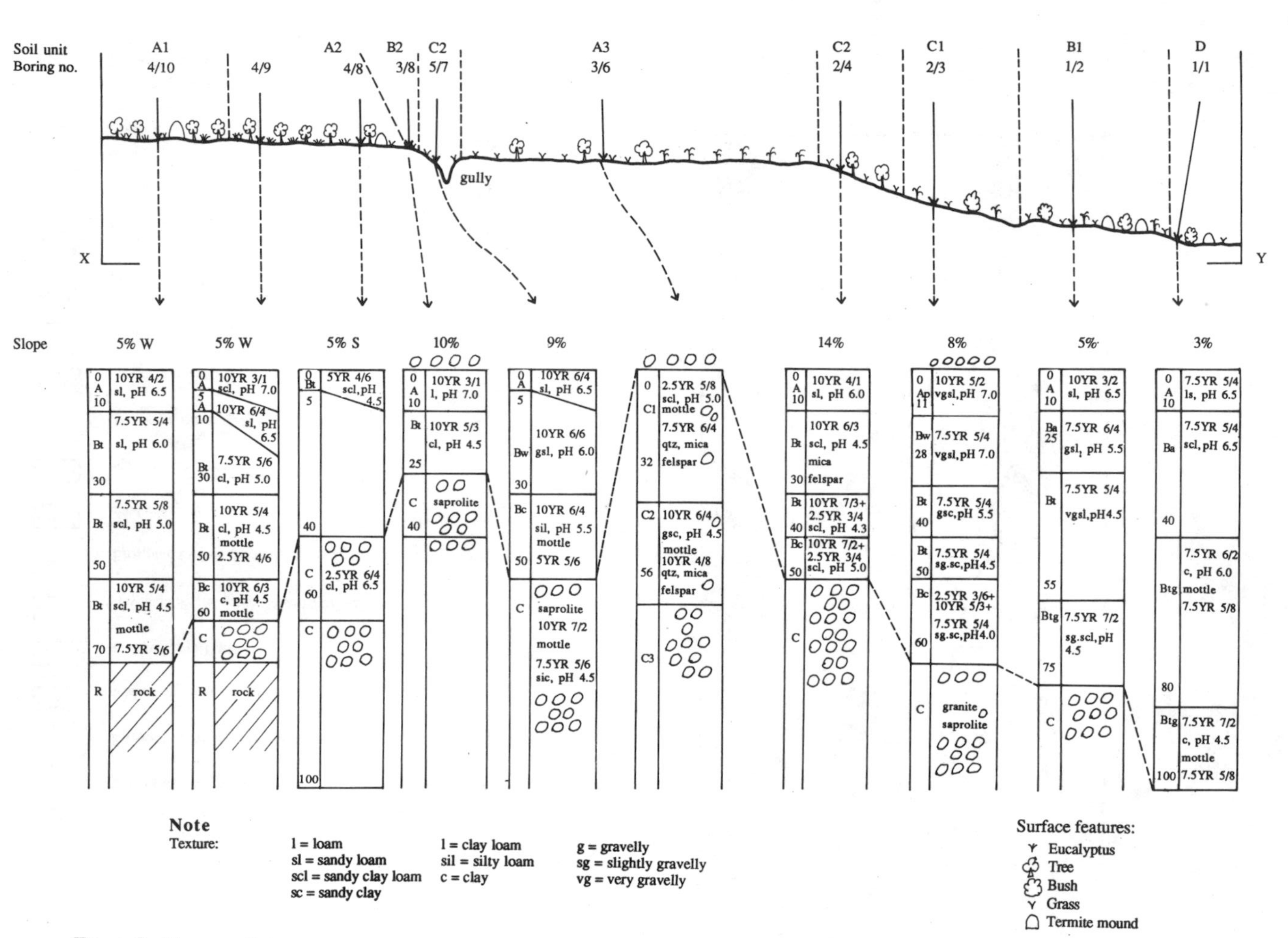

Figure 3. Cross section x-y.

Appendix III

PROGRAMME OF THE WORKSHOP

Monday 7 November

Opening ceremony: Addresses	
- Introduction Thai adviser for IBSRAM	Samarn Panichapong
- Welcome address Deputy governor of Chiang Mai	Ariya Uparamee
- Opening statement Programme officer, IBSRAM	E. Pushparajah
- Formal address Deputy director-general, DLD	Sitilarp Wasuvat
- Inaugural address Deputy minister, Ministry of Agriculture and Cooperatives	H.E. Udorn Tantisunthorn
Technical session I: IBSRAM's network approach	
IBSRAM's role in improving soil management	Marc Latham and E.Pushparajah
Land development and management in Asia	Adisak Sajjapongse
Technical session II: Agroecological approach	
An agroecological system approach for tropical highland research and development	Herman Huizing
Agrosystem analysis for on-farm research in the Chiang Mai valley	Methi Ekasingh
Land evaluation and soil management research	Herman Huizing
Discussion and preparation of guidelines	Herman Huizing

Tuesday 8 November

Technical session III: Farming systems	
Rainfed farming systems in upland soils	Chanuan Ratanawaraha
Cropping system experiments	Adisak Sajjapongse
Socioeconomic evaluation of farming systems	Chatt Chamchong
Physical sustainability evaluation of cropping systems in tropical highlands	Herman Huizing

Excursion to observe different farming systems in northern Thailand

Wednesday 9 November

Technical session IV: Site selection for experiments

Elements of experimental site selection	Samarn Panichapong
Agroclimatic characterization for soil management experiments	Aree Viboonpong
Site selection exercise by groups, using available documents, including landsat images, aerial photos and maps	Herman Huizing

Presentation and discussion of the group exercise

Thursday 10 November

Technical session V: Site characterization (I)

Soil variability on experimental sites	E. Pushparajah
Basic concepts and philosophy of *Soil Taxonomy*	Hari Eswaran
Principles of the FAO legend	F.J. Dent
Soil Taxonomy and agrotechnology transfer	Hari Eswaran

Field exercise (Leaders: Marc Latham and Samarn Panichapong)

Friday 11 November

Field exercise (Leaders: Marc Latham and Samarn Panichapong)

Saturday 12 November

Technical session VI: Site characterization (II)

Physical parameters	C. Valentin
Chemical parameters for site characterization and fertility evaluation	E. Pushparajah
Chemical and physical analysis for monitoring sustainability	Nualsri Kanchanakool
Fertility Capability Classification system: applications and interpretations for crop production planning	Irb Kheoruenromb

Compilation of field exercise results

Sunday 13 November

Free

Monday 14 November

Technical session VII: Erosion and Runoff	
Procedure for monitoring rainfall runoff, and erosion the measurement of land-use effects	C. Ciesiolka & A. Webb
Surface soil properties: crusting, sealing, and plough pans	C. Valentin
Field measurements for soil management studies	C. Ciesiolka & A. Webb

Field visit to observe erosion experiments and carry out soil physics test the field (Leaders: M. Latham and Samarn Panichapong)

Tuesday 15 November

Technical session VIII: Sampling and laboratory analysis	
Sampling soils and plants	J. Gerits
Soil biomass evaluation its quantity and composition in relation to cropping systems	Prosop Virakraphanich
Quality control of plant and soil analysis	J. Gerits
Compilation of results of field exercise and map preparation	(IBSRAM/ITC)
Presentation of results and discussion	

Wednesday 16 November

Technical session IX: Experimental design data processing	
Site variability and experimental design	L. Nelson
Experimental design for land clearing experiments on sloping lands	Dennis E. Sinclair
Statistical package for data processing of yields	L. Nelson
Methods for monitoring agricultural sustainability	C. Ciesiolka and Calvin Rose

Exercise on project proposals by group

Thursday 17 November

Presentation of draft guidelines for site selection and characterization (ITC/IBSRAM/DLD/CMU)

Group discussion

- site selection
- site characterization
- erosion runoff
- monitoring chemical, biological, and physical parameters

Friday 18 November

Presentation of final draft guidelines
Discussion on network implementation

Appendix IV

LIST OF PARTICIPANTS

(in order of country work base)*

AUSTRALIA

Dennis SINCLAIR
University of Newcastle
NSW 2308
Telephone: (049) 685744
Telefax: (049) 674946

Adrian WEBB
Agricultural Research Laboratories
Queensland Department of Primary Industries
Meiers Road, Indooroopilly
QLD 4068

CHINA

Gong ZI-TONG
The Institute of Soil Science
Academia Sinica
P0 Box 821
Nanjing
Telephone: 633318

INDONESIA

Justina Sri ADININGSIH
Agency for Agricultural Research and Development
Centre for Soil Research
Jalan Ir. H. Juanda 98
Bogor 16123
Telephone: (0251)-23012

ISMANGUN
AARD
Centre for Soil Research
Jalan Ir. H. Juanda 98
Bogor 16123
Telephone: (0251)-23012

MALAYSIA

ABDUL WAHAB Nafis
MARDI Research Station
Sg. Baging
Kuantan
Telephone: 09-433430

CHAN Huen Yin
RRIM
PO Box 10150
Kuala Lumpur 50908

GHULAM M. Hashim
Soil Science Unit
Central Research Laboratories Division
MARDI
PO Box 12301
Kuala Lumpur 50774

ZAINOL Eusof
RRIM
260 Jalan Ampang
PO Box 10150
Kuala Lumpur 50908
Telephone: (03) 4567033
Telex: MA 30369
Telefax: 6 (03) 4573512

* Representatives from international organizations are listed separately at the end.

NEPAL

P.L. MAHARJAN
National Agricultural Research and Services Centre
Division of Soil Science and Agricultural Chemistry
Khumaltar, Lalitpur
Kathamandu
Telephone: 5-21149

Ranjit SHAH
Division of Soil Science and Agricultural Chemistry
Khumaltar, Lalitpur
Kathmandu
Telephone: 5-21149

PAKISTAN

Mazhar Iqbal NIZAMI
PARC, Park Road
PO Box 1031
Islamabad
Telephone: 82005341

PHILIPPINES

Alcalde B. CRISOSTOMO
Bureau of Soils and Water Management
Sunvesco Building, Taft Avenue
Metro Manila
Telephone: 50-44-44

Redia ATIENZA
FFSRD-PCARRD
Los Baños
Laguna 4030
Telephone: 50015

Eduardo PANINGBATAN
Department of Soil Science
University of the Philippines
Los Baños
Laguna

SRI LANKA

T. WARUSAWITHAWA
Land and Water Use Division
Department of Agriculture
Peradenya
Telephone: (08) 22440

Anada WICKREMASINGHE
Land and Water Use Division
Department of Agriculture
Peradenya
Telephone: (08) 88355

TAIWAN

Yasuo OTA
AVRDC
PO Box 42
Shanhua
Telephone: 74199

THAILAND

AREE Viboonpong
Faculty of Agriculture
Chiangmai University
Amphoe Muang
Chiang Mai
Telephone: 221275

BOONCHEE Sawatdee
Office of Land Development Region 6
164 Chiang Mai-Fang Road
Dongkaew, Maerim
Telephone: (053) 222694

BOONYONG Phuparuang
Department of Land Development
Phaholyothin Rd., Bangkhen
Bangkok 10900
Telephone: 579-0111

CHAIYASIT Anecksamphant
Office of Land Development Region 7
Amphoe Muang
Nan 55000

CHANUAN Ratanawaraha
Department of Agriculture
Phaholyothin Road
Bangkhen, Bangkok 10900
Telephone: 5790053

CHATT Chamchong
Department of Agricultural Economics
Faculty of Economics and Business
Administration
Kasetsart University
Bangkhen, Bangkok 10903
Telephone: 579-1544

IRB Kheoruenromb
Department of Soils
Faculty of Agriculture
Kasetsart University
Bangkhen, Bangkok 10903
Telephone: 579-2028

KANCHANA Chuenpichai
Office of Land Development Region 7
Amphoe Muang
Nan 55000

METHI Ekasingh
Chiangmai University
Chiang Mai 50002
Telephone: (053) 221275

NUALSRI Kanchanakool
Department of Land Development
Phaholyothin Rd.
Bangkhen, Bangkok 10900
Telephone: 579-5523

PISOOT Vijarnsorn
Department of Land Development
Phaholyothin Rd.
Bangkhen, Bangkok 10900
Telephone: 579-0111

PRASOP Veerakornphanich
Department of Agriculture
Phaholyothin Rd.
Bangkhen, Bangkok 10900
Telephone: 579-7513

PRATEEP Veerapattananirund
Department of Agriculture
Phaholyothin rd.
Bangkhen, Bangkok 10900
Telephone: 579-7516

PRATUMPORN Funnpeng
Department of Land Development
Phaholyothin rd.
Bangkhen, Bangkok 10900

SAHAT Nilapan
Chiang Mai Land Development -
Department Station
Maerim
Chiang Mai
Telephone: (053) 211064

SAMRAN Sombatpanit
Department of Land Development
Phaholyothin Rd.
Bangkhen, Bangkok 10900
Telephone: 579-0111

SORASITH Vacharotayan
Kasetsart University
Bangkhen
Bangkok 10903
Telephone: 579-2028

SONGSAK Sribunjit
Multiple Cropping Center
Chiangmai University
Amphoe Muang
Chiang Mai

SUNTORN Ratchadawong
Office of Land Development region 7
Amphoe Muang
Nan 55000

SUTHAT Julsrigival
Department of Agronomy
Chiangmai University
chiang Mai
Telephone: (053) 221699

THONGLOR Monya
Department of Land Development
Phaholyothin Rd., Bangkhen
Bangkok 10900
Telephone: 579-0111

UDMA Chanphaka
UN Project
Huey Kaew
Chiang Mai
Telephone: (053) 514587

UTIS Tejajai
Office of Land Development Region 7
Amphoe Muang
Nan 55000

USA

David ANDERSON
USDA
Soil Conservation Service
4039 South 80 Street
Lincoln, Nebraska 68506
Telephone: (402) 488-6606

Larry NELSON
Tropsoils-NCSU
Box 7619
Raleigh NC 27695-7619
Telephone: (919) 737-2534

VIETNAM

Thai PHIEN
Institute for Soil and Fertilizers
Hanoi

International Organizations and Institutes

AIT

R.H.B. EXELL
Energy Technology Division
Asian Institute of Technology
PO Box 2754
Bangkok 10501
Thailand
Telephone: (662) 529-0100-13
Telex: 84276 AIT TH
Telefax: (662) 5296374

FAO

F.J. DENT
FAO Regional Office
Maliwan Mansion
Phra Atit Road, Bangkok
Thailand
Telephone: 281-7844
Telefax: 2800445

IBSRAM

Marc LATHAM
CHALINEE Niamskul
E. PUSHPARAJAH
SAMARN Panichapong
ADISAK Sajjapongse
Peter M. AHN
International Board for Soil Research and Management
PO Box 9-109, Bangkhen
Bangkok 10900
Thailand
Telephone: 579-7590, 579-4012, 579-7753
Telex: 21505 IBSRAM TH
Telefax: 66-2-5611230
E.Mail: CGI134

ISRIC

Jan GERITS
International Soil Reference and Information Centre
PO Box 750
6700 AG Wageningen
THE NETHERLANDS
Telephone: 08370-79063

ITC

Herman HUIZING
International Institute for Aerial Survey and Earth Sciences
350 Boulevard 1945 B.P. 6
7500 AA Enschede
THE NETHERLANDS
Telephone: (053) 320330

ORSTOM

Christian VALENTIN
Institut Français de Recherche Scientifique pour le Développement en Coopération
B.P. V-51, Abidjan
COTE D'IVOIRE

IBSRAM

BOARD OF TRUSTEES